IET TELECOMMUNICATIONS SERIES 104

Enabling Technologies for Social Distancing

Other volumes in this series:

Enabling Technologies for Social Distancing

Fundamentals, concepts and solutions

Edited by
Diep N. Nguyen, Dinh Thai Hoang, Thang X. Vu,
Eryk Dutkiewicz, Symeon Chatzinotas
and Björn Ottersten

The Institution of Engineering and Technology

Published by The Institution of Engineering and Technology, London, United Kingdom

The Institution of Engineering and Technology is registered as a Charity in England & Wales (no. 211014) and Scotland (no. SC038698).

The Institution of Engineering and Technology
Futures Place
Kings Way, Stevenage
Hertfordshire, SG1 2UA, United Kingdom

www.theiet.org

British Library Cataloguing in Publication Data
A catalogue record for this product is available from the British Library

ISBN 978-1-83953-490-4 (hardback)
ISBN 978-1-83953-491-1 (PDF)

Typeset in India by MPS Ltd

To all people who have lost their dears
or suffered from COVID. One day we all will rise
again in LOVE and PEACE.

Contents

    Yuris Mulya Saputra, Nur Rohman Rosyid, Dinh Thai Hoang
    and Diep N. Nguyen

    6.1   Ultrasound   169
          6.1.1   Distance among people   170
          6.1.2   Real-time crowd monitoring   172
          6.1.3   Automation   172
    6.2   Inertial sensor   174
          6.2.1   Distance among people   175
          6.2.2   Automation   177
    6.3   Visible light   178
          6.3.1   Real-time crowd monitoring   180
          6.3.2   Automation   181
          6.3.3   Traffic control   183
    6.4   Thermal   184
          6.4.1   Distance among people   185
          6.4.2   Physical contact tracing   186
          6.4.3   Real-time monitoring   187
    6.5   Summary   189
    References   190

7   **Security, privacy and blockchain applications in**
    COVID-19 detection and social distancing   **195**
    Dinh C. Nguyen, Quoc-Viet Pham, Pubudu N. Pathirana, Ming Ding
    and Aruna Seneviratne

    7.1   Introduction   195
    7.2   Security and privacy in COVID-19 detection and
          social distancing   197
    7.3   Blockchain Background   199
          7.3.1   Blockchain Introduction   199
          7.3.2   Potentials of blockchain for intelligent
                  COVID-19 detection   200
    7.4   Blockchain applications in COVID-19 detection and
          social distancing   202
          7.4.1   Outbreak monitoring   203
          7.4.2   User privacy protection   206
          7.4.3   Safe day-to-day operations   207
          7.4.4   Medical supply chain   208
          7.4.5   Social distancing   208
    7.5   Blockchain projects for COVID-19 fighting   209
          7.5.1   Hashlog   209
          7.5.2   Hyperchain   209
          7.5.3   VeChain   210
          7.5.4   PHBC   210

Enabling technologies for social distancing: fundamentals, concepts and solutions

In the presence of contagious diseases such as SARS, H1N1, and COVID-19, social distancing is an effective non-pharmaceutical approach to limit these viral diseases' transmission. Social distancing refers to measures that minimize the disease spread by reducing the frequency and closeness of human physical contacts, such as closing public places (e.g., schools and workplaces), avoiding mass gatherings, and keeping a sufficient distance among people. If implemented properly at the early stages of a pandemic, social distancing measures can play a key role in reducing the infection rate and delay the disease's peak, thereby reducing the burden on the healthcare systems and lowering death rates. In such a context, various technologies have recently been implemented to facilitate social distancing, ranging from supporting people to work from home to monitoring micro- (e.g., individual, the contact tracing apps using Bluetooth) and macro- (e.g., city, tracking the movement/transportation level of a city from the Google map) movement, etc.

Although latest advances in emerging fields like AI, wireless networking, Big Data, and other related technologies have a great potential in facilitating social distancing, implementing them in practical scenarios of social distancing still faces various challenges. First, we need to identify and define these technology-based social distancing scenarios. Different scenarios require different levels of positioning precision, computing power, and hence different technologies (e.g., indoor or outdoor social distancing practices). Second, technologies used for social distancing may compromise or put users' privacy at risk. For that, privacy-preserving solutions are crucial for the success of technology-based social distancing. Emerging technologies like blockchain or privacy-preserving algorithms used in localization/positioning systems, cloud computing, computer vision that have been developed in the literature can together facilitate social distancing to contain the spread of viral diseases.

In this regard, we introduce this book, titled "*Enabling Technologies for Social Distancing: Fundamentals, Concepts and Solutions*," which provides a comprehensive discussion and ideas on how the state-of-the-art technology, e.g., wireless and networking, artificial intelligence (AI), Big Data can enable, encourage, and even enforce social distancing practice. The book begins with a fundamental background on social distancing including basic concepts, measurements, models, and proposing various practical technology-based social distancing scenarios. We then discuss enabling wireless technologies which are especially effective and can be widely adopted in practice to keep distance, encourage, and enforce social distancing in general. After that, other emerging and related technologies such as machine learning/AI, computer vision, thermal, ultrasound, blockchain, and Big Data, are introduced. These technologies offer various solutions and directions to deal with problems in social distancing, e.g., symptom prediction, detection and monitoring quarantined people, and contact tracing. Finally, we provide open issues and challenges (e.g., privacy-preserving, scheduling, and incentive mechanisms) in implementing social distancing in practice. Through this book, we aim to:

- Provide fundamental background on technology-enabled social distancing and how state-of-the-art technologies can enable, encourage, and even enforce social distancing.
- Introduce and discuss emerging social distancing scenarios, ranging from micro- (e.g., individual contacts, indoor environment) to macro-levels (e.g., outdoor environment, social gathering, analysis and control of city's traffic movement or infected movement).
- Provide details on each relevant technology and how the technology can be applied in different social distancing scenarios.
- Highlight and summarize challenges (e.g., privacy, cybersecurity implications on using technologies to facilitate social distancing) and open issues as well as potential research directions to use latest technologies to facilitate social distancing.

Editors:

- Dr. Diep N. Nguyen, University of Technology of Sydney, Australia
- Dr. Dinh Thai Hoang, University of Technology of Sydney, Australia
- Dr. Thang X. Vu, University of Luxembourg, Luxembourg
- Prof. Eryk Dutkiewicz, University of Technology of Sydney, Australia
- Prof. Symeon Chatzinotas, University of Luxembourg, Luxembourg
- Prof. Björn Ottersten, University of Luxembourg, Luxembourg

About the Editors

Diep N. Nguyen is a faculty member and the director of Agile Communications and Computing Group, Faculty of Engineering and Information Technology, University of Technology Sydney (UTS), Australia. His recent research interests are in the areas of 5G/6G, computer networking, wireless communications, and machine learning applications, with emphasis on systems' performance and security/privacy. His work (appeared in more than 150 journal and conference papers, filed patents, technical reports) has been generously supported by various government and industry sponsors, e.g., US National Science Foundation, Raytheon Missile Systems, Australian Research Council, Australia Department of Foreign Affairs and Trade, Intel, US Air Force Research Lab, Australia's Defence Science and Technology Group, Google. He is a senior member of the IEEE. He is also an associate editor, editor of the *IEEE Transactions on Mobile Computing, IEEE Access, Sensors journal, IEEE Open Journal of the Communications Society (OJ-COMS), Scientific Report (Nature's)*. He received his ME and PhD in Electrical and Computer Engineering from University of California San Diego (UCSD) and The University of Arizona (UA), respectively.

Dinh Thai Hoang received his PhD degree from the School of Computer Science and Engineering, Nanyang Technological University, Singapore, in 2016. He is currently a faculty member at the University of Technology Sydney (UTS), Australia. Over the last 10 years, he has made significant contributions to advanced wireless communications and networking systems that is evidenced by around 130 papers published in top IEEE journals and IEEE flagship conferences with more than 8,000 citations to date. Since joining UTS in 2018, he has received several awards including the Australian Research Council Discovery Early Career Researcher Award for his project "Intelligent Backscatter Communications for Green and Secure IoT Networks" and IEEE TCSC Award for Excellence in Scalable Computing (Early Career Researcher) 2021 for his contributions in "Intelligent Edge Computing Systems." Alternatively, he is currently an editor of *IEEE TWC, IEEE COMST* and *IEEE TCCN*. He also served as an editor of *IEEE WCL* and a guest editor for several special issues, e.g., "Deep Reinforcement Learning on Future Wireless Communication Networks" in *IEEE TCCN* and "Internet of Things for Smart Ocean" in *IEEE IoT Journal*.

Thang X. Vu is a research scientist in the Interdisciplinary Centre for Security, Reliability and Trust, University of Luxembourg. His research interests include wireless communications, with particular interests in cache-assisted 5G, cloud radio access networks, machine learning for communications, and resources allocation and optimization. He holds a PhD in Electrical Engineering from the University of Paris-Sud, France.

Eryk Dutkiewicz is a professor and the head of the School of Electrical and Data Engineering, University of Technology Sydney, Australia. He has a joint professor appointment at Hokkaido University, Japan. He is also co-director of the UTS-VNU Joint Research Centres in Hanoi and Ho Chi Minh City, Vietnam. His research interests include 5G communications and IoT systems and networks. He has been the PI on many industry-funded projects with recent ones being *5G Dynamic Spectrum Sharing project* and *Ubiquitous Communications for 5G Networks project* funded by Intel Corporation. He is a co-author of over 350 papers and 19 patent filings.

Symeon Chatzinotas is full professor, chief scientist I and the head of the research group SIGCOM in the Interdisciplinary Centre for Security, Reliability and Trust, University of Luxembourg. His research interests cover multiuser information theory, cooperative/cognitive communications, cross-layer wireless network optimization, and content delivery networks. He has contributed to numerous R&D projects for the Institute of Informatics & Telecommunications, National Center for Scientific Research "Demokritos," the Institute of Telematics and Informatics, Center of Research and Technology Hellas and the Center of Communication Systems Research, University of Surrey. He has co-edited and authored several scientific books. He received his PhD in Electronic Engineering from the University of Surrey, UK.

Björn Ottersten is the director of the Interdisciplinary Centre for Security, Reliability and Trust at the University of Luxembourg. His research interests include signal processing, wireless communications, radar, and computer vision. He has been Digital Champion of Luxembourg, acting as adviser to the European Commission and also member of the Boards of the Swedish Research Council and the Swedish Foundation for Strategic Research. He has served as editor in chief of *EURASIP Signal Processing* and acted on the editorial boards of *IEEE Transactions on Signal Processing*, *IEEE Signal Processing Magazine*, *IEEE Open Journal for Signal Processing*, *EURASIP Journal of Applied Signal Processing*, and *Foundations and Trends in Signal Processing*. He is currently director of technical activities on the Board of EURASIP, and a Fellow of EURASIP and the IEEE and. He has received the IEEE Signal Processing Society Technical Achievement Award and the EURASIP Group Technical Achievement Award. He received his PhD Degree in Electrical Engineering from Stanford University, USA.

Chapter 1
Social distancing and related technologies: fundamental background

*Cong T. Nguyen[1], Dinh Thai Hoang[1], Diep N. Nguyen[1]
and Eryk Dutkiewicz[1]*

COVID-19, with unprecedented consequences on global health and economy in modern history, has dramatically transformed the world's view on pandemics. Starting from Wuhan, China in January 2020 [1], within less than 2 years, there have been more than 261 million people infected, including more than five million deaths [2], in 210 countries and territories around the world [2]. Besides, the dire consequences on global health, COVID-19 has also severely impacted the global economy. Specifically, the global GDP recorded a loss of 4.5% in 2020 (nearly $3.94 trillion) [3]. Thus, there is a pressing need for measures to limit the disease's spread, especially when pharmaceutical solutions are still requiring much more time to be fully implemented.

In the presence of current and previous epidemics and pandemics such as SARS, MERS, H1N1, Ebola, and COVID-19, social distancing has been an effective approach to reduce the disease transmission, especially in the absence of pharmaceutical solutions [4–6]. Particularly, by reducing physical contact among people, the likelihood of disease transmission from an infected to a healthy individual decreases, and thus the disease's spread, as well as its severity, can be significantly reduced. However, there have been many difficulties in social distancing implementation due to its economic repercussions and negative impacts on personal freedom. In such context, technologies can be utilized to ensure proper implementation of social distancing and reduce its negative effects.

In this chapter, we provide a thorough background on social distancing as well as effective technologies that can be leveraged for facilitating social distancing measures. Particularly, we first present an overview and provide fundamental knowledge of social distancing, especially the role of social distancing in the current COVID-19 pandemic. After that, we provide fundamental knowledge on enabling technologies that are particularly effective in the majority of social distancing scenarios.

[1]School of Electrical and Data Engineering, University of Technology Sydney, Sydney, Australia

1.1   Social distancing

1.1.1   Overview

Social distancing refers to a collection of measures that aim to reduce the frequency and closeness of human contacts, thereby minimizing the disease spread. These measures can be applied to public places (e.g., schools, workplaces, and shopping malls), limiting the number of people in buildings, ensuring a safe distance amongst people [4,7], and so on. By reducing physical contact among people, the likelihood of disease transmission from an infected to healthy individual decreases, and thus the disease's spread and severity can be significantly reduced. Studies [4–6] have shown that if social distancing is properly implemented in the early stages of a pandemic, it has the potential to significantly reduce the infection rate as well as to postpone the disease's peak. Figure 1.1 illustrates the impacts of proper social distancing implementation on a pandemic outbreak [8]. As observed in Figure 1.1(a), social distancing can both reduce and delay the outbreak peak [5]. By reducing the peak number of infected cases, social distancing can ensure that the public healthcare capacity can handle the peak number of patients, thereby helping to alleviate pressures off the healthcare system. Moreover, the delay of a pandemic's peak that social distancing brings can be extremely valuable, as it provides more time for preparations and intervention, e.g., improving medical facilities, preparing infrastructure, and developing cures. In addition, as illustrated in Figure 1.1(b) [8], social distancing also helps to reduce the cumulative number of infected individuals [5]. Moreover, early implementation of social distancing can lead to a significantly lower cumulative number of cases.

In the current COVID-19 pandemic, governments around the world have implemented various social distancing measures including border control, travel restrictions, public places closure, and warning their citizens to maintain a safe distance (1.5–2 m) in public [9–11]. However, these measures are often difficult to implement, e.g., borders cannot be closed for too long, some public places cannot be closed, negative impacts on people's freedom and privacy, and people still need to go outside for healthcare, food, or essential work. In these cases, technologies can be utilized to address social distancing's limitation. Due to the complex development of novel strains of viruses and the ever-increasing growth of social interactions, social distancing measures have extended beyond simple physical distancing. For example, in the current COVID-19 pandemic, social distancing also includes many measures such as monitoring, violation detection, and alert. With the rapid development of emerging technologies, e.g., Artificial Intelligence (AI), data analytics, and future communications systems, numerous new solutions have been recently proposed which can encourage or even enforce social distancing.

1.1.2   Measurements and models

1.1.2.1   Basic reproduction number R_o

To decide which level of social distancing and the corresponding measures are suitable, we often rely on the disease's current infection rate. One of the most commonly

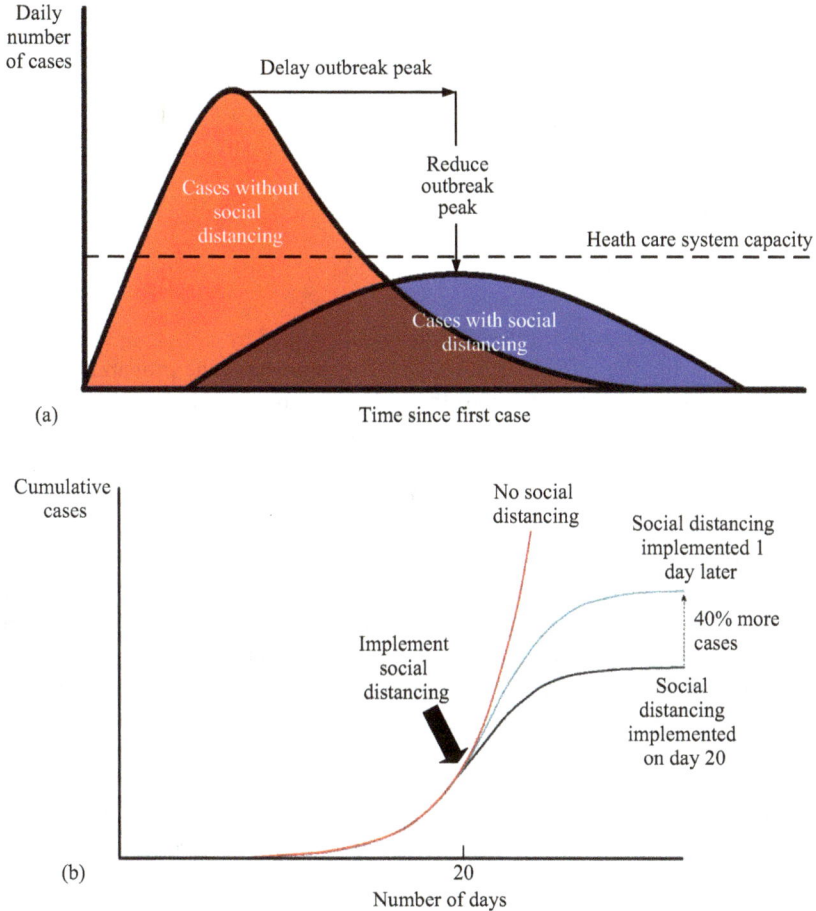

Figure 1.1 *The effects of proper social distancing implementation on a pandemic outbreak: (a) reduce and delay the disease's peak and (b) reduce the total number of infected cases. The earlier social distancing is implemented, the more cases it can reduce [8]*

used metrics for measuring a disease's spread and severity is the basic reproduction number R_o. This number represents the average number of people infected by a case (i.e., an infectious person) over the course of its infectious period [12]. For example, $R_o > 1$ means that on average there will be more than one person infected by a case, and consequently the number of infected people will increase. Conversely, a $R_o < 1$ means that, over its infectious period, each infected individual will infect fewer than one person, and thus the disease will eventually be eradicated in the population under consideration. Since R_o reflects how fast a disease is spreading, R_o has been one of the most important criteria taken into consideration when selecting social distancing measures [5,6]. For example, a high R_o calls for more drastic measures such as travel restriction and public places closure, whereas a lower R_o indicates that more lenient measures can be implemented, e.g., limit the number of people in public places. Note that, drastic measures impose significant consequences on the economy and people's compliance. Therefore, careful selection of appropriate social distancing measures based on the pandemic's infection rate is important, especially when a pandemic lasts for a long time like the current COVID-19 pandemic.

Mathematically, R_o is calculated by

$$R_o = \int_0^\infty b(a)F(a)\mathrm{da}. \tag{1.1}$$

In (1.1), $b(a)$ is the average number of new cases infected by an infectious person per unit of time during the infectious period a. $F(a)$ represents the probability that the infected individual remains infectious throughout the period a [12].

In addition to showing a disease's transmission rate, R_o also provides some intuitive ideas for lowering the rate. There are two main ways to reduce R_o, i.e., to decrease $b(a)$ and $F(a)$, as observed from (1.1). To reduce $b(a)$, approaches that reduce the frequency of human contacts and the transmission probability during these contacts can be employed. For example, there are social distancing measures that reduce the number of contacts made by infected people per unit of time (e.g., travel restriction and limiting the number of people in public buildings) or reduce the likelihood that contacts between the infected and the healthy individuals will result in new cases (e.g., keeping a safe distance and wearing masks). To reduce $F(a)$, pharmaceutical solutions are needed, such as cures and vaccines.

1.1.2.2   Compartmental mathematical model MSEIR

The most commonly used type of model for an infected disease is the compartmental mathematical model MSEIR [13]. In MSEIR models, there are five main classes M, S, E, I, and R, each represents a different type of individual in a population. Specifically, they are:

- M: This class includes infants who have passive immunity inherited from their mothers.
- S: People who can become infected by transmission from an infected case, i.e., susceptible people, belong in this class.
- E: When a susceptible individual of class S becomes infected, the individual begins a latent period. During this period, the individual cannot transmit the disease to a healthy person yet. These individuals constitute class E.
- I: Class I represents the individuals who can transmit the disease to others. After their latent periods are over, individuals from class E are moved to class I.
- R: Individuals who have fully recovered from the disease and acquired immunity are classified by class R.

When choosing a compartmental mathematical model for an infectious disease, there are several factors to be considered. For example, in novel strains of viruses, passive immunity from birth might be non-existent, and thus class M should not be included. Moreover, in cases where the considered period is short, births and deaths might also be omitted from the models. As a result, different types of models, such as MSEIRS, SEIRS, MSEIR, SEIR, SIRS, and SIR, can be created from different combinations of the above classes. From the acronyms of these models, we can deduce which classes are taken into account as well as how individuals move between these classes. For example, in a disease where the infection-acquired immunity is permanent, the MSEIR model can be used. However, if a recovered person can be infected again, the MSEIRS model must be used, as an individual from class R might be moved back to class S again.

The SIR model is the most commonly used among different types of models. Let $S(t)$, $I(t)$, and $R(t)$ denote the number of susceptible, the number of infected, and the number of recovered individuals in a population at time t, respectively. Moreover, let $\frac{1}{\gamma}$ denote the average infectious period and β denote the average number of adequate contacts, i.e., the contacts which infect a new case. Then, the SIR model can be expressed by

$$\frac{dS(t)}{dt} = \frac{-\beta IS}{N}, \tag{1.2}$$

$$\frac{dI(t)}{dt} = \frac{\beta IS}{N} - \frac{\gamma I}{N}, \tag{1.3}$$

$$\frac{dR(t)}{dt} = \frac{\gamma I}{N}, \tag{1.4}$$

where N is the total population. In (1.2), $\frac{\beta IS}{N}$ represents the rate at which new susceptible individuals are infected because β represents the number of adequate contacts made by an infected case and I represents the number of infected cases. Note that newly infected people are moved from class S to class I, and thus there is a minus sign. Moreover, in (1.4), the recovery rate and the infectious period $\frac{1}{\gamma}$ is inversely proportional. Furthermore, the rate of change of class I, i.e., the infected rate, is the difference between the infectious rate and the recovery rate. Finally, $\frac{\beta}{\gamma}$ is the average number of cases infected from an individual because $\frac{1}{\gamma}$ represents the infection time and β represents the number of adequate contacts per unit time. It is worth noting that R_o also represents this average number of cases infected by an individual, and thus $\frac{\beta}{\gamma} = R_o$.

A limitation of the SIR models is that they do not take into account several important aspects of a disease, including people with vital dynamics (birth and deaths) and birth immunity. Therefore, SIR models are only effective for modeling a new type of virus because there is no birth immunity. Moreover, the considered period of SIR models is often short such that births and deaths during the considered period can be neglected. On the other hand, SIR models are well-posed due to their simplicity.

Let $i(t) = \dfrac{I(t)}{N}$, $s(t) = \dfrac{S(t)}{N}$, then if $R_o < 0$, $i(t)$ decrease to zero as $t \to \infty$ [13]. If $R_o > 0$, $i(t)$ reaches its peak, given by

$$i_{\max} = i(0) + s(0) - \frac{1}{R_o} - \frac{\ln(s(0)R_o)}{R_o}, \tag{1.5}$$

then it decreases to zero as $t \to \infty$ [13]. As observed from (1.5), it is possible to reduce the peak infection rate by decreasing $i(0)$ and $s(0)$. Reducing $i(0)$ means reducing the number of infectious cases at the beginning of the considered period. In many cases, it translates to limiting the number infectious people coming from outside the population. This is the aim of social distancing measures that limit the travel of people between different areas, e.g., border control and travel restrictions. For $s(0)$, besides birth immunity, vaccines can be used to reduce this number.

1.1.3   Effectiveness of social distancing

A common approach to evaluate social distancing measures' effectiveness is to measure their effects on the attack rate. This rate is the proportion of infected people in a susceptible population where there is no immunity at the start of the disease outbreak [14]. Since the attack rate represents the disease's severity at a given time, its value varies over the course of the disease. There are different types of attack rates, including the average, peak, and final attack rates. Among them, the peak attack rate is one of the most important. This value is often considered along with the current healthcare capacity to see if the current healthcare facilities can accommodate the highest number of patients or not. After the outbreak is over, the final attack rate is calculated by the total number of infected cases divided by the total population.

Using these attack rate values, many studies have proven the effectiveness of social distancing measures, especially when they are implemented properly [5,14–19]. From these studies, diverse levels of social distancing's effectiveness on the disease spread can be observed, depending on the particular types of measures implemented. The effectiveness of workplace social distancing measures is evaluated in [14] using an agent-based simulation approach. Specifically, six types of strategies, each reducing a different number of workdays, are considered in the study. The findings show that if three consecutive workdays are missed during seasonal influenza ($R_o = 1.4$) outbreak, the final attack rate can be reduced by up to 82%. However, the positive impacts of this strategy are significantly weakened, i.e., 21% decrease in final attack rate, in a pandemic-level influenza where $R_o = 2.0$. Other studies such as [15] and [16] present similar results. In particular, [15] shows that, in a $R_o = 1.4$ setting, social distancing measures applied in workplaces can result in a 39% lower final attack rate. Similarly, [16] shows that the attack rate can be reduced up to 20%, depending on how many times per day the employees have close physical contact with each other.

Besides workplaces, schools closure is also commonly implemented. In [17], the effects of school closure under three varying R_o settings are investigated. The results show that in the $R_o < 1.9$, $2.0 \le R_o \le 2.4$, and $R_o > 2.5$ settings, the final attack rate can be reduced by 20%, 10%, and 5%, whereas the peak attack rate can be reduced by 77%, 47%, and 32%, respectively. Similarly, [18] shows that the final and peak attack rate can decrease by up to 17% and 45%, respectively, when school closure is implemented for a prolonged period in a pandemic.

Isolation of infected and high-risk cases is also commonly implemented during pandemics. In [5], it is shown that with the proper implementation of isolation (less than 10% contact rate) the final attack rate can be reduced by 7% when $R_o = 2$. A similar study [17] shows that when isolation measures are applied in the $R_o < 1.9$, $2.0 \leq R_0 \leq 2.4$, and $R_o > 2.5$ settings, the final attack rate can be reduced by 27%, 7%, and 5%, whereas the peak attack rate can be reduced by 89%, 72%, and 53%, respectively.

Similar to isolation, studies have shown that household quarantines measures are very effective if people can properly follow the rules. The effects of 14-day voluntary household quarantine are examined in [5]. Particularly, the study simulates the cases where 50% of households will comply. In these households, it is assumed that external contact rates are reduced by 75%, while internal contact rate increases by 100%. The results show a reduction of up to 6% and 40% of the final and peak attack rates, respectively. Similarly, [19] studies a case where household quarantines are implemented with a compliance rate of 50% and $R_o = 1.8$. The results show that the final attack rate can be reduced by 31%, and the peak attack rate can be reduced by 68% under the considered setting.

Studies such as [5,17,20] show that social distancing measures can achieve greater effects when they are combined. Particularly, [17] shows that when school closure, isolation, workplace nonattendance, and contact reduction (e.g., avoiding crowds and canceling mass gatherings), are simultaneously in effect, the attack rates can be drastically reduced in all the considered R_o settings. Specifically, the results show that, even with the highest R_o setting, the final attack rate decreases from 65% to only 3%, whereas the peak attack rate can be reduced by nearly 95 times. Similarly, [5] investigates the case where travel restrictions, border control, household quarantines, and workplace closures are applied simultaneously. Simulation results show that both the final and peak attack rates can be reduced up to three and six times, respectively, and the peak of the disease outbreak is postponed by nearly three months. In [20], it is also shown that when household quarantines, school closure, workplace nonattendance, and community contact reduction are implemented at the same time, the final total number of infected cases can be reduced by up to 3–4 times.

On the other hand, there are also studies that investigate the negative effects of social distancing measures. In [21], both benefits and cost of different social distancing strategies are evaluated by simulations. The results show that a favorable result (i.e., benefits higher than costs) can only be achieved if social distancing is applied strictly and properly. Moreover, the results also show that improper implementation of social distancing measures, e.g., prematurely lifting the restrictions, yields even worse results than if social distancing is not implemented. Another study [22] also shows an interesting result. In particular, the effects of social distancing vary according to the values of R_o. For example, when $R_o < 1$, social distancing is not really effective, and social distancing's effectiveness peaks when $R_o \approx 2$.

1.1.4   Social distancing in COVID-19

1.1.4.1   Effectiveness

In the current COVID-19 pandemic, studies have predicted many positive effects of social distancing using modeling and simulation approaches. The results in [23] show that different social distancing measures, such as self-isolation, household quarantine, public place closure, and elderly isolation, have varying effects on the number of severe cases requiring intensive care. Moreover, the study also concludes that implementing all four measures simultaneously achieves the highest effectiveness. Particularly, the peak attack rate can be reduced nearly 4 times, and the disease peak can be postponed by up to three months.

Using the SEIR models, another prediction modeling approach also shows similar results in [24]. In particular, the effects of intermittent and continuous social distancing implementation are investigated and compared in 16 countries. The results show that continuous social distancing can achieve a lower mortality rate than that of intermittent social distancing. However, due to economic constraints, continuous social distancing is very difficult to maintain for a long period of time, especially in low-income countries.

A neural network is developed in [25] to examine the effectiveness of different social distancing measures in Wuhan, South Korea, Italy, and the United States. Specifically, the authors employ the proposed neural network to determine the β and γ parameters. These parameters are then used in the SIR and SEIR models to model the disease spread. The results show that the stricter the social distancing strategies are, the stronger impacts they have on R_o and the disease severity.

Besides modeling and predictions, the effectiveness of social distancing measures, including mandatory masks, quarantine, keeping distance, and traffic restriction, in 190 countries over the period from January 23 to April 13, 2020 are also reported in [26]. The results show that any of the four measures, when implemented individually, can only achieve a maximum attack rate reduction of 34%. However, combinations of three to four measures can achieve a reduction of up to 69%. Nevertheless, it is worth noting that the study cannot take into account factors such as people's compliance and other difficulties in implementation.

Figure 1.2 shows the total number of confirmed COVID-19 cases and the date of social distancing implementation in Australia, France, Italy, and Spain, in the first half of 2020. As observed from Figure 1.2(a), the number of cases started to rise at the beginning of March 2020 in Australia. The authorities then quickly reacted by implementing a nationwide lockdown on March 23 [27]. After around 2 weeks of implementation, the "flatten the curve" effect can be observed, i.e., the number of cases stopped increasing steeply. Similar effects can be observed from Figure 1.2(b)–(d), as the authorities imposed nation-wide lockdown in France, Italy, and Spain, on the 17th, 9th, and 15th of March 2020, respectively [27].

Furthermore, social distancing showed significant positive impacts on COVID-19 fatalities in the above countries during the same time period. Figure 1.3 shows the weekly number of confirmed deaths COVID-19 and the date of social distancing implementation in Australia, France, Italy, and Spain, in the first half of 2020. As

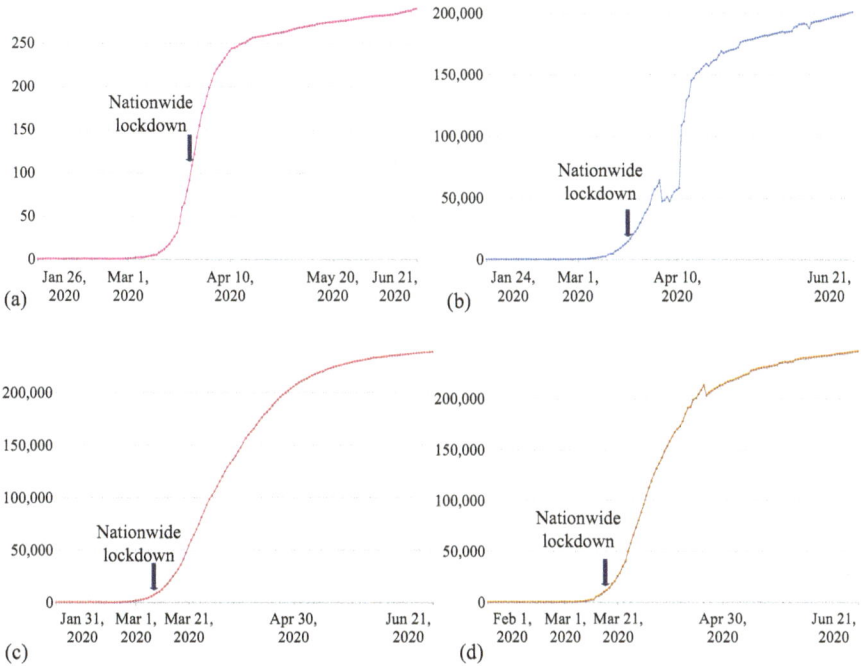

Figure 1.2   Cumulative cases of COVID-19 and date of social distancing implementation in (a) The Australia [28], (b) France [28], (c) Italy [28], and (d) Spain [28]

observed from Figure 1.3, the effects of social distancing on reducing the mortality rate can be observed after around one month of implementation.

1.1.4.2   Second and subsequent waves

Due to their negative impacts, social distancing measures are often very difficult to implement and sustain for a long time. Consequently, it often leads to the premature lifting of restrictions from the authorities as well as reduces people's cooperation [29]. However, such premature termination of social distancing usually has disastrous consequences, such as the second and subsequent waves of the pandemic, in which the attack rate sharply increases again after a period of slowing down.

Evidence of the second waves can be found in the previous pandemics and epidemics, dating back to as early as 1918. Particularly, social distancing measures were lifted or terminated prematurely by the authorities after their initial success in the previous 1918 influenza pandemic (the Spanish flu). Moreover, from the people's side, the perceived risk decreased once the first wave was over. As a result, people resumed their normal daily behaviors, even though there still had not been any effective pharmaceutical solutions. Consequently, a second wave occurred in the 1918 influenza pandemic [30,31]. As shown by data from various geographical locations, this second wave caused even worse consequences than those of the first wave. For example,

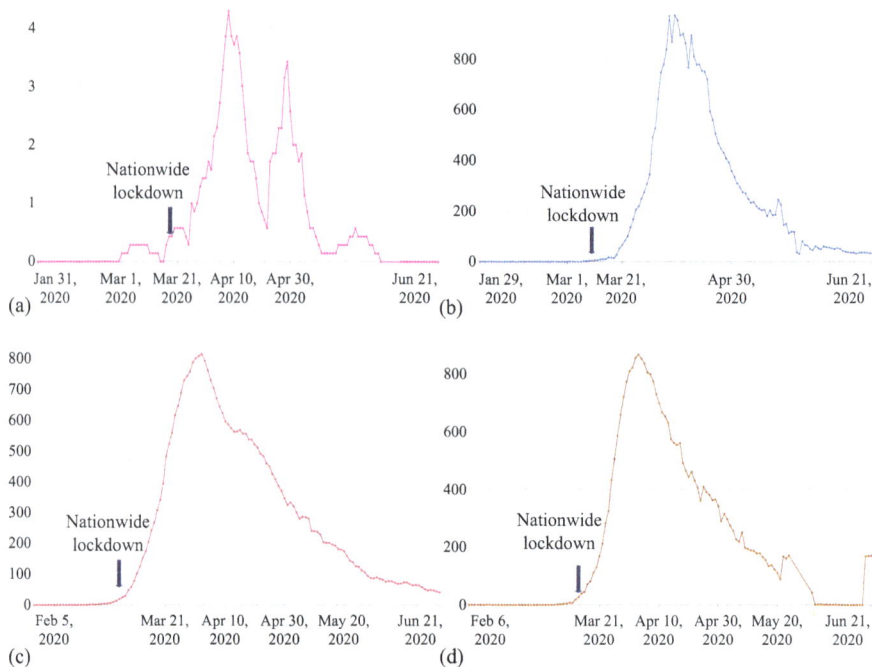

Figure 1.3   *Weekly confirmed deaths of COVID-19 and date of social distancing implementation in (a) The Australia [28], (b) France [28], (c) Italy [28], and (d) Spain [28]*

compared to the first wave, the second wave of the 1918 influenza pandemic that occurred in Sydney has a nearly double mortality rate [30]. In the United States, the second wave of the 1918 influenza pandemic hit 8 out of 16 considered cities, causing higher death rates in 4 cities compared to those of the first wave [31].

By the time of writing, countries around the world have suffered from at least two waves of the current COVID-19 pandemic, as illustrated in Figure 1.4. Many of these waves coincide with the lifting of restrictions in the prior months. For example, as observed from Figure 1.4(a), the United States has been hit by a second wave of the COVID-19 pandemic after the lifting of several restrictions (e.g., partially reopen public places such as bars, restaurants, schools, etc.) in early May 2020. Moreover, in early March 2021, the government lifted several restrictions including mandatory masks in public and reopened businesses, which is followed by a sharp increase in new cases in the following weeks. Furthermore, in early July 2021, several states lifted restrictions, and the subsequent wave reached its peak 2 months later. Similarly, early lifting of restrictions has been followed by new waves of COVID-19 in the United Kingdom, as illustrated in Figure 1.4(b). As the number of new cases dropped, the government ended the lockdown in the United Kingdom in December 2020. One month later, a second wave occurred and reached it's peak. Similar to the United States,

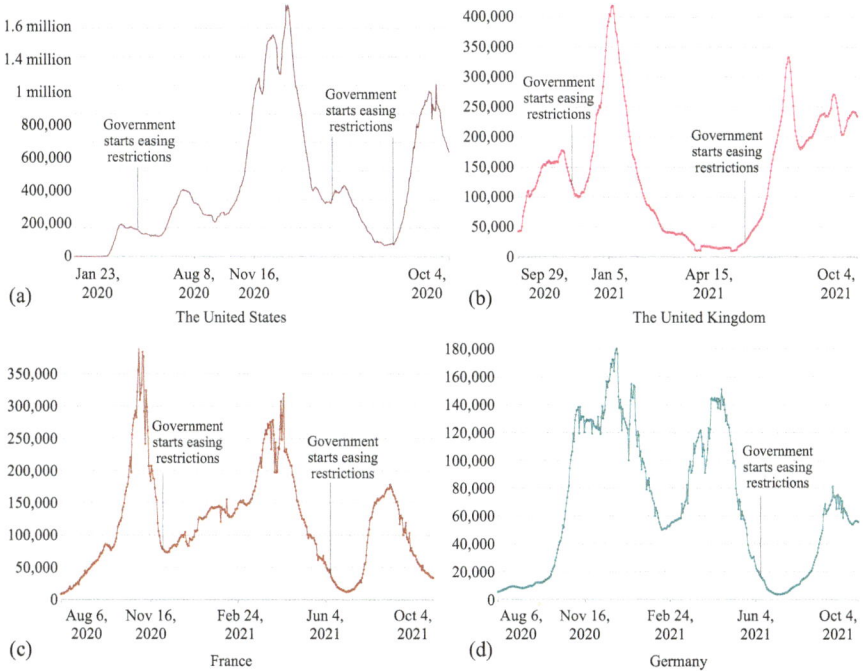

Figure 1.4   Second and subsequent waves of COVID-19 in (a) The United States [28], (b) The United Kingdom [28], (c) France [28], and (d) Germany [28]

the United Kingdom also lifted restrictions again in early July 2021. A new wave then occurred and peaked two weeks later. Similar patterns are reported in France, as illustrated in Figure 1.4(c), and Germany (Figure 1.4(d)).

1.1.4.3   Challenges

Despite its significant potential, it is also shown in the above studies that social distancing is very effective only when implemented properly. Nevertheless, there are many challenges that have been hindering social distancing implementation such as:

- *Negative economic impacts*: Measures such as travel restriction, public places closure, and border control have dramatically disrupted the economy as evidenced by the 4.5% global GDP loss in 2020. This means that strict measures are not sustainable, and consequently they are often lifted prematurely by the authorities.
- *Personal rights violation*: Measures such as isolation, canceling mass gatherings, and quarantines, may cause conflicts with personal freedom as well as religious principles, e.g., many religious facilities are closed during lockdowns all over the world. Moreover, people's privacy is also violated when measures such as contact tracing and movement tracking are implemented, thereby causing noncompliance.

- *Difficulties in changing personal behaviors*: Even with compliance, it is not always easy for a person to follow social distancing instructions. Estimating and maintaining a 1.5–2 m distance with everyone all the time, for example, is difficult, and people must still go outside for healthcare, food, or essential tasks.
- *Difficulties when staying at home*: With the closure of schools and workplaces, many people began to work or study remotely. This causes an overwhelming burden on the current Internet infrastructure and online services. For example, in March 2020, the number of new users of Zoom [32] and Microsoft Teams [33] increased by 1270% and 775%, respectively.

Although there are many difficulties in implementation, social distancing is still the best type of measure available at the beginning of a pandemic, especially when effective pharmaceutical solutions are still not fully developed and available [29]. Thus, for both current and future pandemics, social distancing plays a crucial role in disease's spread mitigation. Therefore, solutions to facilitate social distancing measures are essential. To this end, technologies can be utilized to ensure proper implementation of social distancing and reduce its negative effects.

1.1.5 *Practical social distancing scenarios*

As illustrated in Figure 1.5 [8], the practical implementation of social distancing may involve various scenarios which can be summarized as follows [8,34]:

- *Keeping distance*: Keeping a safe distance is one of the most common scenarios. In these scenarios, often it is hard for a person to be aware of his/her position and distance from others all the time. Therefore, positioning and AI technologies can be employed to assist the person in keeping track of his/her relative position compared to others. Based on that, the person can be alerted when he/she gets too close to another (e.g., by smartphones).
- *Real-time monitoring*: Real-time monitoring, in general, is not a new concept. However, in the context of the current pandemic, many new challenges emerged, such as citizens' privacy, and the unprecedented large scale of implementation. To overcome these challenges, many wireless and related technologies can be utilized. The goal of such monitoring is to gather meaningful data to facilitate social distancing in an efficient manner. Based on the data from monitoring, appropriate measures, such as avoiding crowds, alerting/penalizing violations, limiting access to crowded buildings, can be carried out.
- *Information system*: Due to their omnipresence nowadays, wireless technologies can be utilized to collect the movement and contacts data of the infected individuals. Using this information, susceptible people who had close contact with the infected can be alerted to take cautious actions such as self-isolation and notifying the authorities.
- *Incentive*: Social distancing can negatively impact personal freedom. Therefore, besides enforcing, it is important to incentivize people to comply with social distancing measures. Wireless technologies and economics tools such as contract theory can be leveraged to develop those incentive mechanisms.

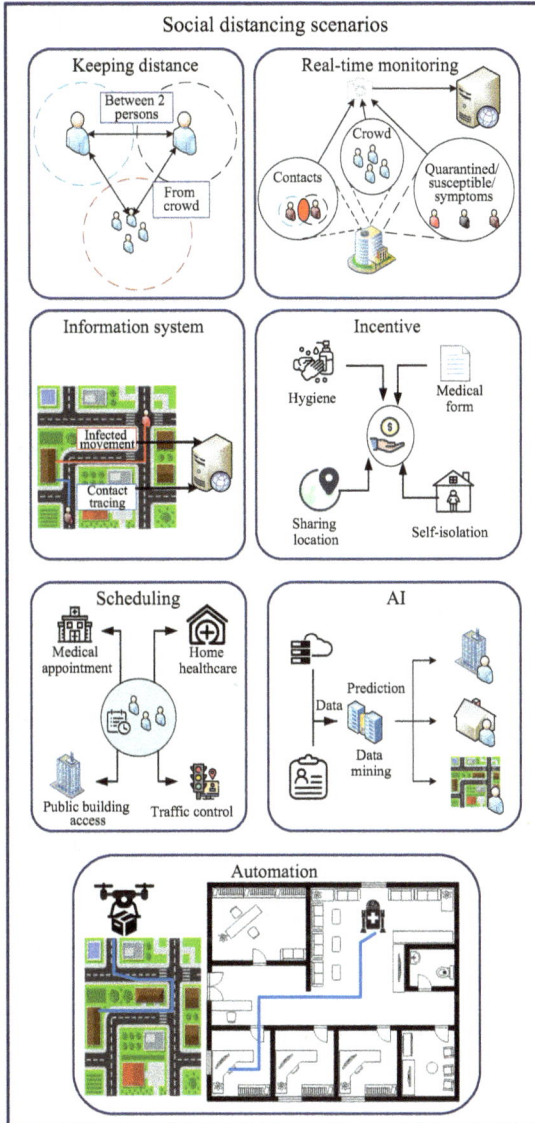

Figure 1.5   Illustrations of the practical social distancing scenarios [8]

- *Scheduling*: Scheduling techniques can be implemented to increase the efficiency of workplaces and hospitals, thereby decreasing the necessary number of people at such places. Moreover, traffic can be scheduled to limit and regulate the number of cars and pedestrians on the road. Furthermore, wireless technologies can be applied to schedule access to public buildings.

- *Automation*: Unmanned aerial vehicles (UAVs) and medical robots are examples of autonomous vehicles that can be used to reduce human presence in essential tasks such as delivery services and medical procedures. Wireless technologies are an essential part of the positioning and navigation systems of these autonomous vehicles.
- *Modeling and prediction*: AI technologies can be employed to mine pandemic-related data. The findings can be very useful to forecast valuable information such as when and where the infected people will go.

1.2   Background of enabling technologies for social distancing

Following the fundamental background of social distancing and its impact on mitigating pandemics, in this section, we identify and briefly discuss enabling technologies for social distancing, such as positioning systems, computer vision, machine learning (ML), operation research, incentive mechanism, and blockchain. More details on how these technologies are actually implemented in their practical social distancing applications will be presented in later chapters.

1.2.1   Positioning-based social distancing solutions

Many of the abovementioned social distancing scenarios involve tracking the positions and locations of people and objects with various levels of accuracy. For example, in the keeping distance scenario, positioning systems can be employed to determine the positions of people inside a room. Using this information, people can be alerted when the distances among them fall below a safe level. Another example is the application of positioning systems to monitor people in quarantines. To this end, wireless technologies with high coverage such as cellular can be employed to determine if a person has left the quarantine area or not.

Figure 1.6 illustrates the process of a typical positioning system [35]. Generally, the aim of such systems is to continuously determine an object's position in real-time [36]. To this end, first, signals (e.g., from sensors) are transmitted from the object to the receivers. Then, in the signal measurement phase, useful properties such as arrival time, signal strength, and signal direction are extracted from the received signals. Using these measured properties, the position of the target can then be determined by various calculation methods in the position calculation phase [35].

In the first phase, methods for measurement can be classified based on which property of the received signals is extracted. The first type is the time-based method which uses the signal's traveling time to determine the distance between the target and the receiving nodes. In the second type of method, namely *Angle-of-Arrival*, the angle of incoming signals can be measured by an array of antennas or directional antennas. The third type is the *Received Signal Strength Indication* method which measures the attenuation introduced during the propagation of the signal from the target to the receiving node [35].

Depending on the measured properties, different methods can be employed to calculate the object's position in the second phase. Among them, *Trilateration* is a

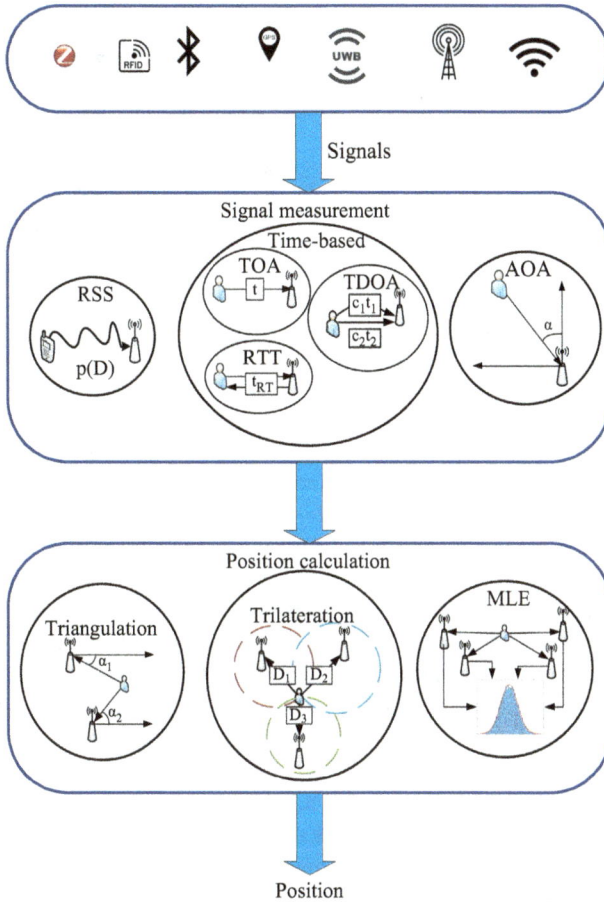

Figure 1.6   General process of positioning systems [8].

common method that calculates the target position based on its distances to three or more different nodes. This type of method is often employed when the signals are measured using time-based methods. When angles of signals are available (from using *Angle-of-Arrival*) instead of distances, the *Triangulation* can be employed to determine the target's position. The third type of method is *Maximum Likelihood Estimation* which is often employed to address the uncertainty in measurements.

1.2.2   AI-based social distancing solutions

1.2.2.1   Computer vision

Computer vision refers to the technology that trains computers to process and extract useful information from visual data, e.g., digital images or videos. With recent advances in AI, particularly in pattern recognition and deep learning, computer

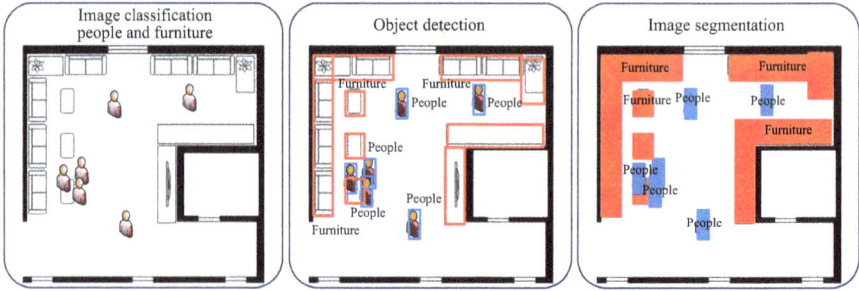

Figure 1.7   Illustrations of computer vision techniques

vision has played a key role in enabling computers to identify and classify objects accurately [37]. With the omnipresence of smartphones and surveillance cameras, computer vision is expected to play an important role in facilitating social distancing. For example, computer vision can enable a surveillance camera to not only detect close contacts among people but also determine if a person is wearing a face mask correctly or not.

As illustrated in Figure 1.7, computer vision can be roughly categorized into three types of techniques. The first one is image classification which refers to the task of automatically classifying images into different predefined classes. With the advent of deep learning, most recent works have been employing deep convolutional neural network (CNN) models for image classification [37].

The second type of computer vision technique is object detection, which aims to identify and localize the position of objects within a picture or video. Similar to the case of image classification, CNN models are the most commonly used technique for object detection. It is worth noting that object detection tasks are often more difficult than the image classification ones, since each object needs to be both identified and localized [37]. Nevertheless, this type of technique is more useful in many social distancing scenarios, such as detecting the distances among people.

Different from image classification and object detection, the main aim of the third type, namely image segmentation, is to partition an image into different segments. The goal of this type of technique is to simplify the image for further adjustment or analysis. Similar to the two other types of methods, CNN is also the most common technique for image segmentation [37].

1.2.2.2   Machine learning

The last 10 years have witnessed a remarkable growth of ML technologies, enabling their applications in many aspects of our lives such as economics, healthcare, automotive, and computer networks [38]. The core idea of ML is that it enables computers to automatically learn and extract useful information from available data. Based on that, important and challenging tasks in social distancing, such as predicting the disease spread, can be done.

Typically, ML techniques can be roughly divided into supervised learning, unsupervised learning, and reinforcement learning. Among them, the supervised learning techniques require data with labels to be available for training. By learning from these labeled data, the machine can then perform tasks such as classification, regression, and estimation [38]. For example, supervised learning is the most common technique used in computer vision for learning visual data. ML algorithms such as Bayesian networks, Support Vector Machine, Hidden Markov Model, Naïve Bayes, and many types of neural networks are often employed for supervised learning [38].

Unlike supervised learning, unsupervised learning techniques do not require the data for training to have labels. Typically, tasks such as clustering and prediction are associated with this type of technique. For example, predictions regarding the number of new cases can be made based solely on historical data, e.g., the number of cases in previous days. Thus, this type of task does not require labels. Algorithms such as K-means, Dirichlet process mixture model, Gaussian mixture model, X-means are commonly applied for unsupervised learning [38].

The third type, namely reinforcement learning, enables machines to learn from feedback received through interactions with an external environment, thereby gradually improving the interactions. Therefore, this type of technique is often employed for decision-making. Typical algorithms used for reinforcement learning are Q-learning, TD learning, R-learning, and Sarsa learning [38].

1.2.3   Other related technologies and research

1.2.3.1   Operations research

Operations research refers to the techniques that improve decision-making processes. Specifically, these techniques aim to find the best decisions to maximize or minimize certain objectives while subjecting to a set of constraints. Since such decision-making processes are ubiquitous, operations research techniques have numerous applications in many areas [39]. In the context of social distancing, these techniques are powerful tools in scenarios including traffic control, workforce planning, and healthcare scheduling.

There is a wide variety of methods applied in operations research, depending on how the problem is modeled and the priority of the decision-makers. For example, if the decisions are only Yes/No, Integer Programming is often employed. In contrast, if the decisions also involve determining a quantity, e.g., number of people allowed inside a building, Mixed integer programming can be employed. Moreover, the selection of methods for solving the problem also depends on the decision-maker priority and available resources. For example, in time-sensitive scenarios, approximate methods such as heuristics are often employed, which can find good (but not optimal) solutions in a limited time. On the other hand, if the optimality of the decision is more important, exact methods, e.g., branch-and-bound algorithm, can be used to find optimal solutions [39].

1.2.3.2   Incentive mechanisms

In large-scale social distancing scenarios such as contact tracing, the authorities' resources alone might be insufficient. In such contexts, incentive mechanisms can be employed to utilize other resources such as buildings' surveillance cameras or

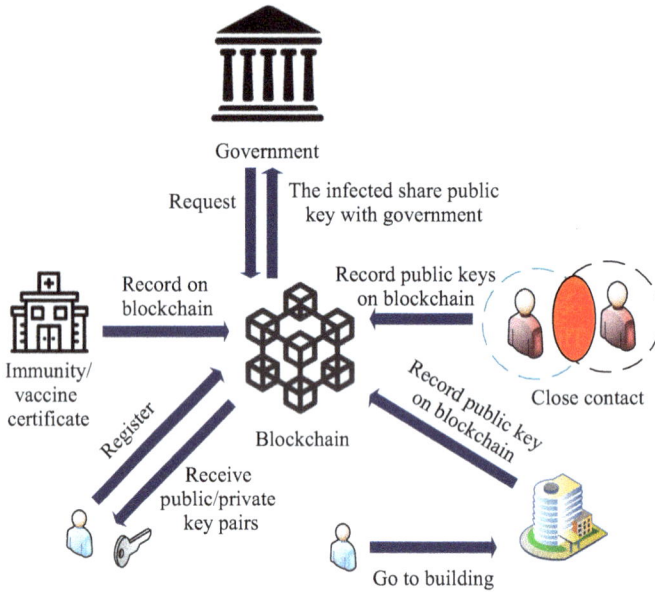

*Figure 1.8   Illustrations of blockchain-based applications for social distancing
practical social distancing scenarios*

people's smartphones. Particularly, individuals can contribute to social distancing implementations in exchange for some forms of incentives.

To design incentive mechanisms, techniques such as game theory and contract theory are often employed. Game theory techniques analyze the rational behaviors and interactions among participants in a system. Based on these analyses, the behaviors of the participants can be anticipated, and thus corresponding actions can be decided accordingly. For example, with the use of Stackelberg game theory, the authority can anticipate how people will react to different levels of incentives, and thus the appropriate level of incentive can be design [40]. Different from game theory, contract theory techniques focus more on how to design contracts such that the employees (people in the above example) are motivated to perform at the desired level [41]. For example, in contact tracing, the authorities can use contract theory to design higher rewards for people who share more of their contacts and movement information.

1.2.3.3   Blockchain

In social distancing scenarios such as contact tracing and infected movement tracking, privacy is one of the most important concerns. In such contexts, blockchain technology is a promising solution. A blockchain is a distributed ledger of information shared across participants in a decentralized peer-to-peer network. Data stored in a blockchain cannot be altered without the majority of the participants' agreement. Moreover, typical mechanisms employed in blockchain networks, such as public-private key pairs, hash functions, and digital signatures, can effectively protect blockchain users' anonymity and identity. Particularly, the public–private key mechanism ensures that the user's blockchain and real-life identities are separated. Moreover, the hash functions and digital signatures can be employed to ensure that private information can only be shared with the user's permission. Thus, as illustrated in Figure 1.8, blockchain technology can be employed to build a database for sensitive information, e.g., contacts, movement information, where users can decide what to share with whom in an anonymized manner.

References

[1]   World Health Organization: *Listings of WHO's response to COVID-19* [online]. 2021. Available from https://www.who.int/news/item/29-06-2020-covidtimeline.

[2]   World Health Organization: *WHO Coronavirus (COVID-19) dashboard* [online]. 2021. Available from https://covid19.who.int/.

[3]   Szmigiera M. Impact of the coronavirus pandemic on the global economy – Statistics & Facts [online]. Statista; 2021. Available from https://www.statista.com/topics/6139/covid-19-impact-on-the-global-economy/.

[4]   Ferguson NM, Cummings DA, Cauchemez S, *et al.* Strategies for containing an emerging influenza pandemic in Southeast Asia. Nature. 2005;**437**(7056): 209–14.

[5]   Ferguson NM, Cummings DA, Fraser C, *et al.* Strategies for mitigating an influenza pandemic. *Nature.* 2006;**442**(7101):448–52.

[6]   Fraser C, Riley S, Anderson RM, *et al.* Factors that make an infectious disease outbreak controllable. *Proceedings of the National Academy of Sciences.* 2004;**101**(16):6146–51.

[7]   European Centre for Disease Prevention and Control. *Considerations relating to social distancing measures in response to COVID-19–second update.* Stockholm: European Centre for Disease Prevention and Control; 2020.

[8]   Nguyen CT, Saputra YM, Van Huynh N, *et al.* A comprehensive survey of enabling and emerging technologies for social distancing—Part I: fundamentals and enabling technologies. *IEEE Access.* 2020;**8**:153479–507.

[9]   Department of Health, Australian Government: *Social distancing for coronavirus (COVID-19)* [online]. 2021. Available from https://www.health.gov.au/news/health-alerts/novel-coronavirus-2019-ncov-health-alert/how-to-protect-yourself-and-others-from-coronavirus-covid-19/physical-distancing-for-coronavirus-covid-19.

[10] Centers for Disease Control and Prevention: *Social distancing, quarantine, and isolation* [online]. 2021. Available from https://www.cdc.gov/coronavirus/2019-ncov/prevent-getting-sick/social-distancing.html.

[11] Public Health England: *Coronavirus: how to stay safe and help prevent the spread* [online]. 2021. Available from https://www.gov.uk/guidance/covid-19-coronavirus-restrictions-what-you-can-and-cannot-do.

[12]   Heffernan JM, Smith RJ, Wahl LM. Perspectives on the basic reproductive ratio. *Journal of the Royal Society Interface*. 2005;**2**(4):281–93.

[13]   Hethcote HW. The mathematics of infectious diseases. *SIAM Review*. 2000;**42**(4):599–653.

[14]   Mao L. Agent-based simulation for weekend-extension strategies to mitigate influenza outbreaks. *BMC Public Health*. 2011;**11**(1):1–10.

[15]   Kumar S, Grefenstette JJ, Galloway D, *et al*. Policies to reduce influenza in the workplace: impact assessments using an agent-based model. *American Journal of Public Health*. 2013;**103**(8):1406–11.

[16]   Timpka T, Eriksson H, Holm E, *et al*. Relevance of workplace social mixing during influenza pandemics: an experimental modeling study of workplace cultures. *Epidemiology and Infection*. 2016;**144**(10):2031–42.

[17]   Milne GJ, Kelso JK, Kelly HA, *et al*. A small community model for the transmission of infectious diseases: comparison of school closure as an intervention in individual-based models of an influenza pandemic. *PLoS One*. 2008;**3**(12):e4005.

[18]   Cauchemez S, Valleron AJ, Boelle PY, *et al*. Estimating the impact of school closure on influenza transmission from Sentinel data. *Nature*. 2008;**452**(7188):750–54.

[19]   Wu JT, Riley S, Fraser C, *et al*. Reducing the impact of the next influenza pandemic using household-based public health interventions. *PLoS Medicine*. 2006;**3**(9):e361.

[20]   Glass RJ, Glass LM, Beyeler WE, *et al*. Targeted social distancing designs for pandemic influenza. *Emerging Infectious Diseases*. 2006;**12**(11):1671.

[21]   Maharaj S, Kleczkowski A. Controlling epidemic spread by social distancing: Do it well or not at all. *BMC Public Health*. 2012;**12**(1):1–16.

[22]   Reluga TC. Game theory of social distancing in response to an epidemic. *PLoS Computational Biology*. 2010;**6**(5):e1000793.

[23]   Ackland GJ, Rice K, Wynne BM, *et al*. The long-term predictions from imperial college CovidSim Report 9. *medRxiv*. 2020.

[24]   Chowdhury R, Heng K, Shawon MSR, *et al*. Dynamic interventions to control COVID-19 pandemic: a multivariate prediction modeling study comparing 16 worldwide countries. *European Journal of Epidemiology*. 2020;**35**(5):389–99.

[25]   Dandekar R, Barbastathis G. Quantifying the effect of quarantine control in Covid-19 infectious spread using machine learning. *medRxiv*. 2020.

[26]   Bo Y, Guo C, Lin C, *et al*. Effectiveness of non-pharmaceutical interventions on COVID-19 transmission in 190 countries from 23 January to 13 April 2020. *International Journal of Infectious Diseases*. 2021;**102**:247–53.

[27]   Moosa IA. The effectiveness of social distancing in containing Covid-19. *Applied Economics*. 2020;**52**(58):6292–305.

[28]   Our World in Data: *Coronavirus (COVID-19) Cases* [online]. 2021. Available from https://ourworldindata.org/covid-cases.

[29]   Kelso JK, Milne GJ, Kelly H. Simulation suggests that rapid activation of social distancing can arrest epidemic development due to a novel strain of influenza. *BMC Public Health*. 2009;**9**(1):1–10.

[30] Caley P, Philp DJ, McCracken K. Quantifying social distancing arising from pandemic influenza. *Journal of the Royal Society Interface.* 2008;**5**(23): 631–9.

[31] Bootsma MC, Ferguson NM. The effect of public health measures on the 1918 influenza pandemic in US cities. *Proceedings of the National Academy of Sciences.* 2007;**104**(18):7588–93.

[32] Williams S. COVID-19: Zoom downloads explode as people work from home [online]. IT Brief Australia. 2020. Available from https://itbrief.com.au/story/covid-19-zoom-downloads-explode-as-people-work-from-home.

[33] Giles M. Microsoft Teams has seen a 775% rise in users in Italy because of COVID-19 [online]. *Forbes*; 2020. Available from https://itbrief.com.au/story/covid-19-zoom-downloads-explode-as-people-work-from-home.

[34] Nguyen CT, Saputra YM, Van Huynh N, *et al.* A comprehensive survey of enabling and emerging technologies for social distancing–Part II: emerging technologies and open issues. *IEEE Access.* 2020;**8**:154209–36.

[35] Zhang D, Xia F, Yang Z, *et al.* Localization technologies for indoor human tracking. *2010 5th international conference on future information technology.* IEEE; 2010. pp. 1–6.

[36] Ferreira AFGG, Fernandes DMA, Catarino AP, *et al.* Localization and positioning systems for emergency responders: a survey. *IEEE Communications Surveys and Tutorials.* 2017;**19**(4):2836–70.

[37] Szeliski R. *Computer vision: algorithms and applications.* Springer Science & Business Media; 2010.

[38] Ertel W. *Introduction to artificial intelligence.* Springer; 2018.

[39] Taha HA. *Operations research: an introduction.* Upper Saddle River, NJ: Pearson/Prentice Hall; 2011.

[40] Han Z, Niyato D, Saad W, *et al. Game theory in wireless and communication networks: theory, models, and applications.* Cambridge University Press; 2012.

[41] Bolton P, Dewatripont M, *et al. Contract theory.* MIT Press; 2005.

Chapter 2

Background on positioning and localization for social distancing

Le Chung Tran[1], Anh Tuyen Le[2], Xiaojing Huang[2], Son Lam Phung[1], Christian Ritz[1] and Abdesselam Bouzerdoum[3]

The recent breakout of coronavirus disease (COVID) has been requiring social distancing to mitigate its spread. Social distancing is a simple-but-effective way to reduce the human-to-human transmission of this deadly virus. A distance of at least 1.5 m between any two nearby people is typically required in many places. However, this social distancing requirement is not always respected, especially in space-confined places. Naturally, it is desired to have an autonomous system that can automatically detect the distance between nearby humans and warn people when the distance requirement is violated. To this end, the human positions are often required to be estimated. Thus, positioning and localization are important techniques to facilitate and enforce the social distancing requirement.

Generally, positioning and localization involve the problem of locating target (or agent) nodes, whose locations are unknown, based on the positions of some known anchors. Since the agents which need to be positioned are usually on the move, by its nature, wireless communications play an important role for navigation and surveillance purposes. Over one century of development, positioning was evolved from navigation aid systems for ships, airplanes, and terrestrial vehicles to many other applications, such as emergency search and rescues, military, location-based marketing, location-based services, tag finding to name a few. Therefore, naturally, wireless communications-based positioning is a promising solution for the social distancing measure to alleviate the outspread of the virus in the COVID-19 pandemic.

Localization schemes can be categorized into range-based and range-free techniques. Range-based localization requires some sort of bearing measurements, such as

[1]School of Electrical, Computer and Telecommunications Engineering, University of Wollongong (UOW), Wollongong, NSW, Australia
[2]School of Electrical and Data Engineering, University of Technology Sydney (UTS), Ultimo, NSW, Australia
[3]Information and Computing Technology Division, College of Science and Engineering, Hamad Bin Khalifa University, Doha, Qatar

the distances and/or angles from anchors to sensor nodes, to calculate the coordinates of the sensors. The distances from anchors to sensors can be estimated by numerous techniques, for example, time-of-arrival (TOA), time-difference-of-arrival (TDOA), receive signal strength indicator (RSSI), and round-trip-time (RTT), while the angles between anchors and sensors can be obtained through angle-of-departure (AOD) or angle-of-arrival (AOA) techniques. Range-free algorithms, however, use the connectivity information or the exchanged multi-hop information between sensors and anchors to acquire indirectly the distances between them. Then, the coordinates of the unknown nodes are estimated from the indirectly obtained distances [1]. Although the nodes in range-free positioning techniques are less complicated as no measurement is taken at those nodes, more nodes are normally required to increase signal coverage and connectivity. Furthermore, the accuracy of the range-free positioning network is poor, compared to the range-based methods. As a result, range-free positioning may not be suitable for social distancing applications. Hence, this chapter only discusses range-based positioning techniques, with the focus on the social distancing context.

The chapter is structured as follows. Section 2.1 provides some background of ranging methods and some most common localization algorithms. Section 2.2 briefly introduces the benchmark to evaluate the performance of the range-based algorithms. Section 2.3 provides some overviews of the global positioning systems (GPS). Sections 2.4 reviews different non-GPS positioning networks, such as cellular, WLAN, Bluetooth, and ultra-wideband (UWB). Finally, the applications of positioning to social distancing are discussed in Section 2.5.

2.1   Fundamental of positioning

2.1.1   Bearing measurement phase

Range-based localization requires two phases, namely the bearing (or range) measurement phase, which is discussed in this subsection, and the localization phase, which will be mentioned in Section 2.1.2. During the bearing measurement phase, either distance or angle (or both) from the target to the anchors is measured. This information is usually obtained by extra hardware deployed at network nodes, except for the case of the received signal strength measurement.

2.1.1.1   Received signal strength

One of the most common measurements to be used in wireless positioning systems is Received signal strength (RSS) due to its readiness and simplicity. In most receivers, the RSS indicator (RSSI) is included to show the quality of the received signal. RSSI is normally required in most wireless standards to support other radio functions, such as handover and resource management. RSSI is normally extracted at an automatic gain control (AGC) amplifier, where an amplitude detector is used to monitor the received signal. This means that RSSI can be obtained without synchronization between transmitters and receivers. It is ready for use without having to deploy extra hardware at the network nodes. Additionally, signal waveforms, modulation schemes, and bandwidth are all unnecessary to estimate a link distance from RSSI. As a result, measuring RSSI is relatively simple and cost-effective.

RSSI is used for the distance estimation due to the common assumption that the received signal power is determined by an exponential decay model. This model is in turn determined by the transmitted power, carrier frequency, propagation loss, small- and large-scale fading, antenna gains of the transmitter and the receiver, and their distance. The small-scale fading can be eliminated by averaging over some time periods or a range of frequencies. The received power, or RSSI (dBm), at the receiver from a distance d is modeled as:

$$P(d) = P(d_0) - 10\gamma \log_{10} \frac{d}{d_0} + n, \tag{2.1}$$

where $P(d_0)$ (dB) denotes the received power measured at the reference distance d_0, γ denotes the path loss exponent (PLE) whose value is typically in the range from two (in free space) to five (in a non-line-of-sight environment), or even six in some serious fading environments. $n \sim \mathcal{N}(0, \sigma)$ is a log-normal shadowing term caused by large-scale fading, which can be modeled as a Gaussian random variable with a zero mean and a variance of σ_n [2]. Since n is normally unknown, the distance d can be calculated from the RSSI as follows:

$$\hat{d} = d_0 10^{\frac{P(d_0)-P(d)}{10\gamma}}. \tag{2.2}$$

The reliability and accuracy of the RSS-based ranging estimation are impacted significantly by the random factors, including the surrounding environment, local network geometry, node positions, and the PLE accuracy. Bayesian filtering and Kalman filtering are often applied to the RSSI signal measurements to reduce the effects of multi-paths and measurement noises [3,4]. Examples of the PLE for different environments are provided in Table 2.1 [5,6]. However, the PLE can change quickly when obstacles present in the propagation paths. Therefore, a technique to track the PLE is required. For example, in [7], a particle Bayesian filter is adopted to

Table 2.1   Typical path loss exponent for various environments

Environment	Typical path-loss exponent
Free space	2
Urban cellular network	2.7–3.5
Shadowed urban cellular network	3–5
Line-of-sight inside a building	1.6–1.8
Non-line-of-sight inside building	4–6
Non-line-of-sight inside factory	2–3
Around human torso (3.1–10.6 GHz)	5.8–6
Along human torso (3.1–10.6 GHz)	3.1
Implant-to-body surface (402–405 MHz, near surface)	4.22
Implant-to-body surface (402–405 MHz, deep tissue)	4.26
Implant-to-implant (402–405 MHz, near surface)	4.99
Implant-to-implant (402–405 MHz, deep tissue)	6.29

estimate the PLE. Particularly, the PLE is considered as random samples or particles of the filter. First, N particles $\gamma_i, i = 1, \ldots, N$, are randomly created a Gaussian distribution over the range [0, 5]. These samples are used to calculate the perceptual model $p(z(t)|\gamma_i(t))$ defined as

$$p(z(t)|\gamma_i(t)) = \frac{1}{\sigma\sqrt{2\pi}} \exp\left(-\frac{[z(t) - P_{r_i}(d)]^2}{2\sigma^2}\right), \tag{2.3}$$

where $\gamma_i(t)$ denotes the state of the ith particle at a given time t, $z(t)$ denotes the observed RSSI value, $P_{r_i}(d)$ is the RSSI calculated by the propagation model (2.1) using $\gamma_i(t)$, σ is the standard deviation obtained from measurements. Then the weights $w_i(t)$ of the filter can be updated as follows

$$w_i(t) = \bar{w}_i(t-1)p(z(t)|\gamma_i(t)), \tag{2.4}$$

where $\bar{w}_i(t)$ is the normalized weight calculated by $\bar{w}_i(t) = w_i(t)/\sum_{i=1}^N w_i(t)$. Finally, the PLE is computed as

$$\gamma(t) = \sum_{i=1}^N \bar{w}_i(t)\gamma_i(t). \tag{2.5}$$

Even though an appropriate PLE estimation is obtained, RSSI-based ranging may still be inaccurate due to the shadowing effects caused by severe multi-path fading and due to a non-line-of-sight environment. Thus, further non-line-of-sight (NLOS) mitigation algorithms, such as [8], or ultra-wideband-based ranging techniques, e.g., [9], could be adopted to improve accuracy. Alternatively, an appropriate combination of RSS-based ranging estimations and other measurements, such as AOA and TOA, could be used.

2.1.1.2 Angle-of-arrival

An important estimation to support beamforming and localization is angle-of-arrival (AOA). A uniform linear antenna array is often used at the receiver to estimate the AOA of one or more incoming signals. Each antenna element in the array is connected to a radio frequency (RF) chain and a signal processing algorithm is needed to calculate the AOA from the signals received by the array. It means that AOA estimations are complicated, compared to RSSI and time-of-arrival (TOA) methods (TOA will be mentioned in the next section). However, the AOA technique has the advantage that only two anchor nodes which perform AOA estimations are sufficient to locate a target moving in the same plane containing these two anchors, while distance-based localization methods require at least three anchors. In addition, AOA is more accurate in in-homogeneous environments, such as water, than TOA and RSSI because, in such environments, transmission speed and propagation loss vary over different water layers.

There exist many methods to estimate the AOA, including the multiple signal classification (MUSIC) technique. This technique first introduced in [10] is the most popular one due to its simplicity. Assuming the received signals at the receiver antenna array comprising of M antenna elements at the time sample t are

$$\mathbf{Y}(t) = \mathbf{A}(\theta)\mathbf{X}(t) + \mathbf{N}(t), \tag{2.6}$$

where $\mathbf{A}(\theta) = [\mathbf{a}(\theta_1), \ldots, \mathbf{a}(\theta_K)]$ denotes the steering matrix, $\mathbf{a}(\theta_k)$ denotes the steering vector with respect to the angle of arrival θ_k of the kth incoming signal ($k = 1, \ldots, K$), i.e., $\mathbf{a}(\theta_k) = [1, \ e^{j(kd_a \sin(\theta_k))}, \ldots, e^{j((M-1)kd_a \sin(\theta_k))}]^T$. d_a denotes the distance between two nearby antenna elements. The time samples $t = 1, \ldots, N_S$, where N_S is the number of signal snapshots. $\mathbf{X}(t)$ denotes the transmitted signal matrix and $\mathbf{N}(t)$ denotes the matrix of additive, independent noise samples, which are assumed to follow the distribution $\mathcal{CN}(0, \sigma^2)$.

The MUSIC algorithm works by taking the correlation of the received signals to estimate the AOA. In particular, the covariance matrix $\mathbf{R}_{yy}$ of $\mathbf{Y}(t)$ can be presented as

$$\mathbf{R}_{yy} = \mathbf{A}\mathbf{R}_{xx}\mathbf{A}^H + \sigma^2\mathbf{R}_{nn}, \tag{2.7}$$

where $(.)^H$ denotes the Hermitian transpose operation, $\mathbf{R}_{xx}$ is the covariance matrix of $\mathbf{X}(t)$, and $\mathbf{R}_{nn}$ is the covariance matrix of $\mathbf{N}(t)$ normalized by σ^2, i.e., $\mathbf{R}_{nn} = E\{\mathbf{N}\mathbf{N}^H\}/\sigma^2$. The notation $E\{.\}$ denotes the expectation operation. Since the noise samples are assumed to be independent, $\mathbf{R}_{nn}$ is an identity matrix $\mathbf{I}$. Note that $\mathbf{R}_{nn}$ might not an identity one in reality, thus this matrix might significantly affect the performance of the MUSIC algorithm as shown in [11].

The correlation matrix $\mathbf{R}_{yy}$ will be then decomposed using the eigenvalue decomposition with the eigenvalues being arranged in an ascending order. The last biggest K eigenvalues are then corresponding to the K incoming signals because these eigenvalues represent the signal powers. The first $M - K$ eigenvalues are corresponding to noises. Therefore, if $\mathbf{U} = [\mathbf{u}_1, \ldots, \mathbf{u}_{M-K}]$ denotes the matrix of the first $M - K$ eigenvectors, $\mathbf{U}$ will represent the noise space [11].

If the received signals are assumed to be independent of noise, then $\mathbf{U}$ and $\mathbf{a}(\theta_k)$ will be orthogonal, i.e., $\mathbf{a}(\theta_k)^H \mathbf{U}\mathbf{U}^H \mathbf{a}(\theta_k) = 0$ for each and every k value, $k = 1, \ldots, K$. In fact, $\mathbf{U}$ and $\mathbf{a}(\theta_k)$ might not be completely orthogonal because of errors. Nevertheless, $\mathbf{a}(\theta_k)^H \mathbf{U}\mathbf{U}^H \mathbf{a}(\theta_k)$ might still be close to zero. In essence, MUSIC is a searching algorithm to estimate the AOA values, which searches for the angles θ_k' that result in the local maxima of the following power spectrum

$$S(\theta_k') = \frac{1}{|\mathbf{a}(\theta_k')^H \mathbf{U}\mathbf{U}^H \mathbf{a}(\theta_k')|}, \tag{2.8}$$

where $S(\theta_k')$ denotes the power spectrum of θ_k' and $\theta_k' \in [-\pi/2, \pi/2]$.

Ideally, $\mathbf{R}_{nn} = \mathbf{I}$ as we assume that noise samples are independent of each others. In this case, the noise subspace and the signal subspace are easily distinguishable, thus the estimations of AOA are often highly accurate. In reality, the above assumption is not always the case, thus it is harder to separate clearly the two subspaces. Therefore, the MUSIC algorithm will be less accurate.

2.1.1.3   Time-of-arrival and time-difference-of-arrival

The distance between a transmitter and a receiver can also be estimated by measuring the TOA or time-difference-of-arrival (TDOA). Both TOA and TDOA methods are based on the time-of-flight (TOF) principle so that the propagation time can be converted to the distance information by multiplying with the propagation speed. For the TOA method, if the transmit time at the transmitter is t_0, and the receiver node i records the received signal at the time t_i, ignoring the noise and measurement errors, the ideal distance between the receiver (anchor node) and transmitter (target node) can be calculated as

$$d_i = c(t_i - t_0), \tag{2.9}$$

where c denotes the propagation speed of the signal, which is the speed of light for propagation in the air. Hence, the clocks of the transmitter and receiver must be synchronized for accurate TOA ranging.

The TDOA method can avoid the requirement of knowing the transmission time t_0 in TOA by using one more anchor node j and taking the difference of the arrival times at the node i and the node j with the assumption that the anchors are synchronized, i.e.,

$$\begin{aligned} d_i - d_j &= c(t_i - t_0) - c(t_j - t_0) \\ &= c(t_i - t_j). \end{aligned} \tag{2.10}$$

The disadvantage of the TDOA technique is that it requires more anchor nodes to perform different distance estimations.

A correlation-based TOF estimation architecture is shown in Figure 2.1 [12]. Assuming the TOF is τ, the received signal at the receiver, denoted as $r(t)$, is modeled as

$$r(t) = s(t - \tau) + n(t), \tag{2.11}$$

where $s(t)$ is the transmitted signal, and $n(t)$ is the receiver noise. To estimate τ, the correlation of $r(t)$ and a reference signal generated by delaying the known transmitted signal takes place. The decision block adjusts the delay amount $\hat{\tau}$ until the maximum output of the correlator is found, i.e.,

$$\tilde{\tau} = \underset{\hat{\tau}}{\operatorname{argmax}} \int r(t)s(t - \hat{\tau})dt. \tag{2.12}$$

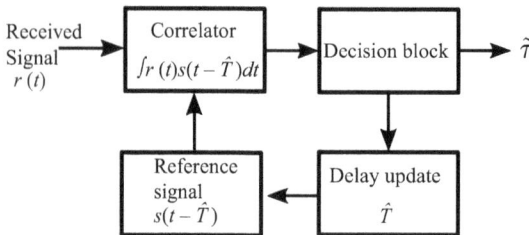

Figure 2.1   Correlation-based TOF estimation

While this technique is straightforward, its accuracy depends significantly on the channel. In severe multipath environments, e.g., urban areas and indoors, the late-arriving components decrease the SNR of the desired line-of-sight (LOS) signal. Hence, the receiver must find the first arriving maximum of the cross-correlation function rather than finding the global maximum peak since the strongest arriving signals may not be the LOS signal [13]. In addition, correlation-based TOF techniques cannot provide a high resolution when multipath signals present closely at the receiver. This is because its resolution depends on the bandwidth of the transmitted signal, which is in turn determined by the minimum pulse width. As an example, if an indoor communications system has a 200 MHz bandwidth, the distance estimation error is about 1.5 m if the LOS path is available [13]. Therefore, more complicated TOA estimation methods have been developed to provide a higher resolution and to mitigate noise as well as multipath effects. Some popular methods include deconvolution, maximum likelihood (ML), and subspace techniques. Deconvolution methods [14,15] convert both received and transmitted signals into the frequency domain, and the channel frequency response is obtained by dividing the corresponding frequency response of the received signal to that of the transmitted signal. ML algorithms [16,17] estimate the multipath channel coefficients and delays by minimizing the difference of the received signal and the convolution of the transmitted signal and the modeled channel. Sub-space methods [18] utilize the orthogonality of noise and signals to decompose the matrix of received signal samples into a signal subspace and a noise subspace. These methods are more complicated than the correlation-based techniques, especially for the case of subspace and ML techniques where matrix decomposition and exhausted search are required, respectively.

The aforementioned TOF estimation techniques are known as two-step methods, i.e., the clock drifts of all the nodes are firstly determined for synchronization before TOF is estimated. However, if clock synchronization performance is poor, the estimated ranging information suffers accelerated errors. Therefore, it has been shown in [19,20] that joint synchronization and ranging can significantly improve the accuracy over the two-step methods. In such joint synchronization and ranging approaches, clock skews and offsets as well as the locations of unknown nodes are estimated simultaneously based on timestamp exchanges. Furthermore, recent advances in interference mitigation technologies, such as in [21–28], have enabled full-duplex communications, which facilitate simultaneous transmission and reception in the same frequency band. A new synchronization and localization algorithm utilizing the advantages of full-duplex radios is proposed in [29,30]. In this scheme, localization and synchronization can be achieved over just two frames of transmission. Due to the full-duplex operation, in the first frame, all nodes simultaneously transmit their signature signals, while receiving signals from other nodes. In the next frame, scrambled versions of the previously received signals or channel parameters estimated based on the received signals will be transmitted. The arrival times of different components of the incoming signals will be then estimated, which allows to determine the clock offsets. Performance evaluations in [29] show that network localization and synchronization deploying full-duplex radios are more efficient than that in the half-duplex counterparts.

2.1.2 Localization phase

Localization phase is the second phase of range-based localization techniques. In the subsections below, some most popular techniques in the localization phase are discussed.

2.1.2.1 Lateration

Lateration methods, or true-range lateration methods, use the distances from the anchors to the target to determine the target's position. Assuming the network includes N anchors whose positions are known at $\mathbf{a}_n = [a_{ix}, a_{iy}]^T, i = 1, \ldots, N$, to locate a targets at $\mathbf{x} = [x_x, x_y]^T$. By using one of the ranging techniques, such as RSSI, TOA, and TDOA, the distances from the target to the anchors can be estimated. The target's position can be found by solving the following set of nonlinear equations of the variables x_x and x_y

$$(x_x - a_{ix})^2 + (x_y - a_{iy})^2 = d_i^2, \text{ for } i = 1, \ldots, N. \tag{2.13}$$

Equation (2.13) means that the target should be on the circle centered at the anchor i with the radius d_i. Therefore, (2.13) is also known as a *circular lateration*. The agent's position can be found by at least two anchors, using the bilateration technique [13], [11], where two circles centered at these two anchors with the distances d_i being their radii are plotted. Then the agent should be at of the two intersections of these two circles. Bilateration can only be used when some constraints are applied so that one intersection can be neglected. Otherwise, at least three anchors are required and the lateration is called trilateration as shown in Figure 2.2. If the distances are perfectly estimated, the intersection of the three circles is a certain point. However, due to the estimation errors caused by shadowing, noise, and multipath effects, the target is in the marked area determined by the intersections of the three circles. Obviously, the

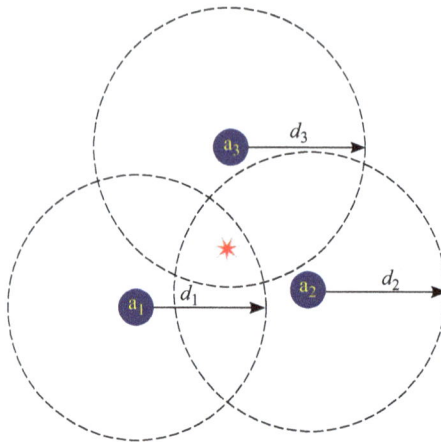

Figure 2.2   Trilateration method

higher the accuracy of ranging estimations is obtained, the smaller the size of this area is.

When the number of anchors $N \geq 3$, the nonlinear equations in (2.13) becomes over-determined. We can first transform them into linear equations, which can then be solved by a linear solution, such as LS and weighted least square (WLS). The latter will be mentioned in the next section. As presented in [31], without loss of generality, the rth anchor is chosen as a reference. Subtracting the other equations to the rth equation, the following set of $(N-1)$ linear equations can be achieved

$$(a_{ix} - a_{rx})x_x + (a_{iy} - a_{ry})x_y = \frac{1}{2}\left[a_{ix}^2 + a_{iy}^2 - a_{rx}^2 - a_{ry}^2 - d_i^2 + d_r^2\right], \qquad (2.14)$$

for $i = 1, \ldots, N$, $i \neq r$.

Equation (2.14) is represented in a matrix form as

$$\mathbf{Hx} = \mathbf{b}, \qquad (2.15)$$

where

$$\mathbf{H} = \begin{bmatrix} a_{1x} - a_{rx} & a_{1y} - a_{ry} \\ \vdots & \vdots \\ a_{Nx} - a_{rx} & a_{Ny} - a_{ry} \end{bmatrix}, \mathbf{x} = \begin{bmatrix} x_x \\ x_y \end{bmatrix}, \mathbf{b} = \frac{1}{2}\begin{bmatrix} K_1^2 - K_r^2 - d_1^2 + d_r^2 \\ \vdots \\ K_N^2 - K_r^2 - d_N^2 + d_r^2 \end{bmatrix},$$

and $K_i^2 = a_{ix}^2 + a_{iy}^2$, $i = 1, \ldots, N$.

The least square (LS) solution attempts to minimize the square of the Frobenius norm (or Euclidean norm), i.e.,

$$||\mathbf{Hx} - \mathbf{b}||^2 \rightarrow \min. \qquad (2.16)$$

Equation (2.16) is minimized when its first derivative over $\mathbf{x}$ is zero. After some simple mathematical manipulations, the LS solution is found as

$$\hat{\mathbf{x}} = (\mathbf{H}^T\mathbf{H})^{-1}\mathbf{H}^T\mathbf{b}. \qquad (2.17)$$

This means that if we know the coordinates of the anchors and the estimated distances from the agent to these anchors, which have been retrieved in the bearing measurement phase mentioned in Section 2.1.1, the position of the agent can be estimated. Knowing the agent's position will help reinforce the social distancing.

2.1.2.2 Weighted least square solution

The solution in (2.17) treats different distance estimates in the same manner. However, in fact, a longer distance tends to cause a larger error in both TOA and RSSI measurements because of the lower SNR and multipath effects [32]. Therefore, the WLS algorithm attempts to minimize the square of the Frobenius norm (or Euclidean norm)

$$\sum_{i=1, i \neq r}^{N} w_i |(H_{i,1}x_x + H_{i,2}x_y) - b_i|^2 = \left\|\mathbf{W}^{1/2}(\mathbf{Hx} - \mathbf{b})\right\|^2 \rightarrow \min, \qquad (2.18)$$

where $\mathbf{W}$ is an $(N-1) \times (N-1)$-sized diagonal matrix of real weighting coefficients w_i, for $i = 1, \ldots, N$ and $i \neq r$, $H_{i,1}$ and $H_{i,2}$ are the $(i, 1)$ and $(i, 2)$ entries of the matrix $\mathbf{H}$, and b_i is the ith entry of the vector $\mathbf{b}$. This weighting matrix is chosen in a way that shorter estimated distances will be weighted more than the longer ones.

Similarly to the LS solution, the WLS solution is found as

$$\hat{\mathbf{x}} = (\mathbf{H}^T \mathbf{W} \mathbf{H})^{-1} \mathbf{H}^T \mathbf{W} \mathbf{b}. \tag{2.19}$$

There are different ways to determine the weighting matrix $\mathbf{W}$. For example, in [32], $\mathbf{W}$ is simply defined by the estimated distances as

$$\mathbf{W} = \text{diag}(\sqrt{w_1}, \ldots, \sqrt{w_N}), \tag{2.20}$$

where $w_i = 1 - \frac{d_i}{\sum_{i=1}^{N} d_i}$. Another way proposed in [33] is based on the covariance matrix $\mathbf{R}_{bb}$ of the vector $\mathbf{b}$ as

$$\mathbf{W}^T \mathbf{W} = \mathbf{R}_{bb}^{-1}. \tag{2.21}$$

$\mathbf{R}_{bb}$ is defined as $\mathbf{R}_{bb} = E\{(\mathbf{b} - E\{\mathbf{b}\})(\mathbf{b} - E\{\mathbf{b}\})^T\}$. Note that $\mathbf{b}$ is a non-zero mean vector. This means that the decision metric in (2.19) is biased.

Assuming that the estimated distances d_i, for $i = 1, 2, \ldots, N$, from the target to the N anchor nodes are independent of each other, we can easily work out

$$\mathbf{R}_{bb} = \frac{1}{4} \begin{bmatrix} \text{Var}\left(d_1^2\right) + \text{Var}\left(d_r^2\right) & \text{Var}\left(d_r^2\right) & \ldots & \text{Var}\left(d_r^2\right) \\ \text{Var}\left(d_r^2\right) & \text{Var}\left(d_2^2\right) + \text{Var}\left(d_r^2\right) & \ldots & \text{Var}\left(d_r^2\right) \\ \ldots & \ldots & \ldots & \ldots \\ \text{Var}\left(d_r^2\right) & \text{Var}\left(d_r^2\right) & \ldots & \text{Var}\left(d_N^2\right) + \text{Var}\left(d_r^2\right) \end{bmatrix},$$

where Var(.) denotes the variance value. When RSSI is used and the channels are log-normal as shown in (2.1), the variance Var(d_i^2) is found in [33] as

$$\text{Var}\left(d_i^2\right) = E\left\{\hat{d}_i^4\right\} - \left(E\left\{\hat{d}_i^2\right\}\right)^2 = e^{4\mu_d}\left(e^{8\sigma_d^2} - e^{4\sigma_d^2}\right), \tag{2.22}$$

where $\mu_d = \ln(d_i)$, $\sigma_d^2 = \sigma^2 \left(\frac{\ln 10}{10\gamma}\right)^2$, γ is the PLE, and σ is the variance of the measurement noise n in (2.1).

Obviously, the WLS algorithm with the covariance matrix would be more accurate than that with the simple weighting coefficient matrix in (2.20). However, the former requires multiple measurements and the estimation of noise variances to be done in advance. Depending on the required accuracy and the required latency in social distancing applications, one might be preferred over the other.

2.1.2.3   Nonlinear approaches

Nonlinear methods find the agent's position by searching the optimized $\hat{\mathbf{x}}$ that minimizes the cost function, denoted as $J(\tilde{\mathbf{x}})$, i.e.,

$$\hat{\mathbf{x}} = \arg\min_{\tilde{\mathbf{x}}} J(\tilde{\mathbf{x}}). \tag{2.23}$$

The cost function is defined as $J(\tilde{\mathbf{x}}) = \sum_{i=1}^{N} (d_i - f(\tilde{\mathbf{x}}))^2$, where $f(\tilde{\mathbf{x}})$ is the ranging function, which is defined by the estimation methods. The functions $f(\tilde{\mathbf{x}})$ for the i-th anchor in the TOA, TDOA, RSS, and AOA ranging-based methods, denoted as $f_{TOA}(\tilde{\mathbf{x}})$, $f_{TDOA}(\tilde{\mathbf{x}})$, $f_{RSS}(\tilde{\mathbf{x}})$, and $f_{AOA}(\tilde{\mathbf{x}})$ respectively, are given below [13]

$$f_{TOA}(\tilde{\mathbf{x}}) = \sqrt{(\tilde{x}_x - a_{ix})^2 + (\tilde{x}_y - a_{iy})^2},$$

$$f_{TDOA}(\tilde{\mathbf{x}}) = \sqrt{(\tilde{x}_x - a_{ix})^2 + (\tilde{x}_y - a_{iy})^2} - \sqrt{(\tilde{x}_x - a_{1x})^2 + (\tilde{x}_y - a_{1y})^2},$$

$$f_{RSS}(\tilde{\mathbf{x}}) = -10\gamma \log_{10}\left(\sqrt{(\tilde{x}_x - a_{ix})^2 + (\tilde{x}_y - a_{iy})^2}\right),$$

$$f_{AOA}(\tilde{\mathbf{x}}) = \tan^{-1}\left(\frac{\tilde{x}_y - a_{1y}}{\tilde{x}_x - a_{1x}}\right). \tag{2.24}$$

Finding the optimum $\hat{\mathbf{x}}$ is challenging because the two-dimensional surface of $J(\tilde{\mathbf{x}})$ may have both local minima and the global minimum. Therefore, optimization algorithms can be used to ensure a global minimum. However, the global search techniques require higher complexity and have slow convergence. Hence, local optimization techniques which use the iterative procedure with an initial position of $\hat{\mathbf{x}}^0$ can possibly find $\hat{\mathbf{x}}$ quickly if $\hat{\mathbf{x}}^0$ is close to the real position $\mathbf{x}$.

Among different local search techniques, the steepest descent method is the simplest, and its iterative procedure at the kth iteration is expressed

$$\hat{\mathbf{x}}^{k+1} = \hat{\mathbf{x}}^k - \mu \nabla(J(\hat{\mathbf{x}}^k)), \tag{2.25}$$

where μ is the step size, which regulates the convergence speed of the algorithm, and $\nabla(J(\hat{\mathbf{x}}^k)) = \begin{bmatrix} \frac{\partial J(\hat{\mathbf{x}}^k)}{\partial x_x} \\ \frac{\partial J(\hat{\mathbf{x}}^k)}{\partial x_y} \end{bmatrix}$ is the gradient vector of $J(\tilde{\mathbf{x}})$ computed at $\hat{\mathbf{x}}^k$. Readers can refer to Section 2.2 for more details of the partial derivatives $\frac{\partial J}{\partial x_x}$ and $\frac{\partial J}{\partial x_y}$ in different ranging methods.

Intuitively, the nonlinear approaches are expected to be more accurate than the linear ones, since they aim to find via iterations the optimal position of the agent to best match the estimated distances d_i, with the cost of more processing complexity and a higher delay.

2.1.2.4   Subspace localization

The core idea of this linear method is that the agent's position can be detected from the eigenvectors of the signal subspace, thus the noise subspace eigenvectors can be removed before the position estimation to improve the accuracy.

Ranging-based subspace localization [34,35] is presented as follows. Consider a wireless sensor network (WSN) where all anchors are assumed to know the positions of one another. Denote $\mathbf{X}$ as the following $N \times 2$ matrix

$$\mathbf{X} = \begin{bmatrix} a_{1x} - x_x & a_{1y} - x_y \\ \vdots & \vdots \\ a_{Nx} - x_x & a_{Ny} - x_y \end{bmatrix}, \tag{2.26}$$

and the matrix $\mathbf{D}$ as

$$\mathbf{D} = \mathbf{X}\mathbf{X}^T. \tag{2.27}$$

Note that the ranks of $\mathbf{X}$ and $\mathbf{D}$ are 2. The $D_{i,k}$ element of $\mathbf{D}$ is

$$D_{i,k} = (a_{ix} - x_x)(a_{kx} - x_x) + (a_{iy} - x_y)(a_{ky} - x_y) = \frac{1}{2}(d_i^2 + d_k^2 - d_{ik}^2), \tag{2.28}$$

where d_i (d_k, respectively) is the true distance between the ith (kth) anchor to the mth agent, and $d_{ik}^2 = \|\mathbf{a}_i - \mathbf{a}_k\|^2$, for $i, k = 1, \ldots, N$. Using the eigenvalue decomposition, $\mathbf{D}$ can be presented as

$$\mathbf{D} = \mathbf{U}\mathbf{\Lambda}\mathbf{U}^T, \tag{2.29}$$

where $\mathbf{\Lambda}$ is a diagonal matrix of the eigenvalues of $\mathbf{D}$, i.e., $\mathbf{\Lambda} = \mathrm{diag}\{\lambda_1, \lambda_2, \ldots, \lambda_N\}$ with $\lambda_1 \geq \lambda_2 \geq \ldots \geq \lambda_N$. The columns of the orthonomal matrix $\mathbf{U}$ are the eigenvectors $\mathbf{u}_n$, $n = 1, \ldots, N$. Because $\mathbf{D}$ has the rank of 2, $\lambda_3 = \lambda_4 = \ldots = \lambda_N = 0$.

Denote $\mathbf{U}_s = [\mathbf{u}_1 \ \mathbf{u}_2]$ and $\mathbf{I}_N$ as the identity matrix of order N. It has been proved in [34,35] that

$$\mathbf{X} = \mathbf{U}_s\mathbf{U}_s^T\mathbf{X}. \tag{2.30}$$

Equation (2.30) reveals that the agent's position $\mathbf{x}$ can be estimated from the eigenvectors corresponding to the signal subspace.

Because $\mathbf{D}$ is in fact not available, its approximate matrix $\hat{\mathbf{D}}$ can be constructed with the $\hat{D}_{i,k}$ element being calculated as

$$\hat{D}_{i,k} = \frac{1}{2}(r_i^2 + r_k^2 - d_{ik}^2), \tag{2.31}$$

where r_i (r_k, respectively) is the estimated distance of d_i (d_k). Hence, (2.30) now becomes an approximation. Thus, we have

$$\mathbf{X} \approx \hat{\mathbf{U}}_s\hat{\mathbf{U}}_s^T\mathbf{X}, \tag{2.32}$$

where $\hat{\mathbf{U}}_s$ represents the signal subspace of $\hat{\mathbf{D}}$. (2.32) indicates that the subspace method suppresses the noise subspace before estimating the positions of the agents.

From (2.26), we can rewrite $\mathbf{X}$ as $\mathbf{X} = \mathbf{A} - \mathbf{1}_N\mathbf{x}^T$, where $\mathbf{A} = \begin{bmatrix} a_{1x} & a_{1y} \\ \vdots & \vdots \\ a_{Nx} & a_{Ny} \end{bmatrix}$, and

$\mathbf{1}_N$ denotes a column vector of N ones. Substituting this expression to (2.32) and rearranging it, we have

$$(\mathbf{I}_N - \hat{\mathbf{U}}_s\hat{\mathbf{U}}_s^T)\mathbf{1}_N\mathbf{x}^T \approx (\mathbf{I}_N - \hat{\mathbf{U}}_s\hat{\mathbf{U}}_s^T)\mathbf{A}. \tag{2.33}$$

Note that $\mathbf{I}_N - \hat{\mathbf{U}}_s\hat{\mathbf{U}}_s^T = \hat{\mathbf{U}}_n\hat{\mathbf{U}}_n^T$, where $\hat{\mathbf{U}}_n$ denotes the noise subspace of $\hat{\mathbf{D}}$. Applying the LS solution to (2.33), the agent's position can be estimated as

$$\hat{\mathbf{x}} = \left[\frac{\mathbf{1}_N^T\hat{\mathbf{U}}_n\hat{\mathbf{U}}_n^T\mathbf{A}}{\mathbf{1}_N^T\hat{\mathbf{U}}_n\hat{\mathbf{U}}_n^T\mathbf{1}_N} \right]^T. \tag{2.34}$$

To demonstrate the performance of the different lateration algorithms presented above, a positioning system is simulated as follows. Five anchor nodes are deployed in an 100 m × 100 m-sized area at known locations of (0,0), (0,100), (100,0), (100,100), and (50,50), while the unknown target locates at (40,70). The ranges estimated by the anchors have the same variance of errors σ^2 (dB). The mean square errors (MSEs) of these algorithms are calculated over 1000 independent runs, i.e., $MSE = \frac{1}{1000} \sum_{i=1}^{1000} \left[(\hat{x}_x - x_x)^2 + (\hat{x}_y - x_y)^2 \right]$. The steepest descent algorithm is simulated with $\mu = 0.1$.

The MSEs of these algorithms are shown in Figure 2.3. In Figure 2.3, "LS" stands for the least square solution in (2.17); "WLS" and "WLS2" are the WLS in (2.19) in which the weighting coefficient matrices are based on the estimated distances (cf. (2.20)) and the covariance errors (cf. (2.21)), respectively; "NLS Steepest Descend" is the non-linear least square solution using the steepest descend local search algorithm in (2.25), and "Subspace" is the subspace solution in (2.34). We can see that, under the same levels of estimation errors, WLS2 has the best performance, while LS is the worst. NLS Steepest descend outperforms the subspace and LS algorithms but the optimized solution is not guaranteed. LS and WLS are the simplest methods, while WLS2 requires a complicated procedure to acquire the covariance of the estimation errors.

2.1.3 Angulation

Angulation algorithms use the estimated AOAs at the anchors to locate the target. In particular, by drawing a baseline started at the anchor position and the AOA as the slope, the target should lie somewhere on this line. Therefore, with as few as two anchors, the target's location is the intersection of the two lines (cf. Figure 2.4(a)).

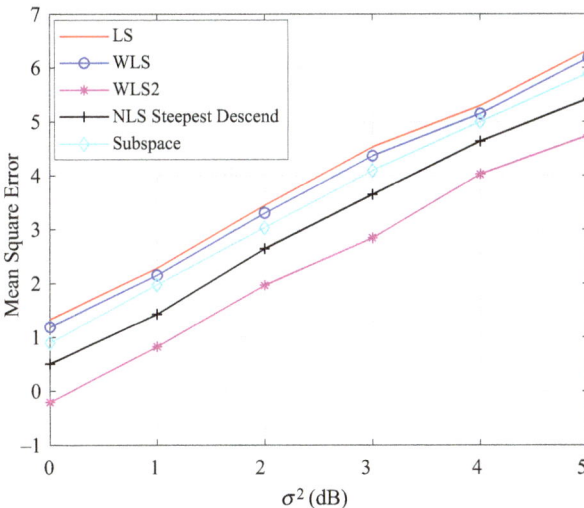

Figure 2.3   Comparison of different lateration localization algorithms

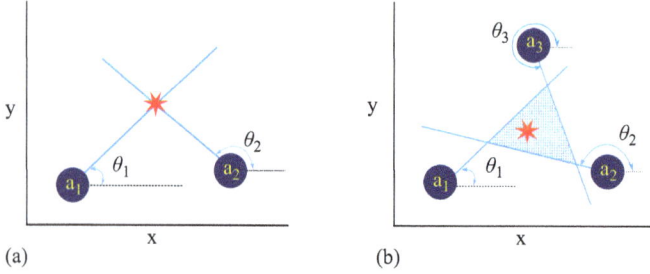

Figure 2.4   Angulation localization. (a) Biangulation and (b) Triangulation

As shown in Figure 2.4(a), the relationship between the DOA estimated by the target's position and the ith anchor is presented as

$$\tan \theta_i = \frac{x_y - a_{iy}}{x_x - a_{ix}}, \quad i = 1, \ldots, N. \tag{2.35}$$

The angulation solution in case of two anchors is given as

$$
\begin{aligned}
\hat{x}_x &= \frac{a_{2x} \tan \theta_2 - a_{1x} \tan \theta_1 - (a_{2y} - a_{1y})}{\tan \theta_2 - \tan \theta_1}, \\
\hat{x}_y &= \frac{(a_{2x} - a_{1x}) \tan \theta_1 \tan \theta_2 + (a_{1y} \tan \theta_2 - a_{2y} \tan \theta_1)}{\tan \theta_2 - \tan \theta_1}.
\end{aligned} \tag{2.36}
$$

From (2.36), it is easy to realize the following relations

$$
\begin{aligned}
\tan \theta_1 \hat{x}_x - \hat{x}_y &= a_{1x} \tan \theta_1 - a_{1y}, \\
\tan \theta_2 \hat{x}_x - \hat{x}_y &= a_{2x} \tan \theta_2 - a_{2y}.
\end{aligned} \tag{2.37}
$$

This equation can be presented in a matrix form as

$$\mathbf{H}_{aoa}\mathbf{x} = \mathbf{b}_{aoa}, \tag{2.38}$$

where $\mathbf{H}_{aoa} = \begin{bmatrix} \tan \theta_1 & -1 \\ \tan \theta_2 & -1 \end{bmatrix}$, and $\mathbf{b}_{aoa} = \begin{bmatrix} a_{1x} \tan \theta_1 - a_{1y} \\ a_{2x} \tan \theta_2 - a_{2y} \end{bmatrix}$.

DOA estimations from more anchors will generally provide better accuracy of the position of the target. When three or more anchors are available, the target is limited in the area defining by three or more lines drawn from the anchors as shown in Figure 2.4(b). The marked area in Figure 2.4(b) becomes smaller. For N anchors, $N \geq 3$, we have an over-determined problem

$$\mathbf{H}_{aoa}\mathbf{x} = \mathbf{b}_{aoa}, \tag{2.39}$$

where $\mathbf{H}_{aoa} = \begin{bmatrix} \tan \theta_1 & -1 \\ \vdots & \vdots \\ \tan \theta_N & -1 \end{bmatrix}$, and $\mathbf{b}_{aoa} = \begin{bmatrix} a_{1x} \tan \theta_1 - a_{1y} \\ \vdots \\ a_{Nx} \tan \theta_N - a_{Ny} \end{bmatrix}$. Since (2.39) is a linear

equation, the LS method can be applied directly. Thus the solution is obtained as

$$\hat{\mathbf{x}} = (\mathbf{H}_{aoa}^T \mathbf{H}_{aoa})^{-1} \mathbf{H}_{aoa}^T \mathbf{b}_{aoa}. \tag{2.40}$$

Clearly, if the coordinates of the anchors and the estimated angle-of-arrival values at these anchors are known, the position of the agent can be calculated.

2.1.4 Hybrid angulation and lateration

Due to measurement errors, standalone lateration and angulation might not be able to locate the target efficiently. The localization accuracy can be significantly improved by the combination of distance measurements and AOA measurements. Figure 2.5 shows that the combination of RSSI and AOA information will significantly narrow down, the possible position of the agent indicated by the red asterisk as shown by the shaded area. This shaded area is smaller than the area of the target's possible position determined separately by either the lateration (three circles) or the angulation (three lines).

With both ranging and AOA information available at each anchor as shown in Figure 2.5, we have

$$x_x = a_{ix} + d_i \cos(\theta_i),$$
$$x_y = a_{iy} + d_i \sin(\theta_i), i = 1, \ldots, N. \tag{2.41}$$

Obviously, with only one anchor node, the position of the target can still be found. When more anchor nodes are available, the over-determined problem can be solved by the LS method as [36]

$$\hat{\mathbf{x}}^{nAnR} = (\mathbf{S}^T\mathbf{S})^{-1}\mathbf{S}^T\mathbf{u}, \tag{2.42}$$

where $\hat{\mathbf{x}}^{nAnR}$ denotes the hybrid AOA/RSSI solution where all N AOA and N distance estimations measured by N anchors are used to predict the position of the target; $\mathbf{S} = \begin{bmatrix} \mathbf{1}_N & \mathbf{0}_N \\ \mathbf{0}_N & \mathbf{1}_N \end{bmatrix}$; $\mathbf{1}_N$ and $\mathbf{0}_N$ denote the $N \times 1$-column vectors of all ones and all zeros, respectively; $\mathbf{u} = [a_{1x} + d_1 \cos(\theta_1), \ldots, a_{Nx} + d_N \cos(\theta_N), a_{1y} + d_1 \sin(\theta_1), \ldots, a_{Ny} + d_N \sin(\theta_N)]^T$.

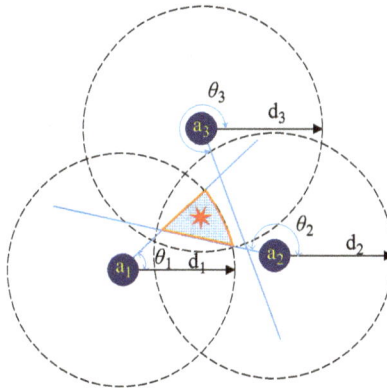

Figure 2.5 Hybrid angulation and lateration

The accuracy of the LS solution can be enhanced by using a weight $w_i = 1 - \frac{d_i}{\sum_{i=1}^{N} d_i}$ for the i-link, $i = 1, \ldots, N$, between the i-anchor and the target [32]. This rationale is that a shorter link often has a smaller error than the longer one. Hence, the WLS equation can be presented as

$$\hat{\mathbf{x}}^{nAnR} = (\mathbf{S}^T \mathbf{W}^T \mathbf{S})^{-1} \mathbf{S}^T \mathbf{W}^T \mathbf{u}, \tag{2.43}$$

where $\mathbf{W} = \mathbf{I}_2 \otimes \mathrm{diag}\{\mathbf{w}\}$, $\mathbf{I}_2$ denotes a 2×2-identity matrix, $\mathbf{w} = \left[\sqrt{w_1}, \ldots, \sqrt{w_N}\right]$, and $\otimes$ denotes the Kronecker operation. The solution in (2.43), denoted as "*nAnR-WLS*" in this chapter, is considered as a benchmark thanks to its simplicity and relatively high accuracy.

Since AOA estimations require an antenna array and signal processing to be equipped at each anchor, conventional hybrid angulation and lateration algorithms lead to a complicated positioning network. To reduce the system complexity, [11] and [37] propose an unbalanced hybrid AOA and RSSI method in which the number of AOA estimations and that of RSSI are unequal. For example, the unbalanced hybrid 1AnR localization technique utilizes one AOA estimation and N RSSI estimations to predict the position of the target. This means that only one anchor, referred to as the master anchor, is required to have an antenna array.

As mentioned in [37], because only one master anchor (e.g., $\mathbf{a}_1$) has both AOA and RSSI estimations while other anchors have RSSI measurements only, the set of N linear equations in (2.41) is no longer available. Instead, it is the mixture of both linear and nonlinear equations. To solve this system of equations, the master anchor is transformed into two virtual anchors $\mathbf{a}_{v1}, \mathbf{a}_{v2}$ with the ranging estimations $d_{v1} = d_1 \sin \theta_1$ and $d_{v2} = d_1 \cos \theta_1$ as shown in Figure 2.6.

Thereby, a new set of $N + 1$ nonlinear equations for $N + 1$ anchors, including the two virtual anchors, based on the RSSI approach is obtained, which can be solved by the LS approach, denoted as 1AnR-LS (cf. (2.17), or the subspace method, denoted as 1AnR-Subspace (cf. (2.34)), as shown in Subsection 2.1.2.1. Simulation results show that the unbalanced hybrid method can achieve equivalent or better accuracy

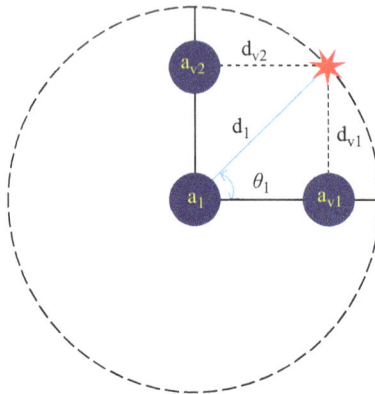

Figure 2.6   Transforming the master anchor into two virtual anchors

than the conventional method when the RSSI estimation errors are small. Readers may refer to [11] and [37] for the more comprehensive performance analyses.

2.2 Performance evaluations and analysis

A popular benchmark to evaluate the accuracy of different localization algorithms is the root-mean square error (RMSE), which represents the deviation of the estimated positions to the real ones. Mathematically, RMSE is calculated by

$$RMSE = \sqrt{E\left\{\frac{1}{M}\sum_{m=1}^{M}(x_{mx} - \hat{x}_{mx})^2 + (x_{my} - \hat{x}_{my})^2\right\}},$$

where $E\{.\}$ is the expectation over M Monte Carlo simulations, and $(x_{mx} - \hat{x}_{mx})^2 + (x_{my} - \hat{x}_{my})^2$ is the square of the distance from the estimated location to the real location of the mth agent. Another benchmark to compare different unbiased estimators is the Cramer-Rao lower bound (CRLB), which provides a theoretical bound that an estimator can ultimately achieve. To calculate CRLB, one needs to derive the Fisher information matrix (FIM), denoted by $\mathbf{I}(\mathbf{x})$. Assuming the estimator has the measurement vector $\boldsymbol{\theta}$, defined as

$$\boldsymbol{\theta} = \mathbf{f}(\mathbf{x}) + \mathbf{n}, \tag{2.44}$$

where $\mathbf{f}(.)$ is a ranging estimation function and $\mathbf{n}$ is an additive zero-mean noise vector. $\mathbf{I}(\mathbf{x})$ is calculated by

$$\mathbf{I}(\mathbf{x}) = E\left\{\frac{\partial^2 \ln p(\boldsymbol{\theta})}{\partial \mathbf{x} \partial \mathbf{x}^T}\right\}, \tag{2.45}$$

where $p(\boldsymbol{\theta})$ is the probability density function of $\boldsymbol{\theta}$ and ln is the natural logarithm. The first derivative term $\frac{\partial \ln p(\boldsymbol{\theta})}{\partial \mathbf{x}}$, referred to as the score, is the gradient (i.e., the rate of change) of the log-likelihood function with respect to each dimension of $\mathbf{x}$. The FIM is based on the derivative of the gradient, thus it provides information of the curvature (i.e., the rate of change of the steepness) of the log-likelihood function along each dimension of $\mathbf{x}$. The sharper (i.e., more curved) the log-likelihood function is, the more confidently the ML estimation of $\mathbf{x}$ found from the measurement data $\boldsymbol{\theta}$ is close to the correct value of $\mathbf{x}$ and vice versa. This also means that the FIM provides the information of the inverse of the variance of the ML estimation of $\mathbf{x}$. The expectation operation guarantees the FIM not to depend on any particular realization of the measurement data $\boldsymbol{\theta}$ because it is averaged over a number of realizations.

When $\mathbf{n}$ is zero-mean Gaussian distributed, $\mathbf{I}(\mathbf{x})$ can be also calculated by

$$\mathbf{I}(\mathbf{x}) = \left[\frac{\partial \mathbf{f}(\mathbf{x})}{\partial \mathbf{x}}\right]^T \mathbf{C}^{-1} \left[\frac{\partial \mathbf{f}(\mathbf{x})}{\partial \mathbf{x}}\right], \tag{2.46}$$

where $\mathbf{C} = \text{diag}\{\sigma_1^2, \ldots, \sigma_N^2\}$ is the covariance matrix of the noise vector $\mathbf{n}$ [38].

For the case of localization in a two-dimensional Cartesian coordinate, the CRLB is obtained as

$$\text{CRLB} = \sqrt{E\left\{\frac{1}{M}\sum_{m=1}^{M}\frac{[\mathbf{I}^{-1}(\mathbf{x})]_{1,1}^{(m)} + [\mathbf{I}^{-1}(\mathbf{x})]_{2,2}^{(m)}}{2}\right\}}, \tag{2.47}$$

where $[\mathbf{I}^{-1}(\mathbf{x})]_{1,1}^{(m)}$ and $[\mathbf{I}^{-1}(\mathbf{x})]_{2,2}^{(m)}$ are the inverse of the $(1,1)$ and $(2,2)$ entries of the FIM $\mathbf{I}(\mathbf{x})$. $[\mathbf{I}^{-1}(\mathbf{x})]_{1,1}^{(m)}$ and $[\mathbf{I}^{-1}(\mathbf{x})]_{2,2}^{(m)}$ represent the variances in the $m$iteration of the maximum likelihood estimations of the x_x and x_y coordinates of the target, respectively.

By stacking the nonlinear ranging functions in (2.24) for N anchors to form the column vectors, denoted as $\mathbf{f}(\mathbf{x})$, for TOA, TDOA, RSSI, and AOA techniques, their corresponding first derivatives with respect to $\mathbf{x}$ can be found as

$$\frac{\partial \mathbf{f}_{\text{TOA}}(\mathbf{x})}{\partial \mathbf{x}} = \begin{bmatrix} \frac{x_x - a_{1x}}{d_1} & \frac{x_y - a_{1y}}{d_1} \\ \vdots & \vdots \\ \frac{x_x - a_{Nx}}{d_N} & \frac{x_y - a_{Ny}}{d_N} \end{bmatrix}, \tag{2.48}$$

$$\frac{\partial \mathbf{f}_{\text{TDOA}}(\mathbf{x})}{\partial \mathbf{x}} = \begin{bmatrix} \frac{x_x - a_{2x}}{d_2} - \frac{x_x - a_{1x}}{d_1} & \frac{x_y - a_{2y}}{d_2} - \frac{x_y - a_{1y}}{d_1} \\ \vdots & \vdots \\ \frac{x_x - a_{Nx}}{d_N} - \frac{x_x - a_{1x}}{d_1} & \frac{x_y - a_{Ny}}{d_N} - \frac{x_y - a_{1y}}{d_1} \end{bmatrix}, \tag{2.49}$$

$$\frac{\partial \mathbf{f}_{\text{RSS}}(\mathbf{x})}{\partial \mathbf{x}} = \begin{bmatrix} -\eta\frac{x_x - a_{1x}}{d_1^2} & -\eta\frac{x_y - a_{1y}}{d_1^2} \\ \vdots & \\ -\eta\frac{x_x - a_{Nx}}{d_N^2} & -\eta\frac{x_y - a_{Ny}}{d_N^2} \end{bmatrix}, \tag{2.50}$$

$$\frac{\partial \mathbf{f}_{\text{AOA}}(\mathbf{x})}{\partial \mathbf{x}} = \begin{bmatrix} -\frac{x_y - a_{1y}}{d_1^2} & \frac{x_x - a_{1x}}{d_1^2} \\ \vdots & \\ -\frac{x_y - a_{Ny}}{d_N^2} & \frac{x_x - a_{Nx}}{d_N^2} \end{bmatrix}, \tag{2.51}$$

where $\eta = \frac{10\gamma}{\ln 10}$ and $d_i = \sqrt{(x_x - a_{ix})^2 + (x_y - a_{iy})^2}$, $i = 1,\ldots,N$. When hybrid lateration and are angulation used, the nonlinear function is a combination of both estimation methods. More particularly, the nonlinear function of the hybrid AOA/RSSI localization, defined by $\mathbf{f}_{AR}(\mathbf{x})$, is written as $\mathbf{f}_{AR}(\mathbf{x}) = [\mathbf{f}_{RSS}(\mathbf{x}), \mathbf{f}_{AOA}(\mathbf{x})]^T$. From this function, the CRLB of the corresponding hybrid method can be calculated.

As an example, the performance of the unbalanced hybrid RSSI/AOA, named as nRSSI/1AOA and denoted as 1AnR, is compared with that of the conventional hybrid AOA/RSSI one, denoted as nAnR in Figure 2.7 [37]. A WSN including N anchors is randomly deployed to locate 1000 agents which randomly appear in an area of 150 m $\times$ 150 m. The first anchor, which is the only anchor providing the AOA estimation for the 1AnR method, is fixed at the origin ($\mathbf{a}_1 = [0, 0]$). Meanwhile, in the conventional AOA/RSSI positioning method, each anchor can provide both AOA and RSSI measurements, i.e., each anchor must be equipped with an antenna array.

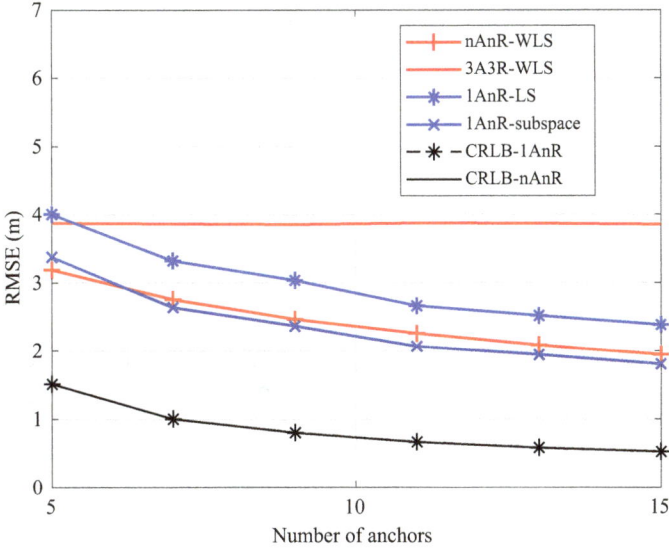

Figure 2.7   RMSE for $\sigma_{m_i} = 5$ degrees, $\sigma_{n_i} = 0.4$ dB

Considering the AOA estimation and RSSI measurement errors at the ith anchor as $\sigma_{m_i} = 5$ degrees and $\sigma_{n_i} = 0.4$ dB, respectively, the RMSEs for 1AnR and nAnR approaches are depicted in Figure 2.7 [37]. It can be seen that the two 1AOA/nRSSI methods, namely 1AnR-LS and 1AnR-Subspace, can achieve a similar accuracy as the 3AOA/3RSSI approach when a small number of simple anchors is added. Especially, the hybrid 1AnR-subspace method surpasses the fully equipped nAnR-WLS method when more than seven simple anchors are employed. Readers may refer to [37] for more comprehensive comparisons.

   In the above example, the average localization errors are around 1–4 m for the considered WSN of size 150 m × 150 m. One would expect that the average localization errors are at a centimeter-level should the localization techniques be applied to the social distancing context, where the typical distances between nearby people are only several meters. From Figure 2.7, it is also clear that there is still room to improve the performance of the current localization and positioning algorithms to reach the CRLB. In addition, the unbalanced hybrid AOA/RSSI algorithms combined with the subspace approach might be worth considering for social distancing due to its comparable performance and the reduced complexity of the anchor nodes, compared to the conventional multi-anchor hybrid AOA/RSSI algorithms.

2.3   Global navigation satellite systems

In the old day before compasses, people used stars as known references to navigate themselves in their adventures. Navigation satellite systems actually operate in the same concept as they are man-made reference points that transmit RF signals for receivers on the Earth's surface to estimate their locations. Global navigation satellite system (GNSS) is a general name for different satellite systems deployed by several governments and unions for the purpose of positioning and tracking. Some well-known GNSSs include global positioning system (GPS) which was originally developed by the US Defense, global navigation satellite system (GLONASS) by Russia, Galileo by the European Union, and BeiDou by China. GPS and GLONASS were originally developed in the 1960s during the Cold War for military purposes. Both are now available for civilian applications but with lower accuracy than their military applications. Galileo and BeiDou have recently been deployed and partially being exploited since 2020. Although these systems have different numbers of satellites and frequencies of transmitted signals, their fundamentals are similar. Particularly, satellites broadcast their clock and positions to users on the Earth so that they can use the TOA technique to estimate the TOF and their clock offset to the satellites. From the estimated ranges to the satellites, the GNSS receiver can calculate its location and tune its clock. Some basic techniques of the GPS system are briefly described as follows [39].

The GPS system includes three major segments, namely, space-, control-, and user-segments. The space segment is the constellation of GPS satellites that transmits signals to users on the Earth. The control segment is a ground network that monitors the health and operations of the space segment. The user segment includes all GPS receivers and software for localization and timing. For a GPS receiver on the ground to estimate its position anywhere and anytime, it needs to "see" at least four satellites. Therefore, 24 active GPS satellites are nominally distributed in six orbital planes at the orbital height of about 20, 200 km. A complete message from a satellite has a duration of 12.5 min and includes 25 data frames. Each data frame of the duration of 30 s is made up of five sub-frames which provide the exact time of transmission, clock, date, time, coordinates of the satellite, and correction data for the receiver to compensate for some errors caused by the parameters of the ionosphere. The transmitted data are modulated with binary phase-shift keying before multiplying with pseudo-random spreading codes so that these satellites can share the same frequency without interfering each other. The modulated data are then transmitted to the user segment at the frequencies of 1575.42 MHz and 1227.60 MHz, which have low attenuation when penetrating through the Earth's atmosphere. This frequency range is also convenient for compact and low-cost GPS receivers.

Unlike the terrestrial TOA estimation presented in Section 2.1.1.3, the GPS receiver is not synchronized to satellites, and it has to simultaneously estimate the TOF from different satellites which transmit signals at different time instants. As a result, the clock bias of the receiver needs to be estimated for accurate positioning. Defining the clock error of the receiver as τ, the pseudo-range, denoted by P^i, from the receiver to the ith satellite is presented as

$$P^i = \sqrt{(a_x^i - x_x)^2 + (a_y^i - x_y)^2 + (a_z^i - x_z)^2} + c\tau, \qquad (2.52)$$

where a_x^i, a_y^i, a_z^i are the coordinates of the ith satellite and x_x, x_y, x_z are the coordinates of the GPS receiver to be determined. Hence, at least four satellites must be seen by the GPS receiver to solve three coordinates $\mathbf{x}$ and the clock bias τ. An iteration method to solve this set of nonlinear equations is described as follows. Defining vector $\mathbf{X} = [x_x, x_y, x_z, c\tau]^T$, and applying Taylor's theorem, the pseudo-ranges can be approximated as

$$P^i(\mathbf{X}) \approx P^i(\mathbf{X}_0) + \frac{dP^i(\mathbf{X})}{d\mathbf{X}}\Big|_{\mathbf{X}_0}(\mathbf{X} - \mathbf{X}_0),$$

$$= P^i(\mathbf{X}_0) + \frac{\partial P^i}{\partial x_x}\Big|_{x_{0x}}(x_x - x_{0x}) + \frac{\partial P^i}{\partial x_y}\Big|_{x_{0y}}(x_y - x_{0y}),$$

$$+ \frac{\partial P^i}{\partial x_z}\Big|_{x_{0z}}(x_z - x_{0z}) + \frac{\partial P^i}{\partial c\tau}\Big|_{c\tau_0}(c\tau - c\tau_0), \tag{2.53}$$

where $\mathbf{X}_0 = [x_{0x}, x_{0y}, x_{0z}, c\tau_0]^T$ is an estimated or guessed vector which is presumed to be near $\mathbf{X}$. Taking the partial derivatives and substituting them into (2.53), we have

$$P^i(\mathbf{X}) = P^i(\mathbf{X}_0) + \frac{(x_{0x} - a_x^i)}{R_0^i}(x_x - x_{0x}) + \frac{(x_{0y} - a_y^i)}{R_0^i}(x_z - x_{0z}),$$

$$+ \frac{(x_{0z} - a_z^i)}{R_0^i}(x_z - x_{0z}) + (c\tau - c\tau_0), \tag{2.54}$$

where $R_0^i = \sqrt{(a_x^i - x_{0x})^2 + (a_y^i - x_{0y})^2 + (a_z^i - x_{0z})^2}$ is the range estimate from the guessed position of the target to the ith satellite. Since (2.54) is a set of linear equations, it can be written in a matrix form as

$$\mathbf{P}(\mathbf{X}) - \mathbf{P}(\mathbf{X}_0) = \mathbf{A}(\mathbf{X} - \mathbf{X}_0), \tag{2.55}$$

where $\mathbf{P}(\mathbf{X}) = [P^1(\mathbf{X}), \ldots, P^N(\mathbf{X})]^T$, $\mathbf{P}(\mathbf{X}_0) = [P^1(\mathbf{X}_0), \ldots, P^N(\mathbf{X}_0)]^T$, and $\mathbf{A}$, called the *design matrix*, for N satellites is

$$\mathbf{A} = \begin{bmatrix} \frac{(x_{0x} - a_x^1)}{R_0^1} & \frac{(x_{0y} - a_y^1)}{R_0^1} & \frac{(x_{0z} - a_z^1)}{R_0^1} & 1 \\ \vdots & \vdots & \vdots & \vdots \\ \frac{(x_{0x} - a_x^N)}{R_0^N} & \frac{(x_{0y} - a_y^N)}{R_0^N} & \frac{(x_{0z} - a_z^N)}{R_0^N} & 1 \end{bmatrix}.$$

The solution for (2.55) can be found in the LS algorithm as

$$(\mathbf{X} - \mathbf{X}_0) = (\mathbf{A}^T\mathbf{A})^{-1}\mathbf{A}^T[\mathbf{P}(\mathbf{X}) - \mathbf{P}(\mathbf{X}_0)]. \tag{2.56}$$

Noted that $\mathbf{X}$ solved from (2.56) is only an approximation since $\mathbf{X}_0$ is the guessed position and the higher-order components in the Taylor series are truncated. Therefore, by denoting this solution as $\mathbf{X}_1$ and replacing $\mathbf{X}_0$ by $\mathbf{X}_1$, we get the new solution $\mathbf{X}_2$ which is closer to the true $\mathbf{X}$. Iterations can continue until the optimal solution is reached.

Due to the availability of GNSS signals and the simplicity of the positioning algorithms, GNSS receivers are widely used to provide tremendous convenience for individuals and businesses. However, GNSS-based positioning may not be suitable for social distancing because its accuracy is relatively low for this application context. For example, GPS positioning accuracy obtained by a mobile phone is about 5 meters [40]. Furthermore, since at least four satellites must be seen by the GNSS receiver, the accuracy of GNSS positioning is even poorer in indoors and dense urban areas where GNSS signals are blocked or reflected by obstacles. Unfortunately, social distancing is most required in such places. Therefore, some other positioning systems without GNSS signaling need to be developed.

2.4   Non-GNSS localization

2.4.1   Localization with UAV

Originally, unmanned aerial vehicles (UAVs), a.k.a. drones, were developed for military purposes, such as reconnaissance, remote surveillance, and armed attacks. Nowadays, due to their flexibility and low cost, UAVs have been widely used in numerous applications, including agriculture, package delivery, aerial videography, and video surveillance [41]. Recently, UAVs are also considered to be deployed as aerial anchors in flexible WSNs for localization services in emergency search and rescue missions [42]. Compared to GNSS systems, UAVs are in shorter distances so that receivers can obtain a higher signal-to-noise-ratio. Compared to terrestrial anchors, UAVs have higher altitudes, resulting in wider ground coverage, less severe shadowing effects, and a higher LOS probability [43]. In such positioning networks, UAVs fly in predetermined trajectories and periodically emit beacons, which include the coordinates of the UAVs, for the unknown nodes to estimate their relative positions.

As shown in Figure 2.8, in a delay-free system, anchors are deployed by N UAVs, while in delay-tolerant network, anchors can be obtained by only one UAV, flying and hovering at N certain positions.

the anchors can be deployed by N UAVs in a delay-free system, or just one UAV to fly and hover at N predetermined locations in a delay-tolerant network.

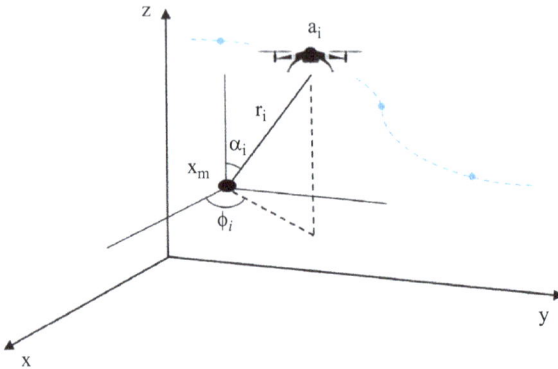

Figure 2.8   UAV-assisted localization using a hybrid TOA/AOA technique

In range-based positioning WSNs, the locations of unknown nodes are derived based on the ranging information estimated from the beacons. The unknown nodes perform ranging techniques, such as TOA, AOA, and RSSI, from the received beacons and then estimate their positions using the localization algorithms. For example, RSSI is used in [44] to estimate the distances from the UAVs to the ground nodes. The PLE and shadowing effects are modeled as functions of the UAVs' altitudes. The optimal altitude of anchors at about 750 m is revealed to minimize the ranging errors in urban areas. However, the localization error is still very high at about 80 m. In [45], the range between the UAV and the target is estimated through RTT measurements. Particularly, a UAV transmits a beacon with a unique identifier to the target, which will send an acknowledgment message back to the UAV for it to calculate the RTT. The UAV then sends another message with the computed distance to the target. After collecting at least three estimated distances to different positions of the UAV, the target can estimate its location using the lateration algorithm as in (2.17). The simulated results show that an accuracy of approximately 1 m can be achieved. In [46], the authors propose a UAV-assisted energy harvesting localization system consisting of a ground station (GS), a UAV, and multiple ground users (GUs), in which both the GS and all the GUs want to know the locations of the GUs, in a spatially correlated shadowing propagation environment. Impacts of some important system parameters, such as the altitude of the UAV, the number of waypoints, energy-harvesting time, and correlated shadowing on both the localization accuracy and the achievable throughput are considered in this joint localization-communications scenario.

An AOA-based technique, named as RFEye, has been proposed to locate the terrestrial transmit source using UAVs [47]. In this technique, the UAV uses an omni-directional antenna combined with a beamforming technique, rather than an antenna array, to conduct AOA estimations. The UAV hovers at random positions within the two *d*-meter apart position areas, each of which is a sphere with a radius of one meter, to measure the AOA of the incoming signal from the transmitter. Collecting signals from these spatially distributed waypoints serves the similar purpose of using an antenna array. After acquiring the AOA estimations at the two spherical areas, the position of the transmitter can be estimated using the angulation method. Experimental results show that the accuracy of RFEye can be approximately 1 m and 2.6 m in two-dimensional (2D) and three-dimensional (3D) spaces, respectively. Although these techniques can achieve good accuracy, their main drawback is the high latency due to multiple messages exchanged between UAVs and targets in [45] and the long total hovering time required for AOA estimations in [47].

Hybrid TOA/AOA localization may also be applied to UAV-assisted WSNs to improve the accuracy [48–50]. Conventionally, to estimate the targets' position in a 3D space, both azimuth and elevation AOA are required to combine with the distance estimation by a TOA technique. However, azimuth and elevation AOA measurements lead to 2D signal processing and an L-shape antenna array to be equipped at each receiver [51]. This requirement leads to more power consumption for signal processing and a larger-sized receiver front-end. To mitigate both computational complexity and hardware complexity of each agent's terminal, a hybrid localization algorithm using TOA combined with elevation AOA, called TOA/1AOA, is proposed in [52].

In this publication, the L-shape antenna array normally required in the conventional TOA/AOA localization can be simplified to just one antenna array. The simulation results show that the TOA/1AOA method can achieve the same or even better accuracy compared to the conventional TOA/AOA techniques when other parameters of the network, including the number of UAVs and their altitude, the signal bandwidth as well as the transmit power from UAVs are controlled properly.

2.4.2   Positioning in WLAN

Wireless fidelity (WIFI) networks or IEEE 802.11 are vital in many indoor settings as well as in metropolitan areas. Positioning using WIFI networks is considered a key technology for enabling location-based services. WIFI localization can be categorized into three major techniques, including TOA, TDOA, and RF fingerprinting.

One-way TOA, where the receiver estimates the TOF by subtracting the receive time to the transmit time, normally requires synchronization between the transmitter and the receiver. Synchronization in IEEE 802.11 standards is ensured by the timing synchronization function (TSF) timers which are maintained to be within 4 μs plus propagation delay. With timer precision of 1 μs, which is equivalent to 300 m in range, one-way TOA is not suitable for positioning in WIFI. Therefore, two-way TOA, or round-trip-time (RTT), which does not require synchronization between transmit and receive timers, is normally used. Since the IEEE 802.11 access protocol is carrier sense multiple access with collision avoidance (CSMA/CA), RTT is easy to be implemented. With the CSMA/CA protocol, when a node has data to send, it senses the channel and only starts transmission when the channel is clear. If the data is received successfully, the receiving node will wait for a short interframe space (SIFS) and then send back an acknowledgment (ACK) message to the transmitter. The SIFS ensures that the ACK message arrives at the transmitter without any collision. Since the transmitter knows the SIFS, from the transmit time of the original message and the receive time of the ACK message, it can calculate the RTT and hence the range between itself and the receiver.

The accuracy of the RTT ranging method depends significantly on the frequency of the chip clock which, in turn, decides the accuracy of clock readings and the SIFS. As an example shown in [53], if the clock frequency has an error of 20 ppm from the nominal and the RTT is 300 μs, then the range estimation error is equivalent to $300 \times 10^{-6} \times 20 \times 10^{-6} \times 3 \times 10^8 = 1.8$ m. The accuracy of RTT ranging methods can be improved by determining the frequency offset between the transmit and receive nodes [54] and/or by averaging over a high number of measurements, e.g., 1000 measurements as in [55]. As a result, high accuracy TOA WLAN positioning requires a modification of hardware for a higher clock rate, or a sufficiently long period of time to measure RTT.

WLAN positioning using RTT can only work if the mobile station (MS) has two-way communication with at least three access points (APs), which is sometimes impractical. The TDOA positioning method can overcome this disadvantage by placing at least three synchronized APs in such a way that they can simultaneously hear the ACK message from the MS. The three APs are synchronized over Ethernet cables and are connected to a location server (LS). When the LS requires the position of an MS, it sends a command through Ethernet cables to these APs so they can make TOA estimations to the MS at the same packet. TOAs estimated by the APs are sent back to the LS. The LS calculates the TDOA of the AP pairs and estimates the position of the MS. With the determined TDOA estimations, WIFI positioning can achieve accuracy around 1 m as reported in [56] in low multipath environments. A disadvantage of TDOA-based WIFI positioning is that it requires the MS to be in the coverage areas of at least three synchronized APs, which is not always available in practice.

RF fingerprinting is one of the most popular techniques used in WIFI positioning networks since it does not require synchronization between APs as TDOA and is less affected by multipath environments, including the multipath correlation problem [57]. This technique bases on the fact that, in a multipath environment, the signal received at each point is a unique combination of different rays. If the signal strengths at all the points of the area are pre-measured and stored in a database, the position of the MS can be estimated by matching the RSS collected at all APs with the database. A typical example of WLAN positioning systems is illustrated in Figure 2.9. Finger-printing localization includes two phases, namely the off-line phase and the on-line phase. In the first phase, a pattern of RSS information obtained by each AP as well as the coordinate of an AP in the floor plan is combined to create an RSS map. Then, in the on-line phase, the RSS sensed by the AP is compared with the RSS map to find the MS's location. Therefore, the density of APs and the number of measurements in the off-line phase are the key factors affecting the accuracy of this technique. As one of the first WIFI positioning systems reported in [58], an experiment was conducted on a single floor of 22 m × 43 m with more than 50 rooms. The area is covered by three APs, and each RSS map is constructed by 70 measurement points. The median 2D positioning error of 2.94 m is reported in this experiment. However, RF fingerprint-ing also has many disadvantages. First, intensive manpower is required to construct

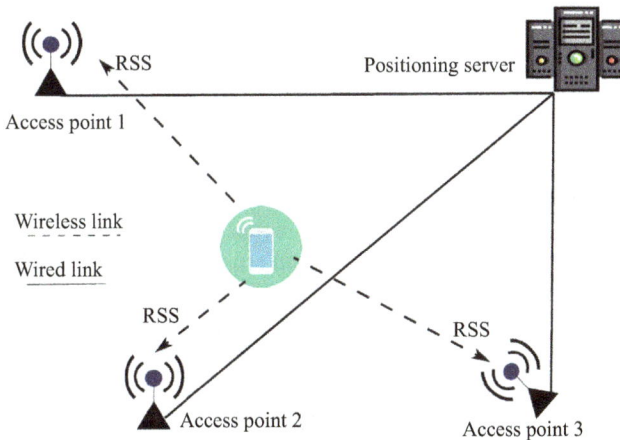

Figure 2.9 A WLAN fingerprinting positioning system

the RSS map and the database during the off-line phase. This database will need to be updated whenever there are some changes in the floor settings. Therefore, the accuracy of this technique deteriorates significantly when the area is crowded by, for example, moving people. Additionally, matching the collected RSSs to the database leads to a long processing time, especially with a large database for higher accuracy or a wider area. Therefore, some additional techniques combined with fingerprinting have been proposed to tackle these problems. For example, in [59], a path-loss model is adopted during fingerprint creation and localization to reduce the workload of the site survey. In [60], a Gaussian process regression model is used to construct and adapt the RSS map frequently. Experimental results over a multi-functional lab with an area of 35 m × 16 m show that an average positioning accuracy of 1.7 m can be obtained with 10 APs.

2.4.3 Positioning in cellular networks

Initially, positioning in cellular networks is mainly used for establishing connections from a Base Station (BS) to Mobile Stations (MSs). The accurate locations of MSs became critical in 1996 when the US Federal Communication Commission (FCC) regulated that wireless operators have to provide the location of emergency calls to the Public Safety Answering Point (PSAP). Since then, with the development of mobile network generations, the accuracy of MS positioning is being enhanced for not only security purposes, but also for optimized mobile network managements, such as handover, power control, and location-based commercial applications.

Positioning in cellular networks can be categorized into network-based and handset-based. Network-based systems use the communication signals received at BSs to estimate the positions of MSs, while in handset-based systems, location estimation is conducted at MSs. Obviously, since BSs are more powerful than MSs, network-based systems can employ more sophisticated positioning algorithms, which are not suitable for implementing in handsets. Additionally, handset-based systems require extra measurements and calculations for positioning tasks which lead to complicated and power-thirsty handsets. Both handset-based and network-based positioning systems use TDOA to estimate the ranges from an MS to at least three BSs, and then lateration algorithms are used to calculate the MS's location [61] as shown in Figure 2.10. In both cases, the accuracy of positioning depends on the number of BSs visible to the MS as well as the reliability of TOA measurements. Therefore, the performance of cellular positioning depends heavily on the density of BSs in the network and the bandwidth of the signals [53]. As studied in [62] over GSM networks by three different operators in the UK, the median error was 246 m in a dense urban area and up to 626 m in a rural setting. A method to improve accuracy in the handset-based systems is to incorporate with a GPS receiver, called assisted GPS (AGPS), which is available in modern mobile phones. As shown in [63], with the third-generation (3G) iPhone, the location estimated using AGPS has the median error of 8 m. Meanwhile, the median error estimated solely by cellular signals in a 3G iPhone is up to 600 m.

This inferior performance has been gradually improved with the development of mobile network generations because a wider bandwidth is used and more specific

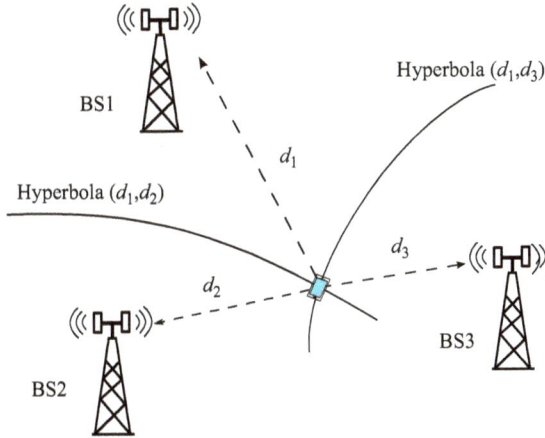

Figure 2.10   Network-based cellular network positioning

solutions are employed. Particularly, the localization error obtained by 2G networks is more than 100 m, while that by 3G and 4G networks is reduced to within 50 m. Hybrid methods using 2G, 3G, and 4G signalings can reduce errors to fewer than 10 m [64]. Fingerprinting methods similar to those for WIFI systems have also been applied for cellular networks in [65–67]. For example, RSS fingerprints were collected from 29 GSM channels and 6 cells in [65,66] to achieve a localization error of below 5 m. Finally, the localization error in 5G networks is expected to be less than 1 m since a higher density of BSs is deployed and a wider bandwidth is used for each MS [68].

2.4.4   Bluetooth positioning

Bluetooth technology was developed in 1994 to originally replace the RS232 cable. Since then, it has been gradually evolved and has replaced cables in a vast number of short-range connections, such as those between a computer and its peripherals, to create a Wireless Personal Area Network (WPAN). Bluetooth transmissions share the 2.4 GHz ISM band with other technologies, such as WIFI, by using an adaptive frequency hopping technique that allows Bluetooth devices to select good channels for communication. In a Bluetooth connection, one device acts as a master which decides the frequency hopping pattern and synchronization clock [69].

Bluetooth technology is developed and managed by the Bluetooth Special Interest Group (SIG). Since 1999, Bluetooth SIG has issued several specifications to enhance the performance of Bluetooth connections. These specifications can be categorized into two groups, namely Bluetooth Classic, which includes core versions (CV) from 1.0 to 3.0, and Bluetooth Low Energy (BLE) in CV 4.0 and 5.0 [70]. A significant improvement of data rates can be witnessed in Bluetooth Classic from 721 kbps in CV 1.0 to 3 Mbps in CV 2.0 and up to 24 Mbps in CV 3.0. The communication range between two devices is also increased from 10 m in CV 1.0 to up to 100 m in CV

3.0. The main reason that CV 3.0 can support such a high data rate and a long-range is that it integrates a WIFI chip for communication through the Bluetooth protocol. However, Bluetooth Classic devices require the peak current of around 30 mA. This makes Bluetooth Classic unsuitable for Internet-of-Things (IoT) applications [71]. On the other hand, BLE is a significant enhancement in terms of power consumption. By implementing a smart host control and an adjustable message length, BLE devices are constantly in a sleep mode except when a connection is initiated. The connection time required in BLE is only a few milliseconds, compared to almost 100 ms in Bluetooth Classic. As a result, the power consumption of BLE devices is reduced by 20 times, compared to that in Bluetooth Classic. This allows BLE devices to operate years-long over a small battery.

The ubiquity of Bluetooth devices combined with the low cost and long battery lifetime of Bluetooth modules make Bluetooth networks to be a promising indoor positioning system. The most popular technique in Bluetooth positioning is to use RSSI for estimating the ranges between the nodes. The coordinates of the nodes are then obtained by lateration or fingerprinting methods. This is because time-based ranging is not accurate in Bluetooth networks as the protocol allows some delay in transferring packets. Furthermore, unlike WLAN with a wide network coverage over multiple room settings, a Bluetooth network is normally limited in a single room. Hence, it is less affected by walls and obstacles. As a result, accurate path-loss models can be developed for RSSI-based ranging.

Initially, RSSI, known as connection-based RSSI in CV 1.0, was simply used to control transmit power levels in a Bluetooth connection. When the received signal power is within an expected range, known as "golden thresholds," RSSI is 0 dB, meaning that no power adjustment is required. RSSI is only related to the real received signal power when the received power is outside of the expected range [72]. Therefore, variable attenuators were adopted in [72] to shift RSSI into the linear region outside the golden threshold. The relationship of RSSI and the communication range is presented as [73]

$$RSSI = -2.28 \ln{(d)} + 5.7, \tag{2.57}$$

where d is the range in meters and ln is the natural logarithm. From (2.57) and using the lateration method, the accuracy of about 2 m is obtained with three anchors to estimate a sensor node's coordinate in a room of size 46 m $\times$ 46 m [73].

Fully functioning RSSI, which shows the relationship between the range and received power, is available from CV 1.2 onward. This inquiry-based RSSI expressed in absolute values (dBm rather than dB as in the connection-based RSSI) is not affected by power control. The main problem of Bluetooth positioning is to construct an accurate path-loss model. As shown in [74], a two-slope model can be used as

$$P(d) = \begin{cases} \beta_1 + \gamma_1 \log_{10}{(d)} & \text{if } d < d_{\text{break}} \\ \beta_2 + \gamma_2 \log_{10}{(d/d_{\text{break}})} & \text{if } d \geq d_{\text{break}}, \end{cases} \tag{2.58}$$

where β_1, γ_1, and β_2, γ_2 are the offset-slope pairs in the near and far regions, respectively, and d_{break} is the breakpoint distance in meters. The values of these parameters need to be determined by the data measured in the off-line training phase. Another method to improve ranging estimation accuracy is presented [75] for BLE devices. Particularly, by using an off-line training phase combined with an active learning mechanism to estimate and adjust propagation parameters while applying a Gaussian filter to smooth the acquired data during the online phase, a localization error of fewer than 1.5 m is achieved at over 80 percent of the test points.

Apart from the lateration algorithm, RSSI-based Bluetooth positioning can also use fingerprinting approaches as in WLAN [76]. After a fingerprint database is built by Bluetooth receivers in the off-line phase, different algorithms can be applied to search for the location of the Bluetooth transmitter in the online phase. Common fingerprinting algorithms include the weighted K-nearest neighbor algorithm (WKNN) [77–79] and the dynamic fingerprinting window (DFW) WKNN [80]. It is shown in [80] that, for a small office area, a Bluetooth fingerprinting network with beacon deployment spacing of 3 m can achieve an average error of 0.7 m and 0.51 m with WKNN and DFW-WKNN, respectively.

A significant enhancement for Bluetooth indoor positioning is issued in CV 5.1 with a new direction-finding feature [81]. Particularly, the conventional low energy (LE) packets transmitted by a Bluetooth tag are modified with an additional frame called Constant Tone Extension (CTE). The CTE frame includes all "1"s so that the antenna receives a constant frequency for this part of the signal. A Bluetooth CV 5.1 indoor positioning network includes several access points or locators, which are equipped with antenna arrays and multiple RF switches, and tags, which are simple single antenna BLE devices. The new direction-finding feature provides two options: angle-of-arrival (AOA) and angle-of-departure (AOD) for positioning. AOA is used when an access point receives signals from a tag, while AOD is utilized when the access point transmits to the tag. As shown in Figure 2.11 for AOA, the signal transmitted from the tag with CTE-enhanced packets arriving at the antenna array of the access point is sampled by the RF switches. With the same signal, both "In-phase" and "Quadrature-phase" samples are obtained, and a pair of the "IQ-samples" is considered to be a complex value containing both phase and amplitude information, which can be an

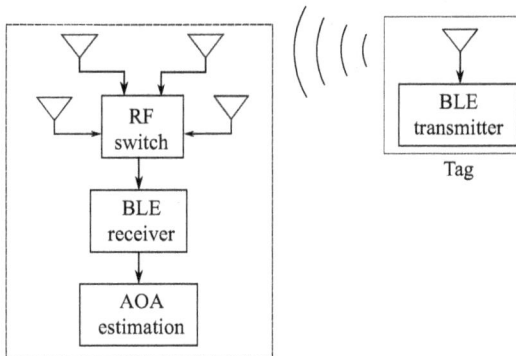

Figure 2.11   AOA feature in BLE 5.1 [81]

input for the AOA estimation algorithm, such as MUSIC. On the other hand, in AOD, the access point sequentially switches the transmitting antenna. Since the tag knows both the switching sequence and the antenna array configuration, it can determine the AOD. Experiments in [82] with a four-element array separated by a half wavelength and with the MUSIC algorithm show that, when the AOA is within 0 to 30°, the RMSE of the AOA estimation is less than 0.5°, while, for the AOA from 30 to 90°, the RMSE is less than 5°. With the direction finding enhancement of Bluetooth 5.1, the indoor positioning network can apply either angulation algorithms or hybrid AOA and RSSI methods as mentioned in Sections 2.1.3 and 2.1.4. As experimented in [83] for a 25 m × 15 m laboratory using four receivers and a 4 × 4 uniform rectangular antenna array, the average positioning error is 0.7 m.

2.4.5   Ultra-wideband positioning

The FCC of the US in its 2002 Part 15 release specified that any signal with an absolute bandwidth wider than 500 MHz or having a fractional bandwidth greater than 20% of the carrier frequency is considered as ultra-wideband (UWB). Additionally, unlicensed access to the 7.5 GHz spectrum from 3.1 GHz to 10.6 GHz is allocated for UWB applications. To ensure that UWB operations do not interfere the existing services, UWB signals must be transmitted with very low power, limited at −41.3 dBm/MHz indoors and 10 dB lower outdoors. This definition makes some variants of conventional modulation strategies, such as wide-band and multi-band OFDM [84–87], and code division multiple access (CDMA), also being considered as UWB. However, commonly, UWB systems are known as impulse radios which utilize ultra-short pulses with very low duty cycles, i.e., the pulse duration is much smaller than the average time between two consecutive pulses.

   UWB signals are fascinating for indoor positioning due to their unique features. Particularly, the use of ultra-short pulses, in the order of a nanosecond, means a very high time resolution. Hence, UWB positioning can achieve accurate distance estimations. In addition, the short pulse can be generated in baseband and, together with a small transmitted power, many UWB systems are carrier-free, i.e., no RF transmit chain is required. As a result, these UWB transceivers are small and power-efficient. These advantages make the UWB technology feasible to be implemented in tags, which are tiny in size and operate on a cell battery.

   Although the large bandwidth of UWB signals results in higher AOA estimation accuracy, this technique might be unsuitable for UWB positioning as the antenna array for AOA estimations results in complicated UWB tags. On the other hand, RSSI seems to be unsuitable for UWB positioning because the very low transmission power makes UWB signals sensitive to the estimation noise. To leverage the advantage of the high time resolution, UWB positioning networks use TOA to estimate the TOF for ranging. However, an accurate TOA measurement in UWB networks is also challenging [12]. In particular, it requires a very strict synchronization between the transmitter and the receiver as even small clock skews and offsets between them will significantly affect the accuracy. Additionally, sampling of the ultra-large bandwidth requires a very high Nyquist frequency (up to several gigahertz), which would lead to expensive and power consuming UWB receivers. Furthermore, the conventional correlation-based TOF estimation as in (2.12) is not applicable because of the exhausting search caused by ultra-small delay time $\hat{\tau}$.

Significant efforts have been devoted to tackle the above challenges and to make UWB positioning affordable. First, the synchronization problem can be solved by using a reference anchor to send the standard clock to other nodes in the network through wired or wireless links [88]. While time transfer via a wired link can achieve great synchronization, it comes with the high cost of infrastructure. On the other hand, wireless timing methods have to deal with high latency, especially in multi-hop networks. A single message protocol was proposed in [89] to achieve synchronization accuracy of 0.46 ns in a single-hop setup and 6 ns in a five-hop scenario.

Regarding the second problem of high-rate sampling, UWB receivers can be designed with an analog correlator, which correlates the received signal with a sine wave [90] or a pulse [91] so that the sampling rate required for the analog-to-digital converter (ADC) can be reduced to the pulse rate only. However, the performance of such architecture is degraded because of circuit mismatches. Therefore, ADC conversion is preferred to perform right after the low noise amplifier (LNA) to make a completely digital UWB receiver. To avoid a high-speed ADC in such receivers, subsampling [92] and frequency-domain channelization techniques [93,94] can be employed.

Some new search strategies have been developed to avoid the exhausted search for TOA in correlation-based receivers because of the very high time resolution of UWB signals. For example, in [95], TOA is estimated using a two-step approach. In the first step, a simple energy detector is used to determine a rough TOA estimation. In this way, the TOA searching space is reduced, and the second step can estimate the fine TOA faster, using statistical change detection approaches.

One of the early UWB positioning systems was demonstrated for military operations in 1997 [96]. The system can track soldiers and vehicles without GPS over a 4 km^2 area. This system was gradually improved to reduce its size and in 2003, after the release of FCC Part 15 rules, a commercial version of this system, named as PAL650, was developed. The PAL650 UWB system includes a set of UWB tags, UWB receivers, and a data processing centre. The UWB tags transmit signals with the pulse width of 2.5 ns at the center frequency of about 6.2 GHz. The pulses are transmitted at an extremely low duty cycle of 0.002%, so that the power consumption of the tags is only about 90 μW, i.e., a tag operating on a 3V 1A-h battery can expect to last for up to 3.8 years. Four UWB receivers conduct TOA measurements and send data to the central processing hub where a lateration algorithm is used to locate the tags. The system can demonstrate an accuracy of better than 30 cm.

Over the last two decades, UWB indoor positioning systems have been well developed and applied in many applications [97]. For example, the Ubisense real-time location system is currently used in automotive plants for tracking every stage of the manufacturing process [98]. Another example is the Humatics rail navigation system, which integrates UWB with other technologies, such as GNSS and Lidar, to seamlessly track and control trains [99]. UWB beacons in this system operating in the 4–4.9 GHz spectrum can reach a LOS range up to 1600 m, and ranging accuracy can

Figure 2.12   Localization of Samsung Galaxy SmartTag+ using UWB

be about 2 cm. Recently, UWB positioning becomes more popular as UWB receiver chips are integrated in smartphones. Button-size tags have been developed to attach on personal gadgets, such as keys and wallets, so that the owners can use their phone to pinpoint their belongings. This UWB feature has been available in IPhone 11 for locating its AirTags and Samsung Galaxy Note 20 for working with its SmartTag+ (cf. Figure 2.12).

2.5   Applications of positioning to social distancing

Social distancing is a simple and effective countermeasure to stop the spread of the SARS-CoV-2 virus in the COVID-19 pandemic. The World Health Organization rec-ommends that a distance of at least 1 m among people is required to avoid infection. The social distancing requirement is expected to become a new norm before a suffi-cient percentage of population is vaccinated. Taking Australia as an example, floor marking signs spaced at 1.5 m to remind people are displayed in buses, trains, and indoor vendors. However, this simple rule is not always respected, or it is easily vio-lated due to the limited available spaces. As a result, an automatic system, which can locate people, measure the distance between them, and send alerts to violating individuals should be deployed in such areas. This system should have the following features:

- The system provides a **complete function** to enforce social distancing, including locating people in the covered area, calculating the distance among them, sending accurate alerts to violating people, and storing necessary information for contact tracing.
- The **accuracy** of the positioning system must be sufficient to ensure a reliable detection of social distancing violations.
- The system must be **robust for different environments** in various application scenarios. That is it should have the capability to mitigate the impacts of dynamic surrounding obstacles and can even work in NLOS conditions, such as in shopping malls.
- The devices equipped for users should be **light weight or wearable**, **relatively convenient**, and have a low power consumption.

Table 2.2   A comparison of RF localization technologies

Technology	Strengths	Weakness	Accuracy
GPS	Low cost, wide outdoor coverage	Low accuracy for indoors	5–10 m
Cellular	Wide coverage in urban areas	Low accuracy for indoors, cooperation of service providers required	1–5 m
UAV	Flexible deployment	Medium accuracy, planning of flight trajectories required	2–4 m
WLAN	Easy to implement, cost efficient	Medium accuracy, modifications of APs required	1.5–4 m
Bluetooth	Low energy consumption, low cost	Error prone to noise	0.5–4 m
UWB	High accuracy in small areas	High cost, complicated receivers	2–1 m

- The system must have **low latency** for measuring the distance between users and for sending alerts when threats are detected so that the alerted users can take actions promptly.
- The system can be **scalable** for the possible increasing number of users.

RF positioning, therefore, is a promising solution to enforce the social distancing. However, not all RF positioning technologies mentioned above are suitable for this application. Table 2.2 explicitly presents the advantages and disadvantages of these techniques.

From the requirements mentioned above, it is obvious that only RF positioning systems with the accuracy of a sub-2-m range might be suitable for the social distancing application. Hence, GPS-based positioning devices are not applicable because of their poor accuracy and unavailability indoors or in urban areas. In contrast, cellular-based positioning has a better coverage for such areas, but the localization accuracy of 3G and 4G networks is insufficient for the social distancing context. Only 5G networks with a higher density of base stations (BSs) can meet such accuracy. However, at the current state, 5G networks are not fully deployed, and the number of 5G mobile handhelds is also limited. UAV-based RF positioning is flexible and achieves higher accuracy than GPS, but such systems are normally dedicated to some special groups of people, such as military troops or emergency responders, who are equipped with compatible UAV transceivers. WLANs are now popular in most buildings and the accuracy of WLAN-based positioning is also sufficient (around 1 m) for this

Figure 2.13   RF components used in [103] for human step length measurements

application [56]. However, this accuracy is only achieved if the mobile user is in the coverage of at least three synchronized access points (APs) for TDOA estimations. Meanwhile, fingerprinting-based WLAN positioning needs a high number of APs for sufficient accuracy and results in a large database, which will cause higher latency for data matching. Hence, Bluetooth and UWB positioning systems are more suitable for the social distancing purpose because of their high accuracy, wide availability, and scalability.

RF positioning for social distancing solutions can be categorized into smart-phone-based and non-smart-phone-based. The former uses the Bluetooth feature available in smartphones to detect the violated distance among people. Since most people have smartphones nowadays, this approach has advantages of wide availability and fast deployment as users are only required to install a mobile application on their phones. For example, in [100,101], RSSI from BLE beacons on Android phones are used for social distancing. Particularly, an RSSI threshold is first set by using the averaged RSSI level received from a distance of 2 m. Then, when a BLE beacon is detected, the RSSI is compared with this threshold to determine if the two people are in a safe or risky zone. If the RSSI falls below the threshold, an alert is sent to the user. Obviously, such simple solution may have wrong alerts or detection failures because the instantaneous RSSI depends on many factors, such as phones' positions and surrounding environments. Another solution is proposed in [102] using a path-loss model and a moving averaged filter to smooth the RSSI. The authors also evaluate the impacts of different factors on the accuracy, such as the distance threshold and mobile phones' positions on the body. Experimental results show that the proposed approach can detect the social distance violation with the accuracy above 80 percent of all cases.

A disadvantage of the smart-phone-based solution is that it requires users to install an additional application on their phone and allow it running most of the time. This leads to the concern of privacy as many users do not want to share their

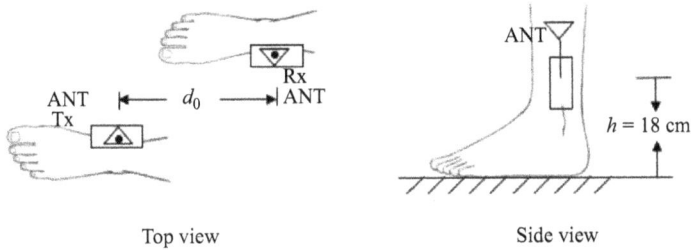

Top view   Side view

Figure 2.14   *Attachment of transceivers to the human ankles for step length measurements in [103]*

information. Therefore, non-smart-phone-based devices, such as bracelets and tags, are useful to monitor the social distancing without requiring any information. Such devices include BLE [104], Zigbee [103], and/or UWB [105] chip-sets combined with a micro-controller to install the data filter and detecting algorithm. The difference between BLE and UWB solutions is that the former consumes a very low power but the accuracy is limited, while the latter can provide higher accuracy but requires more power consumption. In [103] and [106], the authors devise an empirical path-loss model between the two human feet and develop wearable test-beds, using Arduino Uno micro-controllers and Xbee Pro S2C transceivers, to measure the human step length. The transceivers are attached to the human ankles as shown in Figure 2.14. An empirical path loss model between human ankles is proposed and a novel estimation algorithm is developed, which is claimed to be able to estimate the maximum human step length of the person under test with the accuracy of a centimeter level. This technique could be adapted for the social distancing enforcement. In [105], a combination of BLE and UWB low-rate pulses (LRP) is utilized in a wristband to achieve both high accuracy and power saving. Particularly, the BLE chip is always on to detect the proximity of other devices. When a nearby device is detected, the BLE chip wakes the UWB chip up to perform distance estimations. This device is claimed to achieve the distance accuracy within 20 cm and 30 cm in LOS and NLOS environments, respectively. However, more studies should be conducted to evaluate the accuracy in different scenarios, such as the directions and the heights of the devices.

Although many solutions have been proposed for the social distancing applications, to the best of our knowledge, no approach satisfies all the requirements listed above. Smart-phone-based methods are convenient, yet they have low accuracy due to the nature of RSSI in BLE. The concern of privacy also makes social distancing mobile applications have a low installation rate [107]. Not willingly turning on BLE in order for the installed mobile applications to work effectively is another reason for the social distancing enforcement via smart-phone-based methods to be very challenging. Non-smartphone-based devices with the low-rate pulse (LRP) UWB chips are more accurate, but they are more power hungry, thus not being convenient for contact tracing. More efforts, therefore, need to be devoted to developing a complete solution of social distancing enforcement devices.

References

[1]   Han G, Xu H, Duong TQ, *et al.* 'Localization algorithms of wireless sensor networks: a survey'. *Telecommunication Systems*. 2013;**52**(4):2419–36.

[2]   Lee JH, Buehrer RM. Location estimation using differential RSS with spatially correlated shadowing. *In: 2009 IEEE Global Telecommunications Conference*; 2009. pp. 1–6.

[3]   Zafari F, Papapanagiotou I, Hacker TJ. A novel Bayesian filtering based algorithm for RSSI-based indoor localization. *In: 2018 IEEE International Conference on Communications (ICC)*. Kansas City, MO, USA; 2018. pp. 1–7.

[4]   Wang Z, Liu H, Xu S, *et al.* 'A diffraction measurement model and particle filter tracking method for RSS-based DFL'. *IEEE Journal on Selected Areas in Communications*. 2015;**33**(11):2391–403.

[5]   Rappaport T. *Wireless Communications: Principles and Practice*. 2nd ed. USA: Prentice Hall PTR; 2001.

[6]   Yazdandoost KY, Sayrafian K. 'Channel model for body area network (BAN)'. *IEEE P80215 Wireless Personal Area Networks*. 2009.

[7]   Wojcicki P, Zientarski T, Charytanowicz M, *et al.* 'Estimation of the path-loss exponent by Bayesian filtering method'. *Sensors*. 2021;**21**(6). Available from https://www.mdpi.com/1424-8220/21/6/1934.

[8]   Guvenc I, Chong C. 'A survey on TOA based wireless localization and NLOS mitigation techniques'. *IEEE Communications Surveys & Tutorials*. 2009;**11**(3):107–24.

[9]   Zhang J, Orlik PV, Sahinoglu Z, *et al.* 'UWB systems for wireless sensor networks'. *Proceedings of the IEEE*. 2009;**97**(2):313–31.

[10] Schmidt R. 'Multiple emitter location and signal parameter estimation'. *IEEE Transactions on Antennas and Propagation*. 1986;**34**(3):276–80.

[11] Nguyen NM, Tran LC, Safaei F, *et al.* 'Performance evaluation of non-GPS based localization techniques under shadowing effects'. *Sensors*. 2019;**19**(11). Available from https://www.mdpi.com/1424-8220/19/11/2633.

[12] Gezici S, Poor HV. 'Position estimation via ultra-wide-band signals'. *Proceedings of the IEEE*. 2009;**97**(2):386–403.

[13] Zekavat R, Buehrer RM. *Handbook of Position Location: Theory, Practice and Advances*. 1st ed. Wiley-IEEE Press; 2011.

[14] Manickam TG, Vaccaro RJ, Tufts DW. 'A least-squares algorithm for multipath time-delay estimation'. *IEEE Transactions on Signal Processing*. 1994;42(11):3229–33.

[15] Jamalabdollahi M, Zekavat SR. 'High resolution ToA estimation via optimal waveform design'. *IEEE Transactions on Communications*. 2017;**65**(3): 1207–18.

[16] Saarnisaari H. Ml time delay estimation in a multipath channel. *In: Proceedings of ISSSTA'95 International Symposium on Spread Spectrum Techniques and Applications*. vol. 3; 1996. pp. 1007–11.

[17] Lee JY, Scholtz RA. 'Ranging in a dense multipath environment using an UWB radio link'. *IEEE Journal on Selected Areas in Communications.* 2002;**20**(9):1677–83.

[18] Jakobsson A, Swindlehurst AL, Stoica P. 'Subspace-based estimation of time delays and doppler shifts'. *IEEE Transactions on Signal Processing.* 1998;**46**(9):2472–83.

[19] Ahmad A, Serpedin E, Nounou H, *et al.* 'Joint node localization and time-varying clock synchronization in wireless sensor networks'. *IEEE Transactions on Wireless Communications.* 2013;**12**(10):5322–33.

[20] Zheng J, Wu YC. 'Joint time synchronization and localization of an unknown node in wireless sensor networks'. *IEEE Transactions on Signal Processing.* 2010;**58**(3):1309–20.

[21] Le AT, Tran LC, Huang X, *et al.* 'Frequency domain characterization and performance bounds of ALMS loop for RF self-interference cancellation'. *IEEE Transactions on Communications.* 2019;**67**(1):682–92.

[22] Le AT, Tran LC, Huang X, *et al.* 'Analog least mean square loop for self-interference cancellation: a practical perspective'. *Sensors (Switzerland).* 2020;**20**(270):1–15.

[23] Le AT, Tran LC, Huang X, *et al.* 'Analog least mean square loop with I/Q imbalance for self-interference cancellation in full-duplex radios'. *IEEE Transactions on Vehicular Technology.* 2019;**68**(10):9848–60.

[24] Le AT, Tran LC, Huang X, *et al.* 'Beam-based analog self-interference cancellation in full-duplex MIMO systems'. *IEEE Transactions on Wireless Communications.* 2020;**19**(4):2460–71.

[25] Le AT, Tran LC, Huang X, *et al.* 'Analog least mean square adaptive filtering for self-interference cancellation in full duplex radios'. *IEEE Wireless Communications.* 2021**28**(1):12–8.

[26] Huang X, Le AT, Jay Guo Y. 'ALMS loop analyses with higher-order statistics and strategies for joint analog and digital self-interference cancellation'. *IEEE Transactions on Wireless Communications.* 2021;1.

[27] Li J, Tran LC, Safaei F. 'Outage probability and throughput analyses in full-duplex relaying systems with energy transfer'. *IEEE Access.* 2020;**8**: 150150–61.

[28] Li J, Tran LC, Safaei F. 'Throughput analysis of in-band full-duplex transmission networks with wireless energy harvesting enabled sources'. *IEEE Access.* 2021;**9**:74989–5002.

[29] Liu Y, Shen Y, Guo D, *et al.* 'Network localization and synchronization using full-duplex radios'. *IEEE Transactions on Signal Processing.* 2018;**66**(3): 714–28.

[30] Liu Y, Shen Y, Win MZ. 'Single-anchor localization and synchronization of full-duplex agents'. *IEEE Transactions on Communications.* 2019;**67**(3): 2355–67.

[31] Wang Y. 'Linear least squares localization in sensor networks'. *Eurasip Journal on Wireless Communications and Networking.* 2015;**2015**(1):51. Available from https://doi.org/10.1186/s13638-015-0298-1.

[32]   Tomic S, Beko M, Dinis R, *et al.* 'A closed-form solution for RSS/AoA target localization by spherical coordinates conversion'. *IEEE Wireless Communications Letters.* 2016;**5**(6):680–3.

[33]   Tarrio P, Bernardos AM, Besada JA, *et al.* A new positioning technique for RSS-based localization based on a weighted least squares estimator. *In: 2008 IEEE International Symposium on Wireless Communication Systems*; 2008. pp. 633–7.

[34]   So HC, Chan FKW. 'A generalized subspace approach for mobile positioning with time-of-arrival measurements'. *IEEE Transactions on Signal Processing.* 2007;**55**(10):5103–7.

[35]   Lin L, So HC, Chan FKW. 'Multidimensional scaling approach for node localization using received signal strength measurements'. *Digital Signal Processing.* 2014;**34**:39–47. Available from http://www.sciencedirect.com/science/article/pii/S1051200414002243.

[36]   Sayed AH, Tarighat A, Khajehnouri N. 'Network-based wireless location: challenges faced in developing techniques for accurate wireless location information'. *IEEE Signal Processing Magazine.* 2005;**22**(4): 24–40.

[37]   Le AT, Tran LC, Huang X, *et al.* 'Unbalanced hybrid AOA/RSSI localization for simplified wireless sensor networks'. *Sensors.* 2020;**20**(14).

[38]   Chan YT, Ho KC. 'A simple and efficient estimator for hyperbolic location'. *IEEE Transactions on Signal Processing.* 1994;42(8):1905–15.

[39]   Blewitt G. 'Basics of the GPS technique: Observation equations'. *Geodetic Applications of GPS*; 1997.

[40]   Van Diggelen F, Enge P. The world's first GPS MOOC and worldwide laboratory using smartphones. *In: 28th International Technical Meeting of the Satellite Division of The Institute of Navigation (ION GNSS+ 2015)*; 2015. pp. 361–9.

[41]   Matolak DW, Sun R. 'Unmanned aircraft systems: air-ground channel characterization for future applications'. *IEEE Vehicular Technology Magazine.* 2015;**10**(2):79–85.

[42]   Han G, Jiang J, Zhang C, *et al.* 'A survey on mobile anchor node assisted localization in wireless sensor networks'. *IEEE Communications Surveys & Tutorials.* 2016;**18**(3):2220–43.

[43]   Liu Y, Wang Y, Shen X, *et al.* 'UAV-aided relative localization of terminals'. *IEEE Internet of Things Journal.* 2021;**8**(16):12999–3013.

[44]   Sallouha H, Azari MM, Chiumento A, *et al.* 'Aerial anchors positioning for reliable RSS-based outdoor localization in urban environments'. *IEEE Wireless Communications Letters.* 2018;**7**(3):376–9.

[45]   Sorbelli FB, Das SK, Pinotti CM, *et al.* Precise localization in sparse sensor networks using a drone with directional antennas. *In: Proceedings of the 19th International Conference on Distributed Computing and Networking*; 2018. pp. 1–10.

[46]   Le NP, Tran LC, Huang X, *et al.* 'Energy-harvesting aided unmanned aerial vehicles for reliable ground user localization and communications under

lognormal-nakagami-*m* fading channels'. *IEEE Transactions on Vehicular Technology*. 2021;**70**(2):1632–47.

[47] AbdulCareem MA, Gomez J, Saha D, *et al.* 'RFEye in the sky'. *IEEE Transactions on Mobile Computing*. 2020. doi:10.1109/TMC.2020.3038886.

[48] Yu K. '3-D localization error analysis in wireless networks'. *IEEE Transactions on Wireless Communications*. 2007;**6**(10):3472–81.

[49] Taponecco L, D'Amico AA, Mengali U. 'Joint TOA and AOA estimation for UWB localization applications'. *IEEE Transactions on Wireless Communications*. 2011;**10**(7):2207–17.

[50] Li Y, Qi G, Sheng A. 'Performance metric on the best achievable accuracy for hybrid TOA/AOA target localization'. *IEEE Communications Letters*. 2018;**22**(7):1474–7.

[51] Tayem N, Kwon HM. 'L-shape 2-dimensional arrival angle estimation with propagator method'. *IEEE Transactions on Antennas Propagation*. 2005; **53**(5):1622–30.

[52] Le AT, Tran LC, Huang X, *et al.* Hybrid TOA/AOA localization with 1D angle estimation in UAV-assisted WSN. *In: 14th International Conference on Signal Processing and Communication Systems (ICSPCS)*; 2020. pp. 1–6.

[53] Bensky A. *Wireless Positioning Technologies and Applications*. USA: Artech House, Inc.; 2007.

[54] Günther A, Hoene C. Measuring round trip times to determine the distance between WLAN nodes. *In: International Conference on Research in Networking*. Springer; 2005. pp. 768–79.

[55] Ciurana M, Barcelo-Arroyo F, Izquierdo F. A ranging system with IEEE 802.11 data frames. *In: IEEE Radio and Wireless Symposium*; 2007. pp. 133–6.

[56] Makki A, Siddig A, Saad M, *et al.* 'Indoor localization using 802.11-time differences of arrival'. *IEEE Transactions on Instrumentation and Measurement*. 2016;**65**(3):614–23.

[57] Tran LC, Wysocki TA, Seberry J, *et al.* A generalized algorithm for the generation of correlated Rayleigh fading envelopes in radio channels. *In: 19th IEEE International Parallel and Distributed Processing Symposium*; 2005. pp. 1–8.

[58] Bahl P, Padmanabhan V. An in-building RF-based user location and tracking system in IEEE. *In: INFOCOM*; 2000. pp. 775–84.

[59] Zhang J, Han G, Sun N, *et al.* 'Path-loss-based fingerprint localization approach for location-based services in indoor environments'. *IEEE Access*. 2017;**5**:13756–69.

[60] Zou H, Jin M, Jiang H, *et al.* 'Winips: Wifi-based non-intrusive indoor positioning system with online radio map construction and adaptation'. *IEEE Transactions on Wireless Communications*. 2017;**16**(12): 8118–30.

[61] Caffery JJ, Stuber GL. 'Overview of radiolocation in CDMA cellular systems'. *IEEE Communications Magazine*. 1998;**36**(4):38–45.

[62]   Mohr M, Edwards C, McCarthy B. 'A study of LBS accuracy in the UK and a novel approach to inferring the positioning technology employed'. *Computer Communications*. 2008;**31**(6):1148–59.

[63]   Zandbergen PA. 'Accuracy of iphone locations: a comparison of assisted GPS, wifi and cellular positioning'. *Transactions in GIS*. 2009;**13**(s1): 5–25. Available from https://onlinelibrary.wiley.com/doi/abs/10.1111/j. 1467-9671.2009.01152.x.

[64]   del Peral-Rosado JA, Raulefs R, López-Salcedo JA, *et al.* 'Survey of cellular mobile radio localization methods: From 1G to 5G'. *IEEE Communications Surveys Tutorials*. 2018;**20**(2):1124–48.

[65]   Otsason V, Varshavsky A, LaMarca A, *et al.* Accurate GSM indoor localization. *In: International Conference on Ubiquitous Computing*. Springer; 2005. pp. 141–58.

[66]   Varshavsky A, De Lara E, Hightower J, *et al.* 'GSM indoor localization'. *Pervasive and mobile computing*. 2007;**3**(6):698–720.

[67]   Turkka J, Hiltunen T, Mondal RU, *et al.* Performance evaluation of LTE radio fingerprinting using field measurements. *In: International Symposium on Wireless Communication Systems (ISWCS)*. IEEE; 2015. pp. 466–70.

[68]   Dammann A, Raulefs R, Zhang S. On prospects of positioning in 5g. *In: IEEE International Conference on Communication Workshop (ICCW)*; 2015. pp. 1207–13.

[69]   Bisdikian C. 'An overview of the Bluetooth wireless technology'. *IEEE Communications Magazine*. 2001;**39**(12):86–94.

[70]   Jeon KE, She J, Soonsawad P, *et al.* 'BLE beacons for Internet of Things applications: survey, challenges, and opportunities'. *IEEE Internet of Things Journal*. 2018;**5**(2):811–28.

[71]   Collotta M, Pau G, Talty T, *et al.* 'Bluetooth 5: A concrete step forward toward the IoT'. *IEEE Communications Magazine*. 2018;**56**(7): 125–31.

[72]   Bandara U, Hasegawa M, Inoue M, *et al.* Design and implementation of a Bluetooth signal strength based location sensing system. *In: Proceedings. 2004 IEEE Radio and Wireless Conference (IEEE Cat. No.04TH8746)*; 2004. pp. 319–22.

[73]   Feldmann S, Kyamakya K, Zapater A, *et al.* An indoor bluetooth-based positioning system: concept, implementation and experimental evaluation. *In: Proceedings International Conference on Wireless Networks, ICWN '03*; 2003. pp. 109–13.

[74]   Campos RS, Lovisolo L. *RF positioning: fundamentals, applications, and tools*. Artech House; 2015.

[75]   Jianyong Z, Haiyong L, Zili C, *et al.* RSSI based Bluetooth low energy indoor positioning. *In: 2014 International Conference on Indoor Positioning and Indoor Navigation (IPIN)*; 2014. pp. 526–33.

[76]   Faragher R, Harle R. 'Location fingerprinting with Bluetooth low energy beacons'. *IEEE Journal on Selected Areas in Communications*. 2015;**33**(11): 2418–28.

[77] Guo X, Ansari N, Hu F, *et al.* 'A survey on fusion-based indoor positioning'. *IEEE Communications Surveys & Tutorials*. 2019;**22**(1):566–94.

[78] Altintas B, Serif T. Improving RSS-based indoor positioning algorithm via k-means clustering. *In: 17th European Wireless 2011-Sustainable Wireless Technologies*. VDE; 2011. pp. 1–5.

[79] Wang Q, Feng Y, Zhang X, *et al.* 'IWKNN: an effective Bluetooth positioning method based on isomap and WKNN'. *Mobile Information Systems*. 2016, pp. 1–11.

[80] Ruan L, Zhang L, Zhou T, *et al.* 'An improved bluetooth indoor positioning method using dynamic fingerprint window'. *Sensors*. 2020;**20**(24). Available from https://www.mdpi.com/1424-8220/20/24/7269.

[81] Suryavanshi NB, Reddy KV, Chandrika VR. Direction finding capability in bluetooth 5.1 standard. *In: International Conference on Ubiquitous Communications and Network Computing*. Springer; 2019. pp. 53–65.

[82] Toasa FA, Tello-Oquendo L, Peñafiel-Ojeda CR, Cuzco G. Experimental demonstration for indoor localization based on AoA of Bluetooth 5.1 using software defined radio. *In: 2021 IEEE 18th Annual Consumer Communications Networking Conference (CCNC)*; 2021. pp. 1–4.

[83] Pau G, Arena F, Gebremariam YE, *et al.* 'Bluetooth 5.1: an analysis of direction finding capability for high-precision location services'. *Sensors*. 2021;**21**(11). Available from https://www.mdpi.com/1424-8220/21/ 11/3589.

[84] WiMedia Alliance. Multiband OFDM physical layer specification, PHY specification: Final deliverable 1.5; August, 2009.

[85] Sudjai M, Tran LC, Safaei F. Performance analysis of STFC MB-OFDM UWB in WBAN channels. *In: 2012 IEEE 23rd International Symposium on Personal, Indoor and Mobile Radio Communications – (PIMRC)*; 2012. pp. 1710–5.

[86] Tran LC, Mertins A. 'Space–time–frequency code implementation in MB-OFDM UWB communications: Design criteria and performance'. *IEEE Transactions on Wireless Communications*. 2009;**8**(2):701–13.

[87] Tran LC, Mertins A, Wysocki TA. 'Unitary differential space-time-frequency codes for MB-OFDM UWB wireless communications'. *IEEE Transactions on Wireless Communications*. 2013;**12**(2):862–76.

[88] Leugner S, Pelka M, Hellbrück H. Comparison of wired and wireless synchronization with clock drift compensation suited for U-TDoA localization. *In: 2016 13th Workshop on Positioning, Navigation and Communications (WPNC)*. IEEE; 2016. pp. 1–4.

[89] Leugner S, Constapel M, Hellbrueck H. Triclock – clock synchronization compensating drift, offset and propagation delay. *In: 2018 IEEE International Conference on Communications (ICC)*; 2018. pp. 1–6.

[90] Verhelst M, Vereecken W, Steyaert M, *et al.* Architectures for low power ultra-wideband radio receivers in the 3.1-5 ghz band for data rates < 10mbps. *In: Proceedings of the 2004 International Symposium on Low Power Electronics and Design*; 2004. pp. 280–5.

[91]   Terada T, Yoshizumi S, Muqsith M, *et al*. 'A CMOS ultra-wideband impulse radio transceiver for 1-mb/s data communications and /spl plusmn/2.5-cm range finding'. *IEEE Journal of Solid-State Circuits*. 2006;**41**(4): 891–8.

[92]   Chen SWM, Brodersen RW. 'A subsampling radio architecture for ultra-wideband communications'. *IEEE Transactions on Signal Processing*. 2007;**55**(10):5018–31.

[93]   Lee HJ, Ha DS, Lee HS. Toward digital UWB radios: Part i – frequency domain UWB receiver with 1 bit ADCs. *In: International Workshop on Ultra Wideband Systems Joint with Conference on Ultra Wideband Systems and Technologies. Joint UWBST IWUWBS 2004 (IEEE Cat. No.04EX812)*; 2004. pp. 248–52.

[94]   Hoyos S, Sadler BM. 'Ultra-wideband analog-to-digital conversion via signal expansion'. *IEEE Transactions on Vehicular Technology*. 2005;**54**(5): 1609–22.

[95]   Gezici S, Sahinoglu Z, Molisch AF, *et al*. 'Two-step time of arrival estimation for pulse-based ultra-wideband systems'. *EURASIP Journal on Advances in Signal Processing*. 2008;**2008**:1–11.

[96]   Fontana RJ. 'Recent system applications of short-pulse ultra-wideband (UWB) technology'. *IEEE Transactions on Microwave Theory and Techniques*. 2004;**52**(9):2087–104.

[97]   Mahfouz MR, Fathy AE, Kuhn MJ, *et al*. Recent trends and advances in UWB positioning. *In: 2009 IEEE MTT-S International Microwave Workshop on Wireless Sensing, Local Positioning, and RFID*; 2009. pp. 1–4.

[98]   Ward A. In-building location systems. *In: The Institution of Engineering and Technology Seminar on Location Technologies*; 2007. pp. 1–18.

[99]   Humatics rail navigation system datasheet; 2021. Available from https://www.timedomain.com/blog/resource/humatics-rail-navigation-system-datasheet/.

[100] Narvaez AA, Guerra JG. Received signal strength indication—based covid-19 mobile application to comply with social distancing using bluetooth signals from smartphones. In: Data Science for COVID-19. Elsevier; 2021. pp. 483–501.

[101] Kumar S, Gautam V, Kumar A, *et al*. Social distancing using Bluetooth low energy to prevent the spread of Covid-19. *In: 2021 11th International Conference on Cloud Computing, Data Science Engineering (Confluence)*; 2021. pp. 563–7.

[102] Ng PC, Spachos P, N Plataniotis K. 'Covid-19 and your smartphone: BLE-based smart contact tracing'. *IEEE Systems Journal*. 2021. pp. 1–12.

[103] Yang Z, Tran LC, Safaei F. Step length measurements using the received signal strength indicator. *Sensors*. 2021;**21**(2). Available from https://www.mdpi.com/1424-8220/21/2/382.

[104] STMicroelectronics. Social distancing detection using bluetooth® low energy, AN5508; June 2020. Rev. 1.

[105]   Verma N. Turning renesas UWB chips into social distancing solutions —
        phase 1. Available from https://capgemini-engineering.com/us/en/insight/
        turning-renesas-uwb-chips-into-social-distancing-solutions-phase-1/.

[106]   Zhang H, Safaei F, Tran LC. 'Joint transmission power control and relay
        cooperation for WBAN systems'. *Sensors*. 2018;**18**(12). Available from
        https://www.mdpi.com/1424-8220/18/12/4283.

[107]   Barsocchi P, Calabró A, Crivello A, *et al.* 'Covid-19 & privacy: Enhancing of
        indoor localization architectures towards effective social distancing'. *Array*.
        2021;**9**:100051.   Available   from   https://www.sciencedirect.com/science/
        article/pii/S2590005620300369.

Chapter 3

Wireless and networking technologies for social distancing—indoor and outdoor

Jorge Querol[1], Alejandro González[1], Vu Nguyen Ha[1], Sumit Kumar[1] and Symeon Chatzinotas[1]

3.1 Overview

The term *social distancing* describes the limitation of the physical interactions between human beings in the case of a contagious disease outbreak [1]. One may distinguish between individual and crowd/public social distancing measurements. Some examples of public social distancing are travel restrictions, building access limiting, or border control. Examples of individual social distancing are quarantine, trajectory avoidance, and interpersonal distance limitation. Social distancing measures play a crucial role in limiting the severity of the impact of the pandemic, both in terms of the number of infected people and deaths, despite having a negative effect on the freedom of the individuals and the economy. The effectiveness of social distancing measurements has been widely demonstrated [2–7].

Social distancing is based on monitoring, controlling, and reporting the distances between human individuals or crowds. Therefore, wireless and networking technologies able to perform real-time positioning and/or ranging between tracked nodes are of key relevance to the implementation of effective social distancing measures. Some examples of well-known and established technologies to enable social distancing are Wi-Fi, 4G, 5G, and GNSS, which are already integrated and available in smartphones and wearable devices. Figure 3.1 shows the main existing wireless and positioning technologies for social distancing according to their maximum coverage and typical accuracy ranges. In addition, Table 3.1 summarizes the coverage, accuracy, advantages, and drawbacks of the main groups of wireless and technologies described in this chapter.

This chapter is divided as follows. Section 3.2 contains a description of the main positioning technologies that can be potentially used for social distancing applications. Such technologies are presented in decreasing coverage range order. Section 3.3 describes the main social distancing applications that each positioning technology can

[1]Interdisciplinary Centre for Security, Reliability and Trust, University of Luxembourg, Luxembourg

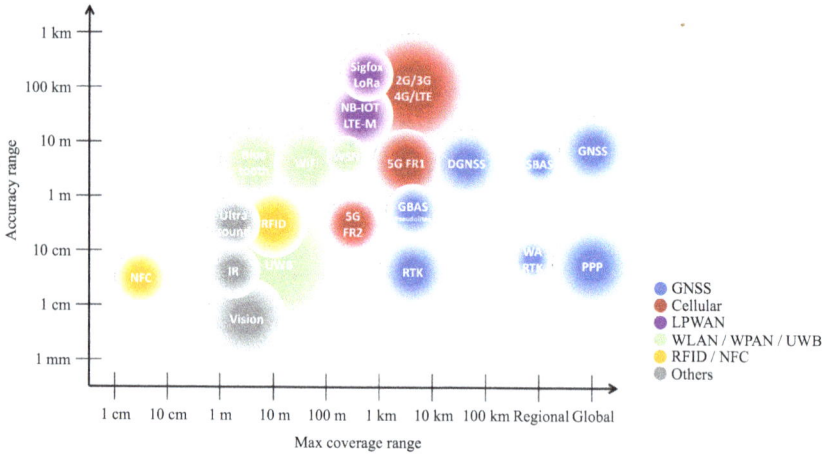

Figure 3.1   Comparison among wireless and networking positioning technologies. The colour code legend indicates the main technology groups from larger to smaller coverage range

Table 3.1   Summary of positioning technologies

Technology	Coverage	Accuracy	Pros	Cons
GNSS standalone	Global	~10 m	– Its global coverage – Unlimited number of users – Free to use	– Need a clear view of sky – Affected by jamming/spoofing – High power consumption to first fix
GNSS with augmentation services	Global Regional	~10 cm	– Very high accuracy	– The receiver needs a communication channel – Higher cost than standalone – Some augmentation systems only have regional coverage
Cellular communications	Regional	~10 m for 5G FR1	– Good coverage indoors – Can provide position and communication at the same time	– Not global coverage – Accuracy in pre 5G systems is poor – Base stations are not deployed thinking in positioning service
Low-power Wide-area Network	Below a kilometre	~1 km	– Low power consumption of the receiver	– Low accuracy – Infrastructure not widely deployed

(continued)

Table 3.1   Summary of positioning technologies (Continued)

Technology	Coverage	Accuracy	Pros	Cons
Wireless Local Area Network	Below a few hundreds meters	~10 m	– Infrastructure already deployed across several buildings – Good indoor coverage – Provides also a communication channel	– Multipath effects – Obstacles – Device heterogeneity
Wireless Personal area Network	Below tens of meters	~10 m	– Includes a communication channel	– Large latency – Low coverage area
Ultra Wideband	Below a few hundreds meters	~1 m	– Good accuracy – Multipath components can be differentiated – Good for indoor environment	– Need a specific infrastructure – Very high bandwidth can interfere with other services
RFID/NFC	Below tens of centimetres	~10 cm	– Low power consumption	– No coverage, just the TAG's position is where the reader is

potentially address. Section 3.4 is devoted to the overall comparison of the technologies and trade-off of the implementation of the different social distancing applications. Section 3.5 presents some future trends in social distancing and new hybrid positioning technologies. Finally, Section 3.6 states the main conclusions of this Chapter.

3.2   Main wireless positioning technologies

3.2.1   Fundamentals of wireless positioning techniques

Wireless localization techniques have been intensively researched and developed for military purposes and also for commercial/professional applications. In the recent pandemic, these techniques have played a very important role in supporting a number of social distancing services. As discussed in the previous chapter, wireless localization techniques are implemented based on geometric measurements, such as distances and angles. Basically, a wireless localization system consists of two components: (1) target nodes (the end devices such as sensors, mobile phones, etc.) and (2) anchors (satellites, base stations, access points, sensors). Both target nodes and anchors in wireless localization systems normally contain the transceiver for transmitting and receiving the RF signal. Extracting the information from the transmitted RF signal, the distances between the wireless nodes can be estimated based on the *'flying time'* of the RF signal, the RF power degradation, the phased variation; the angles of the arriving signals is measured by employing the directional antennas or array of antennas. These geometric measurements suggest the lines, curves (such as circles or hyperbolas), or surfaces (such as spheres or hyperboloids) that the target should locate on. Then, the basic principle of positioning the target nodes is to determine the intersection of these geometric graphics. When the intersection points are not unique, the most potential position can be estimated by employing some advanced techniques like means square error or Kalman filter with considering the impact of environment-related and receiver errors [8]. Recently, the accuracy of wireless localization techniques can be further improved by exploiting the data-driving thanks to the development of machine learning (ML) techniques and the diversification of modern localization scenarios [8]. In what follows, different wireless localization schemes will be introduced.

3.2.2   Global navigation satellite system (GNSS)

The Global Navigation Satellite Systems (GNSS) are the well-known schemes that can provide the Position, Velocity, and Timing (PNT) of devices based on the reception of signals from various satellites. GNSS have been shown several advantages which include but are not limited to the global coverage, the high positioning accuracy of the ranges from a few centimetres to tenths of meters [9,10], and the capacity of supporting a huge number of users. Additionally, all satellites in the same constellation and service share the same transmission frequency (except the legacy GLONASS) using Direct-Sequence Spread Spectrum (DSSS) multiplexing technique; hence, the satellite signals are composed of a pseudorandom code (PRN), used to distinguish each satellite in the same carrier, and a navigation message with information about the satellite position, status and clock, and atmospheric corrections to apply to the signal received [11]. The number of satellites and their orbital positions vary in different constellation designs. However, at least 4 satellites are required for localizing any user. In particular, the signals from these satellites available at the user is able to obtain their own position and also absolute timing.

Generally, there are three main elements in a GNSS architecture:

- *Ground segment*: Composed by the monitoring and the Telemetry, Tracking, and Control (TTC) stations. These stations monitor the health of the satellites and upload the adjustments needed for the satellites to ensure the accuracy requirements at the side of the receiver.
- *Space segment*: Composed by the satellites. They have a very stable clock source and broadcast the navigation message down to the Earth.
- *User terminals*: Composed by the receivers. The user obtains its position and time from the measurement of the distance between itself and the satellites. This information is embedded in the signal transmitted by the satellites. The distance is measured by estimating the signal propagation time, from the satellite transmission to the user device reception, by correlating with the known PRN sequences. Besides, the message that the signal contains, has information about the satellite position and its status. It is needed to solve the equations of positioning and time. One characteristic is that the distances are called pseudoranges because the measurements include the ambiguity corresponding to the clock error of the receiver. The position is obtained using the trilateration method, where the solution is the intersection of 4 spheres, centred each in the satellite and with a radius of the pseudoranges.

In what follows, the different GNSS constellations are briefly introduced which followed by two methods for improving the measurement accuracy in GNSS, the use of the measurements from different constellations at the same time, and the augmentation systems.

3.2.2.1   Constellations description

A GNSS constellation is a complete, independent PNT system, designed to provide a signal that a receiver could use to obtain both, its geographical position and the time with respect to the Universal Time Coordination standard. The four GNSS constellations working globally are GPS/NAVSTAR (USA), Galileo (Europe), GLONASS (Russia), and BeiDdou (China). There are others that cover only a certain region, called Regional Navigation Satellite System (RNSS), like QZSS (Japan) and Navic (India). This subsection is focused on the characteristics of global constellations, not regional ones. For more information about RNSS please refer to [11].

The Table 3.2 contains a summary of all constellation signals. The reader can find a detailed description of each signal in the corresponding constellation Interface Control Document (ICD): GPS [12], Galileo [13], GLONASS [14], and BeiDou [15].

GPS/NAVSTAR
The Global Positioning System (GPS) was the first positioning system with global coverage using satellites. It started with the launch of the Block I satellite in 1978. This constellation design has six equally spaced orbital planes, each one occupied by four satellites at an altitude of 20,200 km. The signal transmitted has two different services, military and civilian, and it is transmitted over three different carriers. The civil signal is open and free to use by any user. It is composed of a coarse and acquisition C/A code. On the other hand, the military signal has an encrypted code.

Galileo
Galileo is a GNSS developed by the EU, and it has five different services defined as:

- *Open*: Free of charge and three carriers offer this service.
- *Public regulated*: It is intended for security authorities. It has enhanced robustness against jamming and spoofing attacks.
- *Commercial*: Provided by two additional signals that are encrypted and have a higher data rate than the other services.
- *Search and Rescue*: Used as a return or uplink channel for distress signals from Earth. The satellites will relay the distress signal to the Rescue Coordination Centre and the user will be informed of its reception.
- *Safety of life*: Used by aviation with increased signal integrity.

Galileo started with the launch of the Galileo In-Orbit Validation satellites in 2005. This constellation design has 27 satellites at an altitude of 23,222 km in three orbital planes. It guarantees to have at least six satellites in view from any point at Earth's surface.

GLONASS
GLONASS was developed by Russia, its design has 24 satellites in three orbital planes with eight satellites in each plane at an altitude of 19,100 km. GLONASS started to be fully operative in 1996. It has, like GPS, two services called:

Table 3.2   GNSS constellations signals summary

Constellation	Band	Carrier (MHz)	PRN code	Modulation	Code rate (Mcps)	Data rate (bps)	Service
GPS	L1	1575.42	C/A	BPSK(1)	1.023	50	Civil
			P	BPSK(10)	10.23	50	Military
			M	BOC sin(10,5)	5.115	N/A	Military
			L1C-I data	MBOC	1.023	50	Civil
			L1C-Q pilot	(6,1,1/11)		–	Civil
	L2	1227.60	P	BPSK(10)	10.23	50	Military
			L2C-M	BPSK(1)	1.023	25	Civil
			L2C-L	BPSK(1)	1.023	–	Civil
			M	BOC sin(10,5)	1.023	N/A	Military
	L5	1176.45	L5-I data	BPSK(10)	10.23	50	Civil
			L5-Q pilot			–	Civil
Galileo	E1	1575.42	E1-A data	BOC cos(15,2.5)	2.5575	N/A	PRS
			E1-B data	MBOC (6,1,1/11)	1.023	125	OS, CS, SoL
			E1-C pilot			–	OS,CS, SoL
	E6	1278.75	E6-A data	BOC cos(10,5)	5.115	N/A	PRS
			E6-B data	BPSK(5)	5.115	500	CS
			E6-C pilot			–	CS
	E5a	1176.45	E5a-I data	BPSK(10)	10.23	25	OS
			E5a-Q pilot			–	OS
	E5b	1207.14	E5b-I data	BPSK(10)	10.23	125	OS, CL, SoL
			E5b-Q pilot			–	OS, CL, SoL
GLONASS	G1	1602.0+ 0.5625k	C/A P	BPSK	0.511 5.11	50 50	Civil Military
	G2	1246.0+ 0.4375k	C/A P	BPSK	0.511 5.11	50 50	Civil Military
BeiDou Phase II	B1	1561.098	B1-I	QPSK(2)	2.046	–	Open
			B1-Q	QPSK		–	Authorised
	B2	1207.14	B2-I	BPSK(2)	2.046	–	Open
			B2-Q	BPSK(10)	10.23	–	Authorized
	B3	1268.52	B3	QPSK(10)	10.23	–	Authorised
BeiDou Phase III	B1	1575.42	B1-C_D	MBOC (6,1,1/11)	1.023	50	Open
			B1-C_P			–	Open
			B1	BOC(14,2)	2.043	50	Authorised
	B2	1191.795	B2-a_D	AltBOC (15,10)	10.23	25	Open
			B2-a_P			–	
			B2-b_D			50	
			B2-b_P			–	
	B3	1268.52	B3	QPSK(10)	10.23	500	Authorised
			B3-A_D	BOC (15,2.5)	2.5575	50	
			B3-A_P	BOC	2.5575	–	

- *SPS*: Standard Positioning Service open and free to use for anyone. This service works in bands G1 and G2 using C/A codes.
- *PPS*: Precise Pointing Service restricted to military users using encrypted codes in the three bands, G1, G2, and G3.

The spectrum management used at the legacy satellites is based on the frequency division multiple access (FDMA) technique, where each satellite transmits at a unique frequency within each band G1 and G2. The legacy satellites have 12 frequency channels, hence the two satellites that transmit at the same time at the same frequency are located in antipodal position. Therefore, any user on Earth will never see these two satellites at the same time. In this case, the PRN sequences for the codes are the same for all satellites, as the satellites are identified by their frequency. However, the latest version of GLONASS allows the use of code division multiple access (CDMA) in bands G1, G2, and G3.

BeiDou
BeiDou was developed by China. It started to be operational in 2011. The total constellation consists of 35 satellites, with 5 GEO, 3 Inclined Geosynchronous Satellite Orbit (IGSO), and 27 in Medium Earth Orbit (MEO) orbit at an altitude of 21,528 km. This system has two different services defined as:

- *Open service* free of charge for all users.
- *Authorized service* where the users have better reliability for positioning, velocity and timing. This service is reserved for authorized users only.

The ranging signals are based in CDMA as the other constellations, sharing all satellites the same carrier frequency and using a PRN code to distinguish each satellite.

3.2.2.2 GNSS performance
The following paragraphs describe different factors that affect the GNSS signal and, therefore, the accuracy of the measurements.

Signal power
Each satellite transmits continuously different signals, usually in two or more frequencies in the spectrum L-band area. The typical output power of a GPS satellite is in the order of +44 dBm, and the free-space path loss is around 170 dB. Therefore, the signal received at the Earth's surface is about −130 dBm. Comparing this value with the noise floor of a receiver with 3 MHz bandwidth (it has a noise power of −109.1 dBm) makes the signal power to be below the noise floor, leading to an intermediate frequency Signal to Noise Ratio (SNR) of −20.9 dB. Values for other GNSS systems are on the same order of magnitude.

The receiver is able to detect the signal by using the spread spectrum technique, plus a coherent integration of the sequences known a priori (the PRN sequences). The expected gain of this process is about +31.1 dB. Therefore, after the coherent integration, the SNR of the signal of a 3 MHz receiver is about +11.7 dB [16].

These values are in ideal outdoor conditions, when the receiver is in an urban canyon, in an area of high multipath, or indoors, the SNR is likely to be below 0 dB, and therefore the receiver is unable to find the signal.

Satellite position

One parameter needed by the user device to find a solution of its position is the localization of the transmitter of the signal. This is a problem in these satellite systems, as most of the constellations use MEO satellites that move fast around the Earth.

The MEO satellites are ideal for global coverage as they orbit around the Earth's surface, whereas the GEO satellites stay in a fixed location (with small deviations). One of the advantages of MEO satellites is that they are closer to the Earth and with better SNR in the reception. The drawback is that using MEO satellites, their positions are a function of time. For that reason, each satellite sends the parameters needed to precisely calculate its position in a precise time t_0. This information is embedded in the navigation message of the signal. The reader interested in the message structure could check the ICD of each constellation and the algorithms involved to get the position in [11].

Once the receiver has the pseudorange measurement and the satellite's position, it can compute its position using the least-squared method for solving the trilateration equations.

One effect that degrades the accuracy is the error in the calculation of the satellite position. This error is improved by the augmentation systems. After the effects that affect the accuracy of the GNSS measurement, a brief description of the different augmentation systems will be made.

Ionospheric/Atmospheric effects

They are one of the biggest sources of error in the pseudorange calculation. This error is due to the changes of the electromagnetic wave produced by the ionization and air molecules at the ionosphere and the troposphere. These effects are time and frequency-dependent. Therefore, in order to remove this error, advanced receivers use signals from different frequency bands. These receivers are called dual-frequency receivers or multi-frequency receivers. As this error is frequency dependent, with the difference of the signals received over different frequencies, the ionospheric and part of the atmospherics effects are mitigated.

However, most of the mass market receivers are single frequency, and therefore, they suffer from these effects. In order to reduce the ionospheric error, the receivers first obtain the parameters of a mathematical model of the channel in the navigation message. Then, via equalization and filtering, the receiver removes in the signal received the effect predicted by this model. There are two-channel models used by different constellations:

- Klobuchar is used in GPS [12], BeiDou [15], Navic [17], and QZSS [18]
- neQuick is used in Galileo [13].

Clocks

For the trilateration methods, all satellite clocks should be synchronized. If there is an offset among them, the position will have an offset too. One of the purposes of the Ground Stations is to measure the offset of each satellite and upload the correction needed to the navigation message. With this information, the receiver is able to compensate for the clock drift.

Some constellations have an extra offset in the navigation message, which is the difference between the constellation time and the UTC time. And it is also sent in the navigation message.

Accuracy of standalone GNSS constellations

Table 3.3 shows the error based on the 95th percentile level of each constellation, using a single frequency receiver for a better comparison between them. These values are the official ones; however, the actual performances exceed these values in an open scenario due to the improvements in signal processing [11].

It is interesting to mention the importance of applying a channel model to the received signal. According to Table 3.3 if the system does not take into account the ionospheric effects, the accuracy is degraded. GLONASS is an example of that lower accuracy than the ones that apply a channel model.

Multi-constellation and multi-frequency systems

There are several situations where a single constellation/frequency receiver can see its performance degraded. Most of these situations (multipath, jamming, ionospheric error) can be greatly improved by the use of a multi-constellation and multi-frequency receiver.

- *Multi-Constellation* receivers' complexity is higher than a single constellation receiver. One of the reasons is that they need to compensate for the time difference between constellations, as each one has its own clock and needs to be compensated for UTC. Another reason is the orbit calculation is different in the case of GLONASS in respect to the other constellations. Also, these receivers have to be able to receive, decode, and process signals from a wider bandwidth. A multi-constellation receiver has a great advantage where the multipath is very high, and where the receiver of a single constellation has just a few satellites in view. If this receiver uses the measurement from the other constellations, the satellite geometry will be much better, and the accuracy of the solution improves as it will have more measurements to find the solution. The algorithm could take the best measurements from different constellations and combine them [19].
- *Multi-frequency* receivers have also advantages over single frequency, as it could discard the ionospheric error because this error depends on the frequency. They also can detect effectively jamming and spoofing attacks, as this kind of attack usually uses a single frequency. The main drawback of these receivers is that the

Table 3.3   Theoretical accuracy of each constellation

Constellation	Coverage	Horizontal (m)	Vertical (m)	Timing (ns)	Service
GPS	Global	13	22	40	SPS
Galileo	Global	15	35	30	Single frequency
GLONASS	Global	28	60	1000	SPS
BeiDou Phase III	Global	10	10	20	Open

bandwidth of the RF front-end should be much higher and also the antenna design should cover all bands. Therefore, the cost is higher than a single band GNSS receiver. In order to overcome this drawback, a more practical approach is using a dual front end with a narrower bandwidth. However, this solution cannot use high-precision algorithms based on the phase relationship between signals.

3.2.2.3   Augmentation services

GNSS augmentation services are methods of improving the performance (i.e. better accuracy, less latency, and less power consumption) of GNSS using external information to retrieve the PNT solution.

D-GNSS

One of the biggest sources of error in the pseudorange measurement is the effect of the ionosphere and atmosphere in the RF signal. One solution to avoid a costly multi-frequency receiver for each user is Differential GNSS (D-GNSS). Originally it was designed for GPS only. It consists of a GNSS receiver fixed in a known position, called the base, which sends to the other receivers, called rovers, its high accuracy observations. Therefore, if the rovers are in the vicinity of the base, the atmospheric errors will be the same, and the rovers could cancel them. These observations are sent to the rovers using another communication channel (e.g. UHF radio or internet). D-GNSS takes advantage of the temporal and spatial correlation of the sources of errors. Therefore, if the baseline between base and rover increases, the error mitigation is less effective.

SBAS

A Satellite-Based Augmentation System (SBAS) is a wide-area or regional augmentation system using geostationary (GEO) satellites, which broadcast the information of the navigation message and other corrections to the receivers. For example, Europe has European Geostationary Navigation Overlay Service (EGNOS), whereas the USA has the Wide Area Augmentation System (WAAS). They are based on the idea of having several monitoring stations around a region and sending the ionospheric corrections to the receivers via geostationary satellites. Although they use the same codes as the conventional GNSS signals, they are not used for ranging. It is just the payload information that is useful for the receiver.

GBAS

The Ground-Based Augmentation System (GBAS), also called *pseudolites*, is an augmentation system used by commercial aviation as part of the Instrument Landing System (ILS). The idea behind this is to have a fixed GNSS station in the vicinity of the airport or another critical infrastructure. These stations send the ionospheric correction to the aircraft receivers. The main difference with SBAS is that the GBAS has a smaller (local) coverage area.

RTK

The Real-Time Kinematic (RTK) technique is designed for high accuracy positioning. It uses the concept of D-GNSS but this time the base sends also the carrier-phase offset of the signal observed in addition to sending its pseudoranges. For example, in GPS L1

C/A, if a receiver could track the PRN-code and the carrier-phase with an accuracy of around 1%, this results in an uncertainty of 3 m for PRN-code but 1.9 mm for carrier-phase. The problem for achieving such high accuracy is that the carrier-phase is a sinusoidal signal and repeats every 19 cm, therefore, the receiver needs to solve this ambiguity before the location can be obtained.

The equations for solving the ambiguity are detailed in [20]. The idea behind them is to perform what is called double differencing. First, find the difference between the measurements of base and rover of one satellite, and then, the difference again between different satellites. This technique reduces the ionospheric, satellite clock error, and tropospheric error in the first instance. After the second difference, it removes the receiver clock error and reduces further the ionospheric and atmospheric errors.

One of the limitations of this approach is the time to the first fix is higher than a receiver without RTK. Another disadvantage is that the coverage of correction measurements is in the vicinity of the base. Furthermore, the base and rover need to have a bidirectional communication channel. Nowadays, with the high availability of different constellations and satellites, the time for the first fix is reduced to a few seconds, opening this system to new applications, like UAVs, self-driving vehicles, etc.

The Wide Area RTK (WARTK) technique is a simplified version of RTK, where just accurate ionospheric corrections are provided and, thus, the distance between the receiver and the reference station can be increased to hundreds of kilometers.

PPP

The Precise Point Positioning (PPP) technique does not use differential observations (no base station needed), but it uses carrier-phase observations from the received signal. Its fundamental features are a highly precise orbit and clock corrections. For example, the International Geospatial Service provides data corrections free of charge, with an accuracy of 5 cm [21]. These measurements are produced independently of GNSS ground control. The GNSS receiver gets a precise ephemeris and a precise satellite clock correction. The accuracy using this technique can be on the order of a few cm compared with the accuracy of a GNSS receiver without any augmentation technique implemented. However, one of the drawbacks is that the receiver needs to be multi-frequency to reduce the ionospheric effects. Furthermore, for a high accuracy solution, it needs much more time for processing. The receiver needs to find the carrier-phase ambiguity iteratively, as there is no double differencing like in the RTK case. The time for the first fix could reach up to 30 min [22].

3.2.3 Cellular communications

Mobile terrestrial networks are characterized by their cellular architecture, where there are several base stations spread across a region, and each base station has an area of coverage. Depending on the population density, this area will be bigger or smaller. For example, in a rural area, the coverage of a single base station could be several km, whereas in an urban area it could be in the range of hundreds of meters, or even less in the case of picocells.

The user equipment (UE) connects to a base station and if it moves outside its coverage, a process called handover takes place, keeping the connection up using the neighbour base station. In this scenario, the network and device need to know where the device is located in order to start the handover process before leaving the coverage area. User localization in the first generations of mobile communications (1G/2G) was based on knowing the location of the cell coverage area, and the accuracy was the size of the cell area.

Specific location services in cellular networks started in the mid-1990s. However, it was not until the LTE release 9 (4G) where the accuracy of the positioning could meet the FCC requirement for emergency calls [23,24]. The accuracy using a legacy mobile network was on the order of hundredths of meters, mainly because of several sources of error, such as multipath, network synchronization and bandwidth of the signal [16].

On the other hand, 5G has a dedicated localization service in its latest review [25]. There are new methods to estimate the device localization, where one of the biggest improvements in 5G is that the base stations are able to send in a broadcast message its position using an extension of LTE Positioning Protocol [26]. This allows the UE to obtain its own position, freeing network resources.

Another major breakthrough in 5G positioning services is the introduction of the mm Wave and the massive MIMO technology for beamforming [27]. This has two advantages, larger bandwidths than the previous technologies, and narrow beam-width used for positioning techniques based on Angle Of Arrival or AOA.

5G release 16 [28] introduces different metrics measuring the accuracy of the positioning, speed, and bearing. Also, it includes three Key Performance Indicators (KPIs) related to the timing, as the latency, time to first fix, and update rate of the position. Moreover, two more KPIs are introduced, power consumption and the energy per fix.

This section will start by describing the different methods used by legacy cellular networks, like 2G/GSM, 3G/UMTS, or 4G/LTE. Then, it will show the improvements that 5G brings to the localization service and the different techniques used by 5G to locate a device.

3.2.3.1   Legacy localization methods

There are different techniques designed to obtain the position of a mobile device. Table 3.4 summarizes the positioning methods in legacy terrestrial cellular networks (GSM, UMTS, CDMA, and LTE). Moreover, the position estimation can happen either network-based, at the base station and network end, or mobile-based, at the UE end.

There is a brief description of each method, for more details of each method, the reader could check the reference [16]:

- *Cell ID (CID)*: Each base station transmits a unique cell ID. The UE can determine its position by obtaining this parameter. When the UE receives the cell ID, the UE assigns its position to the base station with an uncertainty of the area of the

Table 3.4 Positioning methods used by different cellular network generations

Technology	GSM 2G	UMTS 3G	LTE R8 3.9G	LTE-A R9 4G	LTE-A R11	LTE-A pro R12 4.5G	LTE-A pro R13
Cell-ID	x	x					
E-OTD	x	x					
MTA-MOTD	x						
A-FLT		x					
E-CID			x	x	x	x	x
OTDOA	x	x		x	x	x	x
U-TOA	x						
UTDOA		x			x	x	x

cell. This uncertainty could reach from a few dozens of meters in the case of pico and femtocells up to 50 km in rural areas' cells.

It can be improved if the base station takes into account the timing advance TA measurement, as the uplink transmission is based on TDMA. This measurement is used by the network to allow the UE to transmit at a certain time slot, therefore this TA measurement is the time that the signal needs to reach the base station. The resolution of this measurement gives an uncertainty of a few hundredths of meters, and it has a shape of a ring with the base station in its centre and the UE in any location in the ring.

- *Enhanced Observed Time Difference (E-OTD)*: In the E-OTD technique, the UE measures the time that a transmission from a base station is received. It uses a synchronization burst transmission from the cell base station to obtain this measurement. Using TDOA between measurements from different nearby base stations, it computes where the hyperbolas should be located. The problem with this approach is that the base stations in 2G and 3G technology are not required to be synchronized between them. Therefore, the synchronization bursts are sent at different times between base stations, increasing the error. In this system, the position can be calculated by the UE or the UE could send the measurement to the network and the network obtain its position. In case of the position being calculated by the UE standalone, the UE needs to know the position of the transmitter in order to calculate its position as these are parameters in the equations and also the UE needs to know the time difference between base stations called Real-Time Difference (RTD). Another limiting factor is the timing resolution of each measurement, where the smallest step corresponds to a one-time slot of 577 μs corresponding to the uncertainty in the position of hundredths of meters.

- *Uplink TOA (U-TOA)*: U-TOA method was designed for the GSM networks. Here, the UE sends the initial access burst in a specific time slot, and the base station measures the time of arrival of the UE's burst. The network needs two more observations of the ToA, and to obtain them the network initiates a handover procedure for the UE to a neighbour cell. When the network gets a second measurement it continues with the third measurement from another neighbour cell. Once the network has the three measurements, the network can obtain the position of the UE by trilateration. The main drawback of this method is that the UE should be able to reach at least three different base stations. This could be a challenge in a rural scenario where the distance between base stations is high enough for the signal of the UE to not be heard by the neighbour cells.

- *Multilateration Timing Advance (MTA) and Multilateration Observed Time Difference (MOTD)*: In MTA, the position is obtained only between the base station equipment and the UE, the core network is not involved here. MTA starts obtaining the Timing Advance value from nearby different base stations and combining them to obtain the UE position. And MOTD is similar to E-OTD, it uses the TA value of the reference cell, plus the Observed Time difference (OTD) from the neighbour cells and obtains the position as a combination of TOA and TDOA.

- *Advanced Forward Link Trilateration (A-FLT)*: A-FLT technique is used in CDMA networks where the base stations are synchronized. The UE measures the phase of the pilot CDMA signal and calculates the Time of Arrival (ToA) and using trilateration obtain its position. In this case, due to the resolution of the signals, the uncertainty is around 30.5 m. The limitation here is that the UE should be capable of obtaining the phase of the CDMA signal.

- *Enhanced CID (ECID)*: ECID method is similar to Cell-ID with Timing Advance. In this case, this is a network-based method, where the network obtains the round trip time (RTT) between a cell and a UE. The advantage is that clock errors are cancelled out. It is a trilateration method, as the measurement is a range. In this case, the uncertainty is around 9.76 m. However, this method is between a base station and a UE and it needs more measurements from others base stations to solve the position equations. One solution is that the network starts a handover to the next base station, it measures the RTT, changes to other base station and after two changes, it has three range measurements for solving the trilateration equation. This solution has a drawback, the UE should be stationary during the three measurements. Another method available, instead of the RTT is using the Reference Signal Received Power (RSRP) from the neighbours base stations and the distance to these other base stations could be obtained by solving the channel model equation. However, these measurements have a very poor accuracy as the channel is time-varying and the mathematical models are not very accurate.

- *Observed TDOA (OTDOA)*: OTDOA is like E-OTD but instead of using the synchronization bursts, LTE has added a Positioning Reference Signal (PRS) for that purpose. This signal has three parameters, a configuration index (how often the PRS will be transmitted), the bandwidth used by the PRS, and the number of downlink subframes which the PRS will be transmitted. Using this information, the UE knows when the PRS will arrive, and it waits for it to obtain the measurement. The delay measurement is obtained via the correlation between the PRS received and a local copy. The network is responsible to calculate the position, once the measurements are obtained by the UE, it sends them to the network core. The reason for calculating the position in the network is that the UE does not have information about the coordinates of the base stations to solve the equation.

In this case, the theoretical uncertainty under ideal conditions could reach up to 2.44 m.

- *Uplink TDOA (U-TDOA)*: U-TDOA solution is an evolution of the previous method U-TOA from GSM. It relies on the time difference of arrival measurements, and it does not require the handover procedure. Another advantage is that all measurements and complexity are managed by the network. Therefore, it does not use any dedicated signal and benefit from the frequency hopping in GSM to reduce the multipath errors. This method is used in LTE and has an accuracy of 9.76 m.

3.2.3.2 Main sources of error in legacy positioning systems

The main sources of error in terrestrial mobile networks are four [16]:

- One of the major sources of error in the terrestrial networks' positioning system is network synchronization. As has been shown in the previous paragraphs, the uncertainty in TDOA cases depends on the accuracy of the base stations' synchronization within the network timing [29].
- Another important source of errors is the multipath, as the signal bandwidth is small, the resolution to distinguish between different paths is very small, and the receiver could use a reflection instead of the direct path. This effect could be mitigated partially if there are enough measurements and the positioning algorithm detect the worst or faulty measurements and discard them.
- The next source of error is how the base stations' geometry is arranged, this is known as Geometric Dilution of Precision GDP, and only depends on the positions of the base stations. For example, as the variation in elevation is not great, the vertical component of the position has very poor accuracy. For that reason, most of these positioning techniques are used for 2D instead of 3D [29].
- And the last source of error is the database with the base station locations. As this information is needed for solving the equations, if there is an error in the database, the solution will not be valid or will have very high uncertainty.

3.2.3.3 Novel 5G localization methods

The fifth generation of cellular communications (5G) uses natively the legacy localization methods (e.g. ECID, OTDOA now called DL-TDOA and RTT now called multi-RTT), but it also introduces new ones which are power-based (RSS) and angle-based (AOA) solutions.

The main difference between 5G and the others generations is its flexibility in terms of configuration bandwidths, carriers spacings, and so on. Furthermore, the frequency ranges 5G uses are divided into two blocks: FR1 reaches 7.125 GHz with similar properties that LTE but, with a bandwidth up to 100 MHz; FR2 starts at 24 GHz, this high frequency allows angle-based positioning as the beams are very narrow for compensating the higher propagation losses.

One of the main limitations of the previous generations is that the positioning methods UE-based were not possible as the UE does not have information about the position of the base stations, all techniques were UE-assisted where the UE takes the measurements and sent them back to the network to perform the positioning. In

5G, this has been solved by sending the required information to the UE via de LTE Positioning Protocol (LPP). It has been refurbished for this purpose [26]. Therefore, now the UE could get its position based only on the information gathered from the network.

Time-based techniques

Regarding time-based techniques, the basic unit for timing-based positioning methods is based on the maximum supported subcarrier spacing. This value is around 0.5 ns (few centimeter), compared with the same parameter in LTE, where it is around 32.55 ns (few meter).

- *DL-TDOA*: This technique is the evolution of the OTDOA in LTE. It measures the time between the reception of signals from a few neighbouring base stations. In 5G, the PRS has been redesigned, allowing the mobile device to receive this signal from multiple neighbouring base stations. In addition, the PRS is designed to avoid interference between base stations, enabling the UE to detect them independently. These measurements are called RSTD, and they are defined as the difference in time of arrival of the PRS from the neighbouring base station compared to the time of arrival of the PRS of a reference station [30]. The granularity is not yet defined in the standard, but it is proportional to T_c, 1–4 T_c for FR2 and 4–16 T_c for FR1. This is translated into an uncertainty between 60 cm and 2.4 m.
- *UL-TDOA*: The principle is the same as the DL-TDOA, but, in this case, it is the network that does the measurements from different base stations. The main difference with previous generations is that it is not based on the random burst access, it is based on the Sounding Reference Signal (SRS) sent by the UE, which has been extended for positioning purposes [31]. This method is particularly useful for UE tracking services as the measurement and calculation are done on the network side without involving in the UE any calculation.
- *Multi-RTT*: This method is similar to RTT in LTE, with the difference that in 5G it is allowed to get the measurement from the neighbouring base stations. First, the UE synchronizes with the reference base station, adjusting its timing to transmit at the correct time slot. Then, the neighbour cells take the measurement using the SRS signal. After this measurement is done, the neighbour cells respond with the PRS signal for the UE to make the delay measurement. Finally, after both measurements are done, the UE reports them to the network, and the Location Management Function (LMF) in the core network calculates the position.

Power-based techniques

This is another approach to find the position of a mobile, these techniques are much less accurate than timing methods because they need a very good estimation of the channel model [16]. However, the advantage is that there is no need to add extra features to the UE. The localization of the UE can be obtained by measuring the signal power received by different base stations, and with a channel model, the network can estimate the distance to the UE and then, via trilateration, obtain the position.

- *NR ECID*: This method is purely based on power measurement. This is the main difference with ECID from LTE, as NR ECID does not use RTT measurements. It is divided into two groups, using the synchronization signal or using the channel state information reference signal.

- *PRS RSRP* This method is similar to the previous one, with the difference that it uses the PRS signal, and it should be requested by the mobile to the network.

Angle-based technologies

Thanks to the inclusion of the FR2, the angle-based measurements reach an accuracy that could be useful for positioning purposes. The beams in these frequencies should be narrower compared with the typical sectorial approach in the previous generations antenna sites. In that way, the antenna increases its gain, compensating for the higher channel losses in FR2. In a cell, each beam has an identifier that could be used for the positioning of the UE within the cell coverage. There are two different angle measurements:

- *Downlink Angle of Departure* DL-AoD method is based on the measurement of the different beams' ID by the UE from the incoming signals. The measurements are sent back to the network to calculate the UE position. In the case, the network provides the base station positions and beam directions of each beam ID, the mobile could obtain its own position without sending the measurement to the network.
- *Uplink Angle of Arrival* UL-AoA is similar to DL-AoD with the difference that in this case, the base stations obtain the beams' ID that the UE sent to different base stations and the network compute the location based on that information.

This method could increase its accuracy by adding the PRS RSRP measurement from different beams or using RTT measurements and combining all measurements into a single equation.

Other positioning techniques

There are a few more positioning methods that can be used within 5G, but were not included in the specification due to the trade-off between the complexity versus accuracy is not good enough.

One is the carrier-phase-based positioning, which is similar to RTK and PPP in GNSS, that uses the measurement from the positioning reference signal. However, this method needs to dedicate a sinusoidal wave for it, called Carrier-phase PRS or C-PRS. The problem with this method is the synchronization accuracy between base stations should be higher than the current requirement, to reach an accuracy better than the previous methods.

The other method is called Phase Difference of Arrival or PDoA, it is similar to the carrier-phase measurement from the positioning reference signal, but in this case, it measures the difference in phase from multiple subcarriers of the reference signal [32].

And finally, a hybrid positioning, where DL-TDOA and multi-RTT with GNSS measurements could be combined using the weighted least squares algorithm similar to the reference implemented for LTE [33].

3.2.4   Low-power wide-area network (LPWAN)

Low-Power Wide-Area Network (LPWAN) technologies are those wireless communication technologies designed to serve a long-range area with a low data rate connectivity, thus being able to achieve low power consumption. LPWAN technologies are the ones being used by Internet of Things (IoT) service providers since they are able to serve a large number of connected devices in a large area. In some sense, LPWAN can be seen as a range extension of traditional cellular communications. The requirements of LPWAN communications allow designing small, lightweight, and low-cost devices, usually powered by batteries or small solar panels, making them suitable for IoT applications such as asset tracking, people tracking, fleet management, autonomous vehicles or smart manufacturing [34].

Even though multiple communication standards fall in the LPWAN category, the principles for the determination of the position are common. One of the proposed techniques in [35] is based on multilateration by TDOA, commonly used in LoRa and NB-IoT.

3.2.4.1   LPWAN Standards

In the latest decade, a huge number of different LPWAN standards have emerged to address the needs of IoT applications in licensed or license-free frequency bands [36–39]. A non-comprehensive list of LPWAN standards is the following: LoRa [40], Sigfox [41], NB-IoT [42], LTE-M [42], WiFi HaLow [43], Weightless [44], Ingenu RPMA [45], Telensa [46] and Qowisio [47]. Among them, LoRa, Sigfox, NB-IoT and LTE-M are by far the most used ones [8].

LoRa
The Long Range (LoRa) is a networking technology based on a wireless long-range physical layer standard [48] developed by the company Cycleo and later acquired by Semtech in 2012. Since 2015, the LoRa Alliance, with over 500 members, has deployed more than 100 LoRa public networks. LoRa technology is able to serve users in up to 15 km range from the base station thanks to the high sensitivity of its receivers (–135 dBm) [49]. LoRa modules have low-power consumption and are compatible with positioning by multilateration with a reported accuracy of 100 m in a static environment [50].

Lora uses a non-licensed spectrum (typically 433/868/915 MHz [8]), which makes its deployment flexible, particularly, for independent networks such as industrial parks, farms, or remote areas. On the other hand, the use of a non-licensed spectrum makes LoRa not compatible with existing cellular base stations, and it makes it more vulnerable to jamming attacks and inter-operator interference. Despite it, the LoRa standard has an established ecosystem with already working applications in several countries. Moreover, LoRa uses a novel network topology based on a star of stars. Thus making it easier to deploy than other LPWAN technologies.

Sigfox
Sigfox is an LPWAN technology developed by a company of the same name and standardized by ETSI [51], that has been already deployed in over 60 countries [41].

Sigfox also uses free-license frequency bands, but contrarily to LoRa, the Sigfox company is also a network operator providing worldwide low data rate connectivity services.

Sigfox is based on the Ultra NarrowBand (UNB) technology, limiting its data rate as low as 100 bps with a restricted number of messages per day (140 messages/day with 96 bits/message). This allows Sigfox to deploy a low-dense international base station network, which reduces the operational and deployment costs. From the positioning perspective, this is a clear disadvantage with respect to LoRa.

NB-IoT

The 3GPP completed the standardization of NarrowBand Internet of Things (NB-IoT) technology in 2016 [42]. Despite LoRa and Sigfox started their deployment earlier, the NB-IoT community is growing fast with more than 140 operators (mainly cellular communication operators) in 69 countries [8].

NB-IoT uses licensed spectrum, thus making it attractive to cellular communications operators to extend their services by upgrading their existing base stations. It also enables operator-level security and quality assurance. The physical layer interface is designed for integration with 5G and LTE, but low data rate (250 kbps) and low power consumption similar to LoRa. Regarding its localization capabilities, a best-case of 76 m accuracy is reported in a static scenario [8].

LTE-M

LTE-M was also standardized by 3GPP in 2016 [42] and shares feature with NB-IoT. Moreover, LTE-M is also designed for low-data rate cellular communications (up to 1 Mbps) and also supports voice transmission. Its larger bandwidth allows achieving better localization accuracy as compared to NB-IoT.

3.2.4.2 Relation to 5G cellular

5G has introduced innovative solutions for both communications and positioning applications. In particular, they are classified into three groups: Ultra-Reliable and Low Latency Communication (URLLC), Enhanced Mobile Broadband (eMBB), and massive Machine-Type Communication (mMTC) [24]. mMTC has been designed for scenarios with low-cost devices, low data rate, low power consumption, and latency independence, similar to LPWAN applications. NB-IoT and LTE-M are already within the 3GPP standardization body, and the 5G mMTC vertical is expected to merge them in a unified standard.

3.2.5 *Wireless local area network (WLAN)*

A Wireless Local Area Network (WLAN) is a wireless computer networking technique that establishes a communication link between two or more devices within a limited area such as schools, shopping complexes, parking areas, and so on. WLAN connection also provides limited mobility with the indoor environment. Besides the establishment of LAN connection, WLAN, through a gateway, can also provide a connection to the internet. WLAN follows the IEEE 802.11 set of standards [52]. This is commonly known as WiFi*. WiFi operates in the unlicensed Industrial Scientific and Medical (ISM) band (2.4 GHz and 5 GHz) which poses a limit only over the transmit power.

Since 1997, WiFi has evolved in parallel to the evolution in the cellular communication regime. Interestingly, there is an overlap of stakeholders of WiFi [53] and those of cellular communications (3G/4G/5G), hence WiFi has not been seen as a competitor of cellular technologies but it complements and acts as a last-mile solution. Even with the penetration of cellular technologies such as 4G/5G, indoor coverage of cellular technologies still becomes problematic. In such a case, WiFi comes as the last mile solution where the WiFi Access Point (WiFi-AP) is connected to a cellular Base Station or a wired gateway and the end-users connect to WiFi in order to access the internet-based services. Especially inside indoor environments such as restaurants, hotels, airports, train stations people prefer connecting to WiFi hotspots (if available) either through registering or without registration (in public hotspots). This has been standardized in 4G and 5G through the provisioning of evolved Packet Data Gateway (ePDG) [54] and Non-3GPP Interworking Function (N3IWF) respectively [55]. Using ePDG/N3IWF, a user can be connected to the 4G/5G Core using the WiFi-AP. ePDG/N3IWF is useful when the 4G-eNB/5G-gNB is not in the reach of a 4G-UE/5G-UE. A representative diagram of the scheme is shown in Figure 3.2.

Besides, the WiFi chipsets are available in very small form factors and hence can be embedded in almost any device to which the WiFi connection has to be established. As a consequence, WiFi chipsets are by default available on all smartphones. Nowadays, WiFi is omnipresent and hence localization and positioning using WiFi have been an active area of research. Currently, with state-of-the-art technologies, existing WiFi infrastructures can be used for Indoor Positioning System (IPS) with no modification to existing hardware [56–58]. Table 3.5 shows the evolution of WiFi and the main characteristics of every generation.

After IEEE 802.11b which was based on DSSS, FHSS all the new generations of WiFi used OFDM as the physical layer waveform. Additionally, starting from IEEE 802.11n, all the WiFi standards support MIMO (SU-MIMO and MU-MIMO) technology.

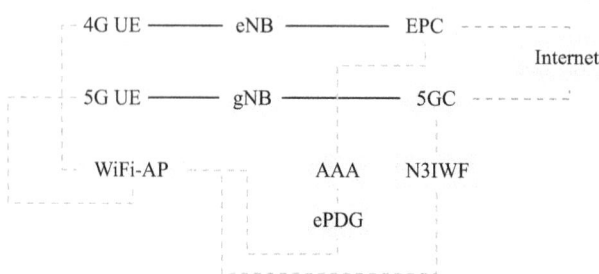

Figure 3.2   Representative diagram showing a cellular 4G/5G user getting access to the internet via either 4G/5G RAN or through WiFi using ePDG/N3IWF

---

*We will use WiFi and WLAN interchangeably in this chapter.

Table 3.5 Evolution of various WiFi standards

Protocol: year	Operating freq (GHz)	Max BW (MHz): max data rate (Mbps)	Outdoor range (m)	Indoor range (m)
802.11b:1999	2.4	22:11	35	140
802.11a:1999	5	20:54	35	120
802.11p:2010	5.9	20:54	NA	1000
802.11g:2003	2.4	20:54	38	140
802.11n:2009	2.4/5	40:600	38	140
802.11ac:2013	5	40:600	38	140
802.11ax:2021	2.4/5/6	80:4804	30	120
802.11ad:2012	60	2160:6757	3.3	NA
802.11ay:2021	60	8000:20, 000	10	100

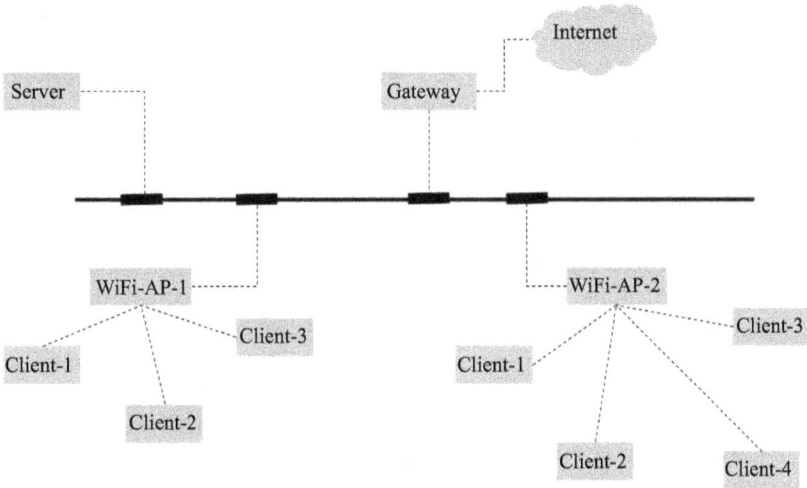

Figure 3.3 A typical architecture of WiFi where multiple WiFi-APs are providing internet access to multiple connected clients

3.2.5.1 Architecture of WiFi

WiFi can operate in either infrastructure mode or ad-hoc mode (also known as peer-to-peer) with the former being the major mode of deployment. In the infrastructure mode, there are two ways for medium access: Distributed Coordinated Function (DCF) and Point Coordinated Function (PCF). In DCF, there is no centralized unit and the users compete with each other to gain access to medium (wireless medium) by using the CSMA-CA technique. While in PCF, there is a central entity that regulates the access of medium. In practice, DCF based deployments are used due to reduced signalling overhead compared to PCF. Figure 3.3 shows the typical architecture of a WiFi network. There is one or more WiFi-AP to which many clients can be connected.

WiFi-APs can be deployed indoors as well as public places in outdoors. Several WiFi-APs can also be connected to each other (through a wired medium) and further connect to the internet using a gateway.

3.2.5.2   WiFi for localization and positioning

Social distancing is required not only outdoors but indoors also. Due to its omnipresence in indoor environments, WiFi technology has always been seen as a potential candidate for localization and positioning indoor, also termed as Indoor Positioning System (IPS). WiFi has emerged as a prominent solution for IPS which can be deployed quickly and managed easily. Considering accuracy, cost, real-time, and reliability, WiFi suitably fits for IPS. This is also supported by the fact that almost every individual today owns a smartphone and almost all public places are equipped with multiple WiFi-APs (if not, then WiFi-APs can be deployed rapidly and easily).

WiFi-based positioning systems can be classified as active and passive positioning systems [59]. In active positioning systems, a mobile WiFi client, for example, a smartphone, actively searches and collects signals from a nearby WiFi-APs and this information is transmitted to the server for processing where the location of the mobile WiFi client (and hence the user) is approximated through localization/positioning algorithms. While in passive positioning systems, the user does not carry any WiFi client, but the WiFi-AP and clients are strategically deployed at known positions, hence, when a user enters the area, the propagation and hence reception of the signal is affected. Based on such changes, the system approximates the position of the user. To enforce social distancing, we consider active positioning systems only as their accuracy is better than that of passive systems [60,61]. Active positioning schemes are further classified into Range-Based and Range-Free methods. In the former, a client approximates its distance from the WiFi-APs (with known locations) using different geometric methods. Range-based methods provide a higher degree of accuracy compared to Range Free methods. Some of the known Range Based methods include: Received Signal Strength Indicator (RSSI) [62], Time of Arrival (TOA) [63], Time Difference of Arrival (TDOA) [63], and Frequency Difference of Arrival (FDOA) [59]. On the other hand, in Range Free methods, the client does not determine its distance from the WiFi-APs directly, instead, it estimates the distance based on hop-count between the WiFi-APs and the client. After that, it uses geometrical methods for the approximation of actual distance. Some of the known Range Free methods include Centroid [64], Amorphous [65], and DV-Hop [66].

3.2.5.3   Challenges and open research issues

The main challenges and open research issues in WLAN-based localization are:

- *Multipath Effects*: Complexity of indoor space is very high compared to outdoors. Apart from direct line-of-sight, many reflections of the signals from WiFi-APs arrive at the client with different phases and cause a multipath effect. This results in signal distortion and has a significant effect on geometric methods [67].
- *Obstacles*: Most of the signal propagation loss models assume line-of-sight between the WiFi-APs and the client, however, in practice this is not the case. There could be several obstacles such as walls, doors, windows which make an

accurate measurement of signal loss almost impossible. With such variance in the models, the accuracy of the positioning methods is affected significantly [59].

* *Device Heterogeneity*: In practice, the client devices, for example, the smart-phones are different from the reference devices used for calibration because different smartphone vendors use different WiFi chipsets, WiFi antennas, hardware drivers, packaging materials, and operating systems. This creates a difference between the strength of the signal received at the reference device and the client device. As many of the algorithms for positioning use RSSI, such difference affects positioning accuracy[68].

Although methods to approximate the position of users using WiFi have matured, there are still some open research issues and challenges while applying such techniques for enforcing social distancing. The indoor environment changes frequently, and the pre-built wireless map may outdate soon. The application of crowd-sources data for this purpose can be researched to solve this problem [69,70]. Further, the WiFi-APs were deployed and positioned for networking purposes and not for positioning/localization. New WiFi-APs can be placed strategically for facilitating the methods of positioning and will lead to an increase in the accuracy of the methods. Besides, positioning in an unknown environment, i.e., an environment whose radio-map and path loss model is not available is still a challenge.

3.2.6 Wireless personal area network (WPAN)

A Wireless Personal Area Network (WPAN) is a network that can be usually set up and dismantled ad-hoc by small devices such as smartphones or laptops. Compared to WLAN, the range of WPANs is shorter, typically between 10 and 15 m. WPANs can be used for both device-to-device (D2D) communication or to connect to a larger network (typically a WLAN). WPANs usually operate in free-licensed bands, and typically in the 2.4 GHz. WPANs are efficiently used for indoor positioning due to the low power and short-range capabilities of their small transceivers [71]. As with any indoor positioning system, their performance suffers more from the complex and changing scenario as compared to outdoor environments.

3.2.6.1 Bluetooth

Bluetooth is a widespread WPAN standard currently embedded in devices such as smartphones, tablets, laptops, and desktop PCs. Therefore, the use of such a technology for positioning purposes does not require additional hardware for its deployment [72].

The Bluetooth Specification v5.3 [73] was released on July 2021 by the Bluetooth Special Interest Group (SIG). The standard defines two types of technology: Bluetooth Basic Rate (BBR) and Bluetooth Low Energy (BLE). The BBR also includes an optional mode named Enhanced Data Rate (EDR). The BBR with EDR extension can reach up to 2.1 Mbps sustained throughput. The BLE technology is designed for lower power consumption, lower complexity, and lower-cost devices as compared to BBR/EDR. On the other hand, BLE offers between 125 kbps and 500 kbps physical layer throughput [73]. BBR/EDR technology splits the 2.4 GHz band (2400–2483.5 MHz) in 79 channels with a 1 MHz spacing, and center frequencies from 2402 MHz to 2480 MHz. Conversely, BLE uses 40 channels, with a 2 MHz spacing, in the FDMA scheme, combined with a TDMA-based polling scheme [73].

As well as in WLAN/WiFi, RSS proximity/fingerprinting and multilateration localization methods can be used in Bluetooth, but with the advantage of lower power consumption [2,74,75]. Bluetooth provides a unique identification number per device in the advertisement packets transmitted by specific beacons. The advertisement interval is defined from 100 to 2000 ms. The typical advertisement interval of BLE beacons used for indoor positioning is 300 ms [74]. In 2013, Apple released its own implementation of indoor localization based on BLE beacons called iBeacons [72]. Indoor localization via Bluetooth offers a clear advantage in buildings without previous infrastructure, enabling battery-powered nodes for months.

However, the use of Bluetooth for positioning has some documented drawbacks. The identification of the device requires executing the device discovery procedure, which may introduce between 10 and 30 s of latency [72]. Despite Bluetooth signal measurements being able to provide accuracy down to 2–3 m, the long latency may make it unsuitable for applications requiring real-time localization [72].

3.2.6.2 WSN

Wireless Sensor Network (WSN) refers to a network of small nodes, equipped with sensors, that collaborate on one or more common tasks. WSN usually allows building ad-hoc networks, without pre-established physical infrastructure or central administration. There exist several WSN standards such as 6LoWPAN, Zigbee, WirelessHART, ISA100.11a, and OCARI [76], defining the higher layer protocol functionalities. However, their physical layer specification is based on the IEEE 802.15.4 standard. IEEE 802.15.4 support several frequency bands: 868.0–868.6 MHz (Europe), 902.0–28.0 MHz (US) and 2.40–2.48 GHz (Worldwide). The standard defines a basic framework of 250 kbps at 10 m distance, with several modulation options using DSSS, Chirp Spread Spectrum (CSS), and Offset Quadrature Phase Shift Keying (O-QPSK) with half-sine pulse-shaping carrier modulation [77].

In WSN, some sensor nodes ('anchors') have a known position. The rest of the nodes ('targets') have to estimate their position using multilateration and/or multiangulation [74]. The RSS received by the anchor nodes is commonly used for positioning [72]. For instance, ZigBee systems consist of a central anchor node and multiple target nodes. The devices can communicate with each other at a range up to 20 m with an unlimited number of hops [2].

3.2.7 Ultra-wideband (UWB)

Ultra-Wideband (UWB), due to the definition of the U.S. Federal Communication Commission, is a wireless transmission technique that spreads information over a wide portion of the frequency spectrum that is greater than 20% of the centre carrier frequency or has a bandwidth greater than 500 MHz [78]. UWB was firstly used by the U.S. Department of Defence before appearing widely in commercial applications in the late of 90s in the previous century [79]. Adopting the unlicensed frequency bands of the 3100–10,600 MHz [78], UWB radio can operate at a deficient energy level for short-range and provide high speed data rate communication with low noise interference thanks to the high bandwidth and extremely short pulses waveforms [80]. Moreover, the low frequency of pulses also allows the UWB signal to go through obstacles, e.g. walls and objects, without accuracy rate degradation. Facilitated by the unlicensed operated frequency band as well as its significant advantages, many academic and commercial projects have exploited UWB radio for several civilian applications which can be catalogued into three main areas: (1) communication and sensors; (2) positioning and tracking; and (3) radar [79,81]. Among these three, UWB-based 'positioning and tracking' applications have attracted much attention for developing suitable social distancing solutions during pandemic times [80]. In comparison to other technologies, UWB has emerged as one of the leading technologies for indoor positioning due to its high accuracy and reliability as well as the low operating cost and latency for localization and tracking.

3.2.7.1 UWB positioning technology

While other technologies like Bluetooth or WiFi need to be re-tooled for location purposes, from the very first implementation, the physical properties of UWB radio were specially designed for achieving real-time, ultra-accurate, ultra-reliable positioning and integrated data transmission [79,81]. Implemented flexibly based on different algorithms or IEEE standards like 802.15.4a and 802.15.4z, UWB positioning technology can deliver the centimetre-level accuracy for distance measurement improve the precision of location estimation on mobiles and IoT devices [78,80]. The localization systems often consist of a number of reference nodes, so-called anchors, which collect the signal from the target nodes, so-called tags, extract the position-related information and then report to the tracking units for place detection purposes. In these systems, the tracking units can be the centralized control units or even the target nodes, and the related information can be the time of arrival (TOA) [82,83], angle of arrival (AOA) [84,85], received signal strength (RSS) [86,87], time difference of arrival (TDOA) [88,89] based on which algorithm has been employed. To increase the accuracy for positioning, multiple localization techniques are employed and integrated while the operating cost may also enlarge [90,91].

3.2.7.2 Advantage and disadvantages of UWB positioning technology

Employing the UWB technologies for localization and distance estimation can bring both advantages and disadvantages.

Advantages
The first advantage of using UWB is operating over the unlicensed frequency band which can be used freely, with the only limitation of radiated power. Due to the wideband, the power spectral density corresponding to the UWB radio propagation is significantly lower than that due to the narrowband signal. Hence, UWB equipment does not interfere with most of the existing radio systems [92]. Then, the power consumption of UWB radio is lower than other positioning technologies, which can prolong the battery life of devices. Propagation over large bandwidth also helps UWB

signals resistant to the multipath problems and interference which results in higher data rate and reliability [79]. In addition, transmitting a signal with a short pulse enables time modulated UWB radio has a very low probability of interception and detection which is very useful for security purposes such as military reconnaissance. Very short pulses offer an advantage in terms of resolvability of multipath components because the reflections from objects and surfaces near the path between the transmitter and receiver tend not to overlap in time for UWB radio signals [78]. Moreover, the carrier-less transmission property of UWB enables much simpler hardware implementation for the transceivers than radiofrequency architecture than other narrowband systems which result in a very less producing cost [78,79].

Disadvantages
Although having many significant strengths for positioning applications, UWB radio also comes with some weaknesses. Operating over an unlicensed band can also bring the possibility of interference with nearby systems in some countries such as World-wide Interoperability for Microwave Access (WiMAX), third-generation 3G wireless systems, and digital TV [92]. In addition, several UWB devices may cause interference to GPS and aircraft navigation radio equipment which raises some serious concerns to flight safety [93]. On another side, UWB is benefited from the very short pulse transmission, but these propagation techniques also require reliable signal acquisition, synchronization, and tracking with very high precision in time relative to the pulse rate at the receiver side which consumes a significant processing time duration [93]. Some techniques have been proposed to deal with this time-consuming challenge such as using a preamble sequence for rapid acquisition [78].

3.2.8   RFID/NFC technologies

This subsection refers to the so-called 'contactless' communications, which includes both RFID and NFC technologies, with a few cm to sub-meter coverage range.

3.2.8.1   RFID

The Radio-Frequency Identification (RFID) systems have two kinds of components: readers and tags. The RFID reader contains a transceiver to transmit RF signals and read the data emitted from the tags. The tags can be passive, getting energy from incoming radio signals, or active [74]. Data stored in the tag consist usually in a univocal identification number, but other information can be stored in a tag depending on its memory size [72].

The RFID technology has different standardized working frequencies: low frequency (LF) at 125–134 kHz, high frequency (HF) at 13.56 MHz, and ultra-high frequencies (UHF) at 860–960 MHz. The limitations of LF and HF RFID systems are the short reading range (few cm) and the inability of reading multiple tags simultaneously. In comparison, the UHF RFID technology is more affected by RF absorption and reflection [74].

The localization based on the RFID technology can be classified in the reader or tag localization. In reader localization, the accuracy of the RFID system is highly dependent on the density of tag deployment and the reading ranges. A mobile RFID reader could obtain its position from the closest tag, either by reading the positioning information or by RSS. Its main disadvantage is the need for the deployment of a large number of RFID tags with prerecorded information. On the other hand, tag localization can be implemented by installing multiple RFID readers in known positions. This method has the same principle as the previous case, but at an increased cost, [74].

3.2.8.2 NFC

Near-Field Communications (NFC) is a short-range wireless networking technology based on magnetic induction rather than electromagnetic radiation. Moreover, it is a half-duplex communication jointly developed by Philips and Sony in 2002 for contactless information exchange [94,95]. NFC is mainly an evolution of high-frequency RFID technologies.

NFC takes place between two devices placed at a few cm of each other. The communications frequency is 13.56 MHz, with transfer rates up to 242 kbps. The device starting the communication is the 'initiator', whereas the respondent is the 'target'. In the communication from an initiator, the Amplitude Shift Keying (ASK) modulation is used, whereas the target responds using a load modulation technique based on a time-variation in the impedance of the listening device generating amplitude or phase changes to the antenna voltage of the polling device [95]. Positioning using NFC can be implemented similarly to RFID by using strategically placed targets.

3.3 Potential social distancing applications

This section presents a number of potential social distancing scenarios and applications, as well as a short discussion on its implementation feasibility per each of the technologies presented in Section 3.2.

3.3.1 Social distancing scenarios

The potential social distancing applications can be classified in several scenarios according to their nature and requirements as follows:

1. *Interpersonal distance*: The main goal is to keep a minimum distance between people. The system increases in complexity when multiple distance ranges have to be calculated and combined. One may distinguish between the *distance between any two people* and the *distance to a crowd*. In both applications, the goal is to warn individuals about when to practice social distancing.
2. *Area monitoring*: The goal is to control or monitor in real-time the number of people within certain areas as a function of the time. This allows enforcing appropriate social distance measurements such as access control or traffic redirection. There are several applications that fit into this category. The most important is *crowd detection*, which is based on determining when and where the density of population is above a threshold. It may refer to either indoor or outdoor public places. Another application is *traffic monitoring* in order to detect violations of a public place lockdown or street movements during confinement periods. One

more application in this category is the *monitoring of quarantined people* to determine compliance with isolation and quarantine requirements.

3.   *Geo-fencing*: Given a safe or non-safe zone, the main goal is to monitor, control, or avoid people trespassing the border of such zone. This may be applied either outdoors (e.g. perimeter confinement) or indoors (e.g. infected homes). For example, *public place monitoring*, or counting the number of people inside/at a public place, falls in this category. Another example is *building access monitoring*.

4.   *Trajectory information*: The main goal is to collect people's trajectory data. Such information can be used to plan alternative routes or to trace back potential contacts. The data can be used to track either the *infected movement* or for *contact tracing*. One more application is *non-essential travel detection*. Using location information to determine if the trip was essential (e.g. medical facilities) or not (e.g. cinemas).

5.   *Automation*: Autonomous machines (e.g. vehicles, robots, drones) can help to reduce the interactions among people in important tasks such as parcel delivery, medical procedures, or transportation. Such autonomy requires reliable and accurate positioning technology.

6.   *Incentive*: Incentive people to stay home when social distancing measures are in place, or to share their location/data to help to enforce social distancing.

The following subsections describe the social distancing applications more suitable to the technologies already described in the previous section. Table 3.6 summarizes the social distancing applications versus the technologies under consideration.

3.3.2   *GNSS-based applications*

The main applications enabled by the GNSS technology for social distancing are in open areas, as it needs a clear view of the sky. This will give the user the best accuracy of its position. However, the main drawback is that the user device will need a communication link to send its position to a server and receive the position of other persons [2].

- *Crowd detection and public place monitoring*: A central server gathering the GNSS position of people in a certain area can estimate its density. Therefore, warnings to new people arriving in this area can be sent to avoid this area or to use protective measures before entering there.
- *Monitoring of quarantined people and traffic monitoring*: GNSS can be used to track the movement of infected people if they wear a compatible device. A simple GNSS device (i.e. single-band code-based tracking with no augmentation techniques) embedded in a smartphone or a wearable can be an efficient solution to track if a person leaves a quarantine zone.
- *Infected movement data and non-essential travel detection*: GNSS can be used to track the movement path of the infected and non-infected people, especially outdoors, which can be useful to estimate the contact between them. In addition,

Table 3.6   Summary of the social distancing applications versus the wireless and networking technologies under consideration

Scenarios	Applications	Description	Technologies
Interpersonal Distance	Distance between any two people	Measure and track the distance between two people	Bluetooth, UWB
	Distance to crowds	Alert when approaching a crowd	UWB
Area Monitoring	Crowd detection	Crowd detection in public places	GNSS, Cellular, WiFi, Bluetooth, WSN, UWB, RFID
	Traffic monitoring	Find violations of a public place lockdown or street movements during confinement periods	GNSS, Cellular, LPWAN
	Monitoring of quarantined people	Compliance with the isolation/quarantine requirements	GNSS, Cellular, LPWAN
Geo-fencing	Public place monitoring	Count the number of people in a public place	GNSS, Cellular, WiFi, WSN, RFID, NFC
	Building access monitoring	Count number of people in a public building	WiFi, WSN, UWB, RFID, NFC
Trajectory Information	Infected movement	Monitor infected people trajectories and notify people at the same places	GNSS, Cellular, LPWAN
	Contact tracing	Trace the contacts that an infected individual made	Bluetooth, UWB
	Non-essential travel detection	Determine if a trip is essential or not using the location information	GNSS, Cellular
Automation	Robot-assisted social distancing	Enhance robot positioning and navigation	GNSS, Cellular, UWB
	Autonomous delivery systems	Diminish number of people outside their homes	GNSS, Cellular
Incentive	Location sharing encouragement	Incentive people to share location history	Bluetooth, NFC
	Stay-at-home encouragement	Incentive people to stay home	Cellular, WiFi

the trajectory time series can be used to detect non-essential trips during lockdown events.

- *Automation*: This application is useful in order to keep the distance between infected people and delivery services. These delivery services can use automated UAVs, robots, or self-driving cars to avoid physical contact with the infected.

3.3.3   Cellular-based applications

The main advantage of this technology for positioning is that at the same time you get the user's positions, it offers a bidirectional communication link. Another advantage is that most people have a smartphone in their pockets, meaning that tracking their position is straightforward for a network operator. And there is no need to deploy any infrastructure extra for positioning. On the other hand, the disadvantage here is that the accuracy is not as good as other technologies, like GNSS. However, it is compensated with the fact that it can be used indoors.

The main social distance applications for cellular positioning are [2]:

- *Crowd detection*: The density of people can be determined by the estimation of the network usage by a single cell or a cluster of them. In that case, the crowd and gatherings can be detected/predicted in an area, the authorities can take actions to avoid it.
- *Monitoring of quarantined people and traffic monitoring*: The network operator could gather the historic movement data from a user. This information can be used to track people infected or in quarantine. This history could be used to investigate how the infected could spread the virus with its movements.
- *Public place monitoring*: Cellular communications can be used by the government, who require it, to ensure that people remain at home. Also, it can be used to obtain how the crowd moves and take decisions based on that information. This is particularly useful to detect a large number of people around a public area and trigger social distancing measurements.
- *Infected movement and non-essential travel detection*: As people usually carry a smartphone, it can be used to track their movement. The new uplink measurements based on 5G are particularly useful for this purpose since it is the network that is obtaining these measurements without the knowledge of the user.
- *Stay-At-Home encouragement*: Due to the increase of work from home during the pandemic, the videoconference calls have seen a huge increase in users. 5G network could provide the service requested by these people who live where landlines are not capable enough.

3.3.4 LPWAN-based applications

LPWAN technologies are suitable for those applications that are not sensitive to latency and have a large number of connected nodes [8].

- *Monitoring of quarantined people and traffic monitoring*: LPWAN technologies can be very efficient to determine the location of a user in a city-level wide area.
- *Infected movement*: If a number of IoT base stations are deployed around an area, it is possible to track the movements of infected people in that area. This may be particularly useful for public areas such as parks, or other open areas.

3.3.5 WLAN-based applications

WLAN technology is suitable for applications where the penetration of satellite-based technologies is very difficult (or their performance is very bad). Besides, WiFi-APs are almost ubiquitous and hence either no dedicated infrastructure is needed or can be quickly deployed on-demand for the purpose of indoor positioning. The main WLAN-based applications are:

- *Crowd Detection*: With WiFi-based positioning systems, the formation of crowds inside indoor spaces can be detected and forced to disperse or maintain safe distances. The recent advancements in WiFi-based positioning and localization systems can also provide accuracy as good as satellite-based systems, for example, accuracy up to 0.6 m can be obtained. WiFi access points (WiFi-AP) become the primary point of data connection inside airports, train stations. Besides, the deployment of WiFi-AP in crowd-prone places is not very hard. Once, the connection to a particular WiFi-AP exceeds a predetermined threshold, authorities can be alerted and appropriate measures can be taken. For example, if the number of people inside the supermarket is very high, a restriction/recommendation can be placed for new coming customers to divert towards other supermarkets or visit at another time.
- *Building and public place access monitoring*: WiFi technology can also be applied to control/schedule the extent of people presents inside a supermarket, university, etc. With various WiFi access points implemented inside the building, the number of people currently inside the building can be estimated based on the number of connections from user devices to the access points. The number of persons currently inside a building can be estimated through the number of established connections on the WiFi-AP. Based on the messages exchanged between the WiFi-AP and the users, several people within the coverage area of the WiFi-AP can be extracted and the information can be used to take measures such as forcing people to form the queue (before entering) or maintain a safe distance. In addition, people who plan to enter the building complex can also be notified in advance or encouraged/forced to stay at home.
- *Stay-at-home encouragement*: In another interesting application of WiFi technology, people can be encouraged to stay at home. This can be based on their frequency of outside movement or can be based on time (some particular time during the day). When a user connected to a WiFi-AP, moves away from it, this phenomenon can be detected through the decreased RSSI and the user can be notified/encouraged for staying at home.

3.3.6   Bluetooth-based applications

The main social distancing applications enabled by Bluetooth are:

- *Interpersonal distance*: In this case, each device (e.g. smartphone, smartwatch, etc.) can estimate the distance to the nearby Bluetooth devices by obtaining the RSS levels [96]. If the proximity threshold is attained, the devices can trigger an alarm to the users to practise social distancing. Bluetooth has the capability to connect to multiple devices simultaneously [2]. Therefore, it can track the distance to multiple people in the coverage area.
- *Crowd detection*: Bluetooth can also be used for indoor crowd detection [97]. A centralized network controller can use the Bluetooth signals from nearby devices to estimate and/or predict the position of the users/crowd. If a crowd is present, the building administrator can enforce different policies such as moving people out and/or prohibiting the entrance to the building.

- *Contact tracing*: Bluetooth can be particularly effective for contact tracing. The main principle is to use the Bluetooth identifier to identify and keep track of the persons in the vicinity. In case of an identified positive case, the identifier records can be centralized by the health authorities to trace back his/her close contacts. Those persons can be notified and put into quarantine as soon as possible thanks to the records obtained from the Bluetooth device.
- *Location sharing encouragement*: The capabilities of Bluetooth to measure inter-personal distances can be used to encourage people to share their trajectory data for contact tracing or infected movement reporting. This can be done based on the number of people and their proximity to the smartphone or wearable owner through a full day.

3.3.7 *WSN-based applications*

WSN is a promising candidate in several applications to enable social distancing. The most relevant are:

- *Crowd detection*: WSN is seen as a potential technology for the detection of indoor users by using the RSS levels of the received signal from multiple nearby devices. The central WSN node can retrieve the information reported by the devices and inform the building manager in case a crowd is present. There are several studies in the literature showing how the accuracy of WSN can be improved for social distancing. For instance, in [98], a ZigBee-enabled system is reported to achieve high accuracy (4–7 m) with low-power devices. In [99], the 'drift phenomenon' was considered in order to enhance the localization of ZigBee-enabled devices. The effect takes place when the users are moving from one place to another in indoor environments. The proposed method is reported to increase the accuracy of WSN localization up to 60% with respect to traditional methods.
- *Building and public place monitoring*: WSN can be used efficiently to control the number of people in an indoor environment. For example, a mobile node, such as an access token, enters a building and connects to the central node. The central node counts the number of people inside. If a threshold is attained, the central node can report it to the building manager, which can control the access to the building of new people.

3.3.8 *UWB-based applications*

Despite some minor drawbacks, the positioning possibilities with significant advantages of UWB technology have opened the door to a wide range of applications. Specifically, it is deployed in several applications with different verticals, such as mobile inventory and locator beacons for emergency services, improving operational efficiencies or improving personnel safety in manufacturing [100], flight control for drones or even swarms of hundred drones, indoor people navigation for the smart home of smart health-care systems [101], and military reconnaissance. Finally, it is worth mentioning the applications of UWB technology in pandemic times. Regarding contact tracing, which has been a topic of intense research and development during the COVID-19 pandemic, UWB technology has been considered as a solution for user location tracking, proximity detection and distance alerts to avoid infection [80,102]. In this direction, multiple commercial solutions for ensuring the maintenance of social distancing in industrial environments relying on UWB have emerged [103–105]. In particular, NXP company offers an UWB-based device for social distancing which is optimized for IoT trackers with remarkably precise positioning performance including motion tracking based on MEMS accelerometer technology, with a low-power wake feature [103]. Renesas has been turned UWB chips for social distancing which aims to COVID-19 prevention. These chips can function on a wearable device that can detect nearby other wearing users. The chip then alerts if the distance between them is not safe or collects the data for contact tracing purposes [104]. A similar solution for social distance is also provided by Noccela [105] with UWB-based badges and a contact tracing tool platform. The badges record intruding incident information including ID/time/duration which can be updated to the service stored in the remote memory for tracing. To collect the information, only one UWB-beacon needed to receive the contact traces from badges when one enters or leaves the facility. The solution can be adapted later on to personnel tracking or added with an item tracking feature. When updating badges to the RTLS solution, a beacon network should be installed as an addition. Otherwise, the badges can be updated remotely.

3.3.9 RFID-based applications

The RFID technology has also the potential to be used in social distancing applications. However, it has not been well adopted in practice because of its deployment complexity [2]. In particular, RFID tags are not commonly available such as other technologies like WiFi or Bluetooth. Nevertheless, the main potential applications are the following:

- *Crowd detection*: For this application, each person has to be equipped with an RFID tag. An RFID strategically placed in the indoor environment can estimate the position of the person based on RSS measurements. If an excessive number of people are detected in the same area, the system can notify the authorities to take the corresponding actions. One example of the state-of-the-art RFID technology is described in [106], where the use of a moving robot is proposed to enhance the positioning accuracy. The robot is able to perform SLAM, and thus it can continuously interrogate all RFID tags in its area. Then, based on passive RFID tags at known locations, the system can estimate the location of target tags by properly manipulating the measured backscattered power.
- *Building and public place monitoring*: An RFID reader can be installed at the main entrance of a building to control the number of people inside. The user RFID tags will be read by the RFID reader and their IDs will be used to control and count the number of people inside. If a threshold is attained, the building manager will be notified. The RFID technology can be deployed in workplaces where the worker has an access ID card equipped with an RFID tag.

3.3.10   NFC-based applications

NFC is a cm-level range technology, which can suitable for some of the following applications that require access control and identification of the users:

*   *Building and public place monitoring*: NFC technology is commonly used for gate access control [95]. For instance, in [107], an NFC identification and attendance service were tested and implemented to simplify attendance monitoring and tracking in school classrooms. The students use their smartphones to register their attendance themselves.
*   *Location sharing encouragement*: In [108], cities like Oulu (Finland) and Frankfurt (Germany) have NGC tagged places enabling smartphone users to get location-specific data. The NFC tags conform to an infrastructure useful to get information about local services. The same principle can be used to encourage users to share their location data to fight against pandemics.

3.4   Comparison and trade-off analysis

As discussed in the previous sections, there is a large variety of wireless and networking positioning technologies that can be used for social distancing. For most of them, their use for positioning and localization purposes is secondary, and therefore, their performance is limited. Table 3.7 shows a comparison among the technologies for social distancing under consideration. The comparison indicators are coverage and accuracy ranges, (which depend on different factors such as type of signal, geometry, environment, etc.), cost, complexity and power consumption, and environment (indoor, outdoor or, both). Other technologies such as vision, infrared (IR), and ultrasound have been added for comparison purposes.

GNSS technology is the only one that has been specifically designed for (outdoor) localization purposes. It provides the largest coverage, from several km to global in the case of conventional GNSS, with accuracy ranging from sub-meter to few meters. This makes GNSS suitable for area monitoring, geo-fencing trajectory information, and automation in outdoor environments. Moreover, GNSS receivers are already integrated into smartphones, making them ubiquity, and the satellite infrastructure is already deployed and working. However, GNSS receivers have a high cost due to their high complexity and their power consumption is usually high as compared to other technologies because the received signal is continuously tracked. Relaxing positioning update rates and target accuracy requirements helps to reduce the power consumption of GNSS receivers. Besides, GNSS does not perform well in indoor environments and deep urban canyons. The hybridization of GNSS technology with other communications standards, such as 5G or WiFi, is the first step towards a unified indoor/outdoor positioning technology.

Traditional cellular communications show poorer accuracy as compared to GNSS, although they show better performance in indoor scenarios due to the larger power and lower frequency bands. The new capabilities introduced by 5G will make them a serious candidate to replace the use of GNSS in urban scenarios. Moreover, the

Table 3.7   Comparison among wireless and networking technologies for social distancing [2,8,9,11,72,74,109,110]

Technologies		Coverage range	Accuracy range	Cost	Complexity power cons.	Environment
GNSS	GNSS	Worldwide	5–50 m	Medium/high	High	Outdoor
	SBAS	Regional	2–10 m	High	High	outdoor
	GBAS/ Pseudolites	1–20 km	20–2 m	High	High	outdoor
	WARTK	500–900 km	5–20 cm	High	High	outdoor
	D-GNSS	10–200 km	1–10 m	High	High	outdoor
	RTK	1–20 km	1–5 cm	High	High	outdoor
	PPP	Worldwide	1–30 cm	High	High	outdoor
Cellular	Pre-5G	1–30 km	10–1 km	Medium/low	Medium	Indoor/ Outdoor
	5G FR1	1–30 km	1–20 m	Medium	Medium	indoor/ outdoor
	5G FR2	100–1 km	10–1 m	High	High	indoor/ outdoor
LPWAN	NB-IOT/ LTE-M	1–10 km	10–1 km	Low	Low	Outdoor
	LoRa/ Sigfox	1–30 km	100–1 km	Low	Low	Outdoor
WLAN	WiFi	20–130 m	1–10 m	Medium/low	Medium	Indoor/ outdoor
WPAN	Bluetooth	1–75 m	1–10 m	Low	Very low	Indoor/ outdoor
	WSN/ ZigBee	100–300 m	1–10 m	Low	Low	Indoor/ outdoor
	UWB	1–50 m	1–1 m	Medium	Very low	Indoor/ outdoor
Contactless	RFID	1–30 m	10–3 m	Medium	Low	Indoor
	NFC	1–10 cm	1–10 cm	Low	Low	Indoor
Others	Ultrasound	2–10 m	1–1 m	Medium	Low	Indoor
	Infrared	1–5 m	1–2 m	Medium/high	Low	indoor
	Vision	1–10 m	0.1–10 cm	High	High	indoor/ outdoor

combination of cellular and WiFi technologies is an effective solution for stay-at-home encouragement.

LPWAN technologies are also suitable for outdoor social distancing applications, showing an accuracy range of hundreds of meters with a very low cost and low power consumption. This makes them a good candidate for monitoring rural areas. In addition, they can be combined with GNSS for providing augmentation data or to send GNSS measurements to a central server for post-processing.

The omnipresence of WiFi-APs in indoor environments gives WiFi-based IPS a distinct advantage over other indoor IPS. However, due to low coverage (maximum

130 m), it is not as effective (or accurate) compared to GNSS in outdoor scenarios. Further, by combining the data from other WiFi-APs and other technologies, the indoor positioning accuracy of WiFi-based IPS can be made comparable to what is achievable through GPS and GNSS outdoors. On the downside, WiFi-based IPS consume more power compared to any other system deployed for IPS, hence probably not suitable for battery operated indoor scenarios.

WPAN technologies (Bluetooth and WSN) have coverage and accuracy ranges similar to WiFi, but with lower cost, lower complexity, and low power consumption. However, WLAN infrastructure is already in place and it is highly used for data communications, whereas WSN needs dedicated deployment. Bluetooth is widely used in smartphones and it is an efficient solution for interpersonal distance, contact tracing, and location sharing applications.

The UWB technology is the most promising one in the IPS category, showing the best accuracy, with good indoor coverage, and low power consumption. However, its slow adoption rate, due to the issues of co-existence and interference with other radio-based technologies, is translated in a medium cost of the technology. Nevertheless, its adoption in the future in smartphones and its hybridization with other technologies such as 5G or GNSS can make UWB the best candidate for interpersonal distance, crowd detection, trajectory information, and indoor automation.

Eventually, contactless technologies (RFID and NFC) have limited performance due to their short range. However, their use is widely spread in a public place and building access monitoring due to their low complexity and low power consumption.

3.5   Future trends and technologies

COVID-19 waves with a number of variants have caused significant changes across all aspects of life as well as the behaviour of society. The people need new tools and services which are useful for protecting their health, while the national governments require highly efficient technology-based social distancing protocols/solutions for protecting their people. In this circumstance, developing new technologies and applications to improve the quality of human life, such as, safer, healthier, and more comfortable is a significant trend of the near future [8,111], especially, in an era of the IoT with significant advancements in computation and communication networks and tremendous growth of data traffic over the wireless communication networks [112]. Hence, wireless localization, one of the key technologies for various social-distancing applications, is expected to be more intelligent and provide long-term location accuracy [113]. The constantly growing demand for effective social distancing solutions has strengthened the important role of wireless localization but also required the new dynamical features from this technology. Particularly, it must be flexible and suitable for multiple services as well as applications in various environments, such as indoor or outdoor with different communication ranges. Additionally, it should be capable to deal with challenges caused by signal attenuation and interference in the dense wireless network [8]. To cope with these requirements, some future trends of evolving wireless positioning technologies can be listed as follows.

3.5.1 Cooperative localization

The cooperative localization trend suggests scaling up the wireless localization networks by emerging available positioning schemes in both categories, equipment and technologies [8]. In particular, connecting multiple nodes or multiple sets of nodes, such as positioning sensors, tag devices, smartphones, smartwatches, and smart glasses can help the system collect and extract the local information more accurately and cope with the loss of signal strength at high distance range, NLoS or the interference problems. The cooperation is not limited to the positioning systems but also includes other parts of the future communication networks such as computation components, e.g. cloud computing and mobile edge computing (MEC) [111], and the physical-layer equipment, like intelligent reflection surfaces (IRS) [114]. This equipment-sense cooperative solution can be possible in dense networks with a huge number of transmitters and receivers [8,115]. Another cooperative localization is to integrate different wireless positioning technologies such as GNSS, 5G, WiFi, UWB in one system. This trend not only can increase the locating accuracy but also enlarge the effective positioning range of the systems [115].

3.5.2 Sensor fusion

In order to increase the positioning accuracy, employing adaptive and online methods for calibrating the localizing results instantaneously have been implemented in several localization frameworks. To fulfil this requirement of the future wireless positioning schemes, sensor fusion can be considered as an excellent solution since installing and aggregating a huge number of sensors is a foreseeable trend in the coming years [116]. The basic idea of integrating sensor fusion techniques in wireless positioning schemes is to combine the measurements from different sensors installed for various purposes and services to improve the final accuracy, resilience, and reliability. For example, in autonomous cars, the positioning engine uses data from a wide range of sensors (e.g. IMU, Lidar, GNSS, RADAR) and combines them to precisely locate the vehicle and get information about its environment. There are three main sensor aggregation categories that are separated based on the abstraction level, i.e. how to combine the measurements from different sensors: data-level, feature-level, and decision-level fusion [117–119].

3.5.3 Machine learning/artificial intelligence

In the future wireless dense networks which require a huge computation effort, enhance big data analytics, and improve security and data privacy, the role of ML tools will become more important. Then, exploiting ML techniques in wireless positioning is an expected trend which can be proven by a number of works on localization using ML or deep learning [118,120]. How to combine the existing geometrical and ML techniques to achieve three goals: high localization accuracy, low maintenance cost, and delightful deployment ubiquity will be a significant direction for future wireless localization [120].

3.6   Conclusions

This chapter reviews the suitability of the most relevant wireless and networking communications technologies (e.g. GNSS, cellular, WiFi, Bluetooth and RFID) for the implementation of social distancing applications such as interpersonal distancing, real-time area monitoring, or geo-fencing.

Technologies with a larger coverage range (e.g. GNSS, cellular and LPWAN) are suitable for outdoor social distancing applications where wide areas have to be monitored such as traffic monitoring, quarantine people monitoring, or infected movement data gathering. Conversely, technologies with a shorter coverage range (e.g. WiFi, WSN, UWB and NFC) are effectively used in indoor social distancing applications such as building access monitoring and indoor crow detection. Moreover, technologies with peer-to-peer positioning (e.g. Bluetooth and UWB) are suitable for interpersonal distance monitoring. Furthermore, highly accurate technologies (e.g. GNSS, 5G FR2, UWB) enable the automation of robots and UAV-based delivery. Eventually, several technologies help to incentive social distancing such as stay-at-home encouragement (e.g. Cellular, WiFi) or location sharing encouragement (e.g. Bluetooth and NFC).

Although most of the mentioned technologies are already well matured and widely used, there are future trends to improve their accuracy, reliability, and resilience. The most relevant are cooperative localization, sensor fusion, and AI-powered localization.

References

[1]   Kelso J, Milne G, Kelly H. 'Simulation suggests that rapid activation of social distancing can arrest epidemic development due to a novel strain of influenza'. *BMC Public Health*. 2009;**9**:117.

[2]   Nguyen CT, Saputra YM, Huynh NV, *et al*. 'A comprehensive survey of enabling and emerging technologies for social distancing – part I: fundamentals and enabling technologies'. *IEEE Access*. 2020;**8**:153479–507.

[3]   Ferguson NM, Cummings DAT, Fraser C, *et al*. 'Strategies for mitigating an influenza pandemic'. *Nature*. 2006;**442**(7101):448–52.

[4]   Kumar S, Grefenstette JJ, Galloway D, *et al*. 'Policies to reduce influenza in the workplace: impact assessments using an agent-based model'. *American Journal of Public Health*. 2013;**103**(8):1406–11.

[5]   Mao L. 'Agent-based simulation for weekend-extension strategies to mitigate influenza outbreaks'. *BMC Public Health*. 2011;**11**(1).

[6]   Milne GJ, Kelso JK, Kelly HA, *et al*. 'A small community model for the transmission of infectious diseases: comparison of school closure as an intervention in individual-based models of an influenza pandemic'. *PLoS One*. 2008;**3**(12):e4005.

[7]   Timpka T, Eriksson H, Holm E, *et al*. 'Relevance of workplace social mixing during influenza pandemics: an experimental modelling study of workplace cultures'. *Epidemiology and Infection*. 2016;**144**(10):2031–42.

[8]   Li Y, Zhuang Y, Hu X, *et al.* 'Toward location-enabled IoT (LE-IoT): IoT positioning techniques, error sources, and error mitigation'. *IEEE Internet of Things Journal.* 2021;**8**(6):4035–62.

[9]   Frederick D Moorefield J. Global Positioning System. Standard Positioning Service Performance Standard; 2020. *ANNEX B.* Available from: https://www.navcen.uscg.gov/pdf/gps/geninfo/2020SPSPerformanceStandar dFINAL.pdf.

[10] Zumberge JF, Heflin MB, Jefferson DC, *et al.* 'Precise point positioning for the efficient and robust analysis of GPS data from large networks'. *Journal of Geophysical Research: Solid Earth.* 1997;**102**(B3):5005–17. Available from: https://agupubs.onlinelibrary.wiley.com/doi/abs/10.1029/96JB03860.

[11] J Sanz Subinara JMJZ, Hernández-Pajares M. *GNSS Data Processing. Volume I: Fundamentals and Algorithms.* 1st ed. ESA Communications; 2013.

[12] Anthony Flores CT. Navstar GPS Space Segment/Navigation User Interfaces. Available from: https://www.gps.gov/technical/icwg/IS-GPS-200L.pdf.

[13] Union E. Signal-In-Space Interface Control Document. Available from: https://www.gsc-europa.eu/sites/default/files/sites/all/files/Galileo_OS_SIS_ ICD_v2.0.pdf.

[14] Russian Space Systems J. General Description of Code Division Multiple Access Signal System. Available from: http://russianspacesystems.ru/wp-content/uploads/2016/08/ICD-GLONASS-CDMA-General.-Edition-1.0-201 6.pdf.

[15] Office CSN. Interface Control Document. Open Service Signal B2b. Available from: http://en.beidou.gov.cn/SYSTEMS/ICD/202008/P02020080 3539206360377.pdf.

[16] Adrián Cardalda García AP Stefan Maier. *Location-Based Services in Cellular Networks from GSM to 5G NR.* 1st ed. Artech house; 2020.

[17] Organization ISR. IRNSS Signal-In-Space ICD for SPS v1.1. Available from: https://www.isro.gov.in/sites/default/files/irnss_sps_icd_version1.1- 2017.pdf.

[18] Union E. Quasi-Zenith Satellite System. Interface Specification. Centimeter Level Augmentation Service. Available from: https://qzss.go.jp/en/ technical/download/pdf/ps-is-qzss/is-qzss-l6-001.pdf.

[19] Shabnam M, Chowdhury IH, Tushar ZH, *et al.* Performance evaluation of GNSS receiver in multi-constellation system. In *2017 International Conference on Electrical, Computer and Communication Engineering (ECCE);* 2017. pp. 610–14.

[20] Blewitt G. Basic of the GPS Technique: Observation Equations. In *Geodetic Applications of GPS. Swedish Land survey;* 1997.

[21] Adavi SS, Nisha MS. GNSS positioning analysis of different positioning modes using open- source tools – a comparative study. In *2021 5th International Conference on Computer, Communication and Signal Processing (ICCCSP);* 2021. pp. 199–203.

[22] Teunissen PJG. In *GNSS Precise Point Positioning;* 2021. pp. 503–28.

[23]   ETSI TR. Functional stage 2 description of Location Services (LCS) in GERAN. Available from: https://www.3gpp.org/DynaReport/43059.htm.

[24]   del Peral-Rosado JA, Raulefs R, López-Salcedo JA, *et al.* 'Survey of cellular mobile radio localization methods: From 1G to 5G'. *IEEE Communications Surveys & Tutorials*. 2017;**20**(2):1124–48.

[25]   ETSI TR. Summary of Rel-16 Work Items. Available from: https://www. 3gpp.org/DynaReport/21916.htm.

[26]   ETSI TR. LTE Positioning Protocol LPP. Available from: https://www. 3gpp.org/DynaReport/37355.htm.

[27]   Guerra A, Guidi F, Dardari D. 'Single-anchor localization and orientation performance limits using massive arrays: MIMO vs. beamforming'. *IEEE Transactions on Wireless Communications*. 2018;**17**(8):5241–55.

[28]   ETSI TR. GPP TR-2 2.872: study on positioning use cases. Available from: https://www.3gpp.org/DynaReport/22872.htm.

[29]   Weiss MA, Barry C. 'Hyperbolic positioning accuracy issues: measurement noise, geometric dilution of position, and synchronization errors'. In *2021 IEEE International Symposium on Precision Clock Synchronization for Measurement, Control, and Communication (ISPCS)*; 2021. pp. 1–7.

[30]   ETSI TR. Physical layer measurements. Available from: https://www. 3gpp.org/DynaReport/38215.htm.

[31]   ETSI TR. Physical channels and modulation. Available from: https://www. 3gpp.org/DynaReport/38211.htm.

[32]   LG. Discussions on DL only based positioning. Available from: https://portal. 3gpp.org/ngppapp/CreateTDoc.aspx?mode=view&contributionUid=R1-190 3346.

[33]   Cardalda García A. *Hybrid localization algorithm for LTE combining satellite and terrestrial measurements*. University of Oviedo; 2015.

[34]   Zhou X, *Flora E. Comparison of Performance and Power Consumption Between GPS and Sigfox Positioning Using Pycom Modules*; 2018.

[35]   Marceli M, Sandri B, Jeleni J, *et al.* Determining Location in LPWAN using Multilateration. In *2019 2nd International Colloquium on Smart Grid Metrology (SMAGRIMET)*; 2019. pp. 1–4.

[36]   Mekki K, Bajic E, Chaxel F, *et al.* 'A comparative study of LPWAN technologies for large-scale IoT deployment'. *ICT Express*. 2019;**5**(1):1–7.

[37]   Raza U, Kulkarni P, Sooriyabandara M. 'Low power wide area networks: an overview'. *IEEE Communications Surveys & Tutorials*. 2017;**19**(2):855–73.

[38]   Haxhibeqiri J, De Poorter E, Moerman I, *et al.* 'A survey of LoRaWAN for IoT: from technology to application'. *Sensors*. 2018;**18**(11):3995.

[39]   Ikpehai A, Adebisi B, Rabie KM, *et al.* 'Low-power wide area network technologies for internet-of-things: a comparative review'. *IEEE Internet of Things Journal*. 2019;**6**(2):2225–40.

[40]   LoRa Alliance; 2021. Available from: https://lora-alliance.org/.

[41]   Sigfox.com; 2021. Available from: https://www.sigfox.com/en.

[42]   Flynn K. Release 13; 2021. Available from: https://www.3gpp.org/release-13.

[43] Wi-fi.org; 2021. Available from: https://www.wi-fi.org/discover-wi-fi/wi-fi-halow.

[44] Openweightless.org; 2021. Available from: https://www.openweightless.org/.

[45] Ingenu; 2021. Available from: https://www.ingenu.com/technology/rpma/.

[46] Telensa; 2021. Available from: https://www.telensa.com/.

[47] Qowisio; 2021. Available from: https://www.qowisio.com/en/qowisio-english/.

[48] Alliance L. *LoRaWAN What is it? A Technical Overview of LoRa and LoRaWAN LoRa Alliance*; 2015.

[49] Fargas BC, Petersen MN. GPS-free geolocation using LoRa in low-power WANs. In *2017 global internet of things summit (Giots)*. IEEE; 2017. pp. 1–6.

[50] GHADIRZADEH M. *GPS free geolocation in Lora networks*; 2018.

[51] ETSI. Technical characteristics for Low Power Wide Area Networks Chirp Spread Spectrum (LPWAN-CSS) operating in the UHF spectrum below 1 GHz. System Reference document (SRdoc); 2018. Available from: https://www.etsi.org/deliver/etsi_tr/103500_103599/103526/01.01.01_60/tr_103526v010101p.pdf.

[52] IEEE 802; 2021. Available from: https://www.ieee802.org/11/.

[53] WiFi Alliance; 2021. Available from: https://www.wi-fi.org/membership/member-companies.

[54] Naik GI. LTE WLAN interworking for Wi-Fi hotspots. In *2010 Second International Conference on COMmunication Systems and NETworks (COMSNETS 2010)*. IEEE; 2010. pp. 1–2.

[55] Kunz A, Salkintzis A. 'Non-3GPP Access Security in 5G'. *Journal of ICT Standardization*. 2020; pp. 41–56.

[56] Zou H, Jin M, Jiang H, *et al.* 'WinIPS: WiFi-based non-intrusive indoor positioning system with online radio map construction and adaptation'. *IEEE Transactions on Wireless Communications*. 2017;**16**(12):8118–30.

[57] Toh C, Lau SL. Indoor localisation using existing WiFi infrastructure—a case study at a university building. In *2016 22nd International Conference on Virtual System & Multimedia (VSMM)*. IEEE; 2016. pp. 1–5.

[58] Retscher G, Moser E, Vredeveld D, *et al.* Performance and accuracy test of a WiFi indoor positioning system. *Journal of Applied Geodesy*. 2007;**1**:103–10.

[59] Liu F, Liu J, Yin Y, *et al.* 'Survey on WiFi-based indoor positioning techniques'. *IET Communications*. 2020;**14**(9):1372–83.

[60] Pirzada N, Nayan MY, Subhan F, *et al.* 'Comparative analysis of active and passive indoor localization systems'. *AASRI Procedia*. 2013;**5**:92–7.

[61] Deak G, Curran K, Condell J. 'A survey of active and passive indoor localisation systems'. *Computer Communications*. 2012;**35**(16):1939–54.

[62] Wang P, Luo Y. Research on WiFi indoor location algorithm based on RSSI Ranging. In *2017 4th International Conference on Information Science and Control Engineering (ICISCE)*. IEEE; 2017. pp. 1694–8.

[63] Makki A, Siddig A, Saad M, *et al.* 'Survey of WiFi positioning using time-based techniques'. *Computer Networks*. 2015;**88**:218–33.

[64] Kluge T, Groba C, Springer T. Trilateration, fingerprinting, and centroid: taking indoor positioning with bluetooth LE to the wild. In *2020 IEEE 21st International Symposium on "A World of Wireless, Mobile and Multimedia Networks"(WoWMoM)*. IEEE; 2020. pp. 264–72.

[65] K Sheshadri R, Arslan MY, Sundaresan K, *et al*. Amorfi: amorphous wifi networks for high-density deployments. In *Proceedings of the 12th International on Conference on emerging Networking Experiments and Technologies*; 2016. pp. 161–75.

[66] Xiao H, Zhang H, Wang Z, *et al*. An RSSI based DV-hop algorithm for wireless sensor networks. In *2017 IEEE Pacific Rim Conference on Communications, Computers and Signal Processing (PACRIM)*. IEEE; 2017. pp. 1–6.

[67] Sen S, Lee J, Kim KH, *et al*. Avoiding multipath to revive inbuilding WiFi localization. In *Proceeding of the 11th Annual International Conference on Mobile Systems, Applications, and Services*; 2013. pp. 249–62.

[68] Zou H, Huang B, Lu X, *et al*. Standardizing location fingerprints across heterogeneous mobile devices for indoor localization. In *2016 IEEE Wireless Communications and Networking Conference*. IEEE; 2016. pp. 1–6.

[69] Ma L, Fan Y, Xu Y, *et al*. Pedestrian dead reckoning trajectory matching method for radio map crowdsourcing building in WiFi indoor positioning system. In *2017 IEEE International Conference on Communications (ICC)*. IEEE; 2017. pp. 1–6.

[70] Li Z, Zhao X, Liang H. Automatic construction of radio maps by crowdsourcing PDR traces for indoor positioning. In *2018 IEEE International Conference on Communications (ICC)*. IEEE; 2018. pp. 1–6.

[71] Lim JM, Jeong WM, Sung TK. Development of a WPAN-based Self-positioning System for Indoor Flying Robots. *Journal of Institute of Control, Robotics and Systems*. 2015 May;**21**(5):490–5. Available from: https://doi.org/10.5302/j.icros.2015.15.0018.

[72] Mainetti L, Patrono L, Sergi I. A survey on indoor positioning systems. In *2014 22nd International Conference on Software, Telecommunications and Computer Networks (SoftCOM)*; 2014. pp. 111–20.

[73] SIG B. Bluetooth Core Specification v5.3; 2021. Available from: https://www.bluetooth.org/DocMan/handlers/DownloadDoc.ashx?doc_id=5 21059.

[74] Subedi S, Pyun JY. 'A survey of smartphone-based indoor positioning system using RF-based wireless technologies'. *Sensors*. 2020;**20**(24):7230.

[75] Subedi S, Pyun JY. 'Practical fingerprinting localization for indoor positioning system by using beacons'. *Journal of Sensors*. 2017;**2017**:1–16.

[76] Labeau F, Agarwal A, Agba B. Comparative study of wireless sensor network standards for application in electrical substations. In *2015 International Conference on Computing, Communication and Security (ICCCS)*; 2015. pp. 1–5.

[77] Gutierrez JA, Callaway EH, Barrett R. *IEEE 802.15.4 low-rate wireless personal area networks: enabling wireless sensor networks*. USA: IEEE Standards Office; 2003.

[78] Alarifi A, Al-Salman A, Alsaleh M, *et al.* 'Ultra wideband indoor positioning technologies: analysis and recent advances. *Sensors.* 2016;**16**(5). Available from: https://www.mdpi.com/1424-8220/16/5/707.

[79] Ghavami M, Michael LB, Kohno R. *Ultra Wideband Signals and Systems in Communication Engineering.* 2nd ed. John Wiley & Sons, Ltd; 2007. Available from: https://onlinelibrary.wiley.com/doi/abs/10.1002/9780470060490.fmatter.

[80] Shubina V, Holcer S, Gould M, *et al.* 'Survey of decentralized solutions with mobile devices for user location tracking, proximity detection, and contact tracing in the COVID-19 era'. *Data.* 2020;**5**(4). Available from: https://www.mdpi.com/2306-5729/5/4/87.

[81] Siwiak K, McKeown D. *Ultra-wideband Radio Technology.* 2nd ed. John Wiley & Sons, Ltd; 2005.

[82] Alsindi NA, Alavi B, Pahlavan K. 'Measurement and modeling of ultra-wideband TOA-based ranging in indoor multipath environments'. *IEEE Transactions on Vehicular Technology.* 2009;**58**(3):1046–58.

[83] Gifford W, Dardari D, Win M. 'The impact of multipath information on time-of-arrival estimation. *IEEE Transactions on Signal Processing.* 2020;1.

[84] Tan Z, Zhu X, Zhao Z, *et al.* UWB-AOA estimation method based on a spare antenna array with virtual element. In *2018 IEEE International Conference on Computational Electromagnetics (ICCEM)*; 2018. pp. 1–3.

[85] Smaoui N, Heydariaan M, Gnawail O. Single-antenna AoA estimation with UWB radios. In *2021 IEEE Wireless Communications and Networking Conference (WCNC)*; 2021. pp. 1–7.

[86] Barral V, Escudero CJ, García-Naya JA. NLOS classification based on RSS and ranging statistics obtained from low-cost UWB devices. In *2019 27th European Signal Processing Conference (EUSIPCO)*; 2019. pp. 1–5.

[87] Sourya A, Dutta S, Chandra A, *et al.* Find my car: simple RSS-based UWB localization algorithms for single and multiple transmitters. In *2020 IEEE Latin-American Conference on Communications (LATINCOM)*; 2020. pp. 1–6.

[88] Zhao P, Zhu X, He L, *et al.* UWB-RTK Positioning System Based on TDOA. In *2019 UK/China Emerging Technologies (UCET)*; 2019. pp. 1–4.

[89] Djaja-Josko V, Kolakowski J. A new method for wireless synchronization and TDOA error reduction in UWB positioning system. In *2016 21st International Conference on Microwave, Radar and Wireless Communications (MIKON)*; 2016. pp. 1–4.

[90] Choliz J, Eguizabal M, Hernandez-Solana A, *et al.* Comparison of algorithms for UWB indoor location and tracking systems. In *2011 IEEE 73rd Vehicular Technology Conference (VTC Spring)*; 2011. pp. 1–5.

[91] Gunia M, Lu Y, Joram N, *et al.* On the Precision of Common Individual or Hybrid Positioning Systems. In *2019 International Conference on Localization and GNSS (ICL-GNSS)*; 2019. pp. 1–6.

[92] Abtahi M, Magné J, Mirshafiei M, *et al.* 'Generation of power-Efficient FCC-Compliant UWB Waveforms Using FBGs: Analysis and Experiment'.

Journal of Lightwave Technology. 2008;**26**(5):628–35. Available from: http://jlt.osa.org/abstract.cfm?URI=jlt-26-5-628.

[93] Aiello R. 3 – Interference and Coexistence. In: Aiello R, Batra A, editors. *Ultra Wideband Systems*. Burlington: Newnes; 2006. p. 53–71. Available from: https://www.sciencedirect.com/science/article/pii/B9780750678933500043.

[94] Thilak N, Braun R. Near field magnetic induction Communication in Body Area Network. In *2012 International Conference on Devices, Circuits and Systems (ICDCS)*; 2012. pp. 124–5.

[95] Coskun V, Ozdenizci B, Ok K. 'The survey on near field communication'. *Sensors*. 2015;**15**(6):13348–405.

[96] Wang Y, Yang X, Zhao Y, *et al.* Bluetooth positioning using RSSI and triangulation methods. In *2013 IEEE 10th Consumer Communications and Networking Conference (CCNC)*; 2013. pp. 837–42.

[97] Wang Y, Ye Q, Cheng J, *et al.* RSSI-Based Bluetooth Indoor Localization. In *2015 11th International Conference on Mobile Ad-hoc and Sensor Networks (MSN)*; 2015. pp. 165–71.

[98] Luoh L. 'ZigBee-based intelligent indoor positioning system soft computing'. *Soft Computing*. 2013;**18**(3):443–56.

[99] Chu CH, Wang CH, Liang CK, *et al.* High-accuracy indoor personnel tracking system with a ZigBee wireless sensor network. In *2011 Seventh International Conference on Mobile Ad-hoc and Sensor Networks*; 2011. pp. 398–402.

[100] Barral V, Suárez-Casal P, Escudero CJ, *et al.* 'Multi-Sensor Accurate Forklift Location and Tracking Simulation in Industrial Indoor Environments. *Electronics*. 2019;**8**(10). Available from: https://www.mdpi.com/2079-9292/8/10/1152.

[101] Raza U, Khan A, Kou R, *et al.* Dataset: indoor localization with narrow-band, ultra-wideband, and motion capture Systems. In *Proceedings of the 2nd Workshop on Data Acquisition To Analysis. DATA'19*. New York, NY, USA: Association for Computing Machinery; 2019. pp. 34–6. Available from: https://doi.org/10.1145/3359427.3361919.

[102] Social distancing wristwatch uses Ultra-Wideband Chipset; 2020. Available from: https://www.eenewseurope.com/news/social-distancing-wristwatch-uses-ultra-wideband-chipset.

[103] UWB for social distancing: take the guesswork out of staying safe. Available from: https://www.nxp.com/company/blog/uwb-for-social-distancing-take-the-guesswork-out-of-staying-safe:BL-UWB-SOCIAL-DISTANCING.

[104] Turning Renesas UWB chips into social distancing solutions – Phase 1. Available from: https://capgemini-engineering.com/us/en/insight/turning-renesas-uwb-chips-into-social-distancing-solutions-phase-1/.

[105] Social distancing solutions – Noccela. Available from: https://social distancing.noccela.com/.

[106] Megalou S, Tzitzis A, Siachalou S, *et al.* Fingerprinting localization of RFID tags with real-time performance-assessment, using a moving robot. In *2019 13th European Conference on Antennas and Propagation (EuCAP)*; 2019. pp. 1–5.

[107]   Sánchez I, Cortés M, Riekki J, *et al*. NFC-based interactive learning environments for children. In *Proceedings of the 10th International Conference on Interaction Design and Children. IDC '11*. New York, NY, USA: Association for Computing Machinery; 2011. pp. 205–8. Available from: https://doi-org.proxy.bnl.lu/10.1145/1999030.1999062.

[108]   Prinz A, Menschner P, Altmann M, *et al*. inSERT – An NFC-based self reporting questionnaire for patients with impaired fine motor skills. *Third International Workshop on Near Field Communication*. 2011; pp. 26–31.

[109]   Dwivedi S, Shreevastav R, Munier F, *et al*. *Positioning in 5G networks*; 2021.

[110]   Wang Z, Yang Z, Dong T. 'A review of wearable technologies for elderly care that can accurately track indoor position, recognize physical activities and monitor vital signs in real time. *Sensors*. 2017;**17**(2):341.

[111]   Pham QV, Fang F, Ha VN, *et al*. 'A survey of multi-access edge computing in 5G and beyond: fundamentals, technology integration, and state-of-the-art. *IEEE Access*. 2020;**8**:116974–7017.

[112]   Cisco Annual Internet Report – Cisco Annual Internet Report (2018–2023) White Paper. Available from: https://www.cisco.com/c/en/us/solutions/collateral/executive-perspectives/annual-internet-report/white-paper-c11-74 1490.html.

[113]   What to watch out for in 2021: Localization Trends. Available from: https://www.welocalize.com/what-to-watch-out-for-in-2021-localization-trends/.

[114]   Elzanaty A, Guerra A, Guidi F, *et al*. Towards 6G Holographic Localization: Enabling Technologies and Perspectives. *CoRR*. 2021;abs/2103.12415. Available from: https://arxiv.org/abs/2103.12415.

[115]   Chukhno N, Trilles S, Torres-Sospedra J, *et al*. 'D2D-based cooperative positioning paradigm for future wireless systems: a survey'. *IEEE Sensors Journal*. 2021; p. 1.

[116]   LiDAR in Non-Automotive Applications to Drive Sensor Installations to 16M Units by 2030. Available from: https://www.eetasia.com/lidar-in-non-automotive-applications-to-drive-sensor-installations-to-16m-units-by-2030/.

[117]   Tamar Peli RKKEFB Mon Young. *Feature level sensor fusion*; 1999.

[118]   Howard SM. *Deep learning for sensor fusion*. Case Western Reserve University; 2017.

[119]   Khaleghi B, Khamis A, Karray FO, *et al*. 'Multisensor data fusion: a review of the state-of-the-art'. *Information Fusion*. 2013;**14**(1):28–44. Available from: https://www.sciencedirect.com/science/article/pii/S1566253511000558.

[120]   Li D, Xu J, Yang Z, *et al*. Train once, locate anytime for anyone: adversarial learning based wireless localization. In *IEEE INFOCOM 2021 – IEEE Conference on Computer Communications*; 2021. pp. 1–10.

Chapter 4

Computer vision technologies for social distancing

Tran Hiep Dinh[1], Nguyen Linh Trung[2] and Chin-Teng Lin[3]

Many technological applications have been developed and implemented in the last two years to fight the COVID-19 pandemic via social distancing. Despite its importance in response to the coronavirus, the remaining challenge is the limitation of human resources to monitor and provide timely warnings to maintain appropriate activities, such as keeping distance between each other, wearing face masks, or complying with curfew restrictions. To tackle this problem, computer vision researchers have proposed numerous approaches for autonomous object detection and distance measurement, which will be summarised in this chapter. First, vision-based applications in intelligent surveillance systems for social distancing monitoring as well as masked face detection, are introduced. Then, core classical and modern neural network-based methodologies for these applications are analysed. A simple masked face detection is developed to verify its effectiveness and limitations, followed up by remarks and discussions on open problems.

4.1 Introduction

4.1.1 Fundamental background of computer vision

Computer vision researchers have been working on mathematical techniques to reconstruct objects' shape and appearance in imagery, understand the mechanism of the human visual system, and develop algorithms for the computer to interpret images automatically [1]. In contrast to the effortless process of image understanding and

[1] University of Engineering and Technology, Vietnam National University, Hanoi (VNU-UET), Joint Technology and Innovation Research Centre (JTIRC), a partnership between the University of Technology Sydney and the VNU-UET, Hanoi, Vietnam
[2] Advanced Institute of Engineering and Technology (AVITECH), University of Engineering and Technology, Vietnam National University, Hanoi, Vietnam
[3] Australian Artificial Intelligence Institute, School of Computer Science, Faculty of Engineering and Information Technology, University of Technology Sydney, Sydney, Australia

interpreting in humans, a complete solution to the understanding of the visual system has yet to be developed [2]. As a result, computer vision remains an unsolved challenge, even after decades of development [3].

Despite the aforementioned difficulties, tremendous advances have been made in computer vision to solve a wide range of real-world applications, such as optical character recognition, medical imaging, surveillance and monitoring, face detection, and so on, among which object detection is one of the most fundamental problems. Object detection can be defined as a task of scene analysis for object localisation and recognition. As mentioned in [4], traditional approaches follow a three-stage strategy, where a multi-scale sliding window is employed to exhaustively search for interested objects, the features of which are then represented via descriptors such as Haar wavelet [5], Histogram of Oriented Gradients (HOG) [6], Scale-invariant feature transform (SIFT) [7], or SURF [8]. In the final stage, objects are classified into different categories using classifiers, among which the most popular ones are Support Vector Machine [9], AdaBoost [10], and Deformable Part-based Model [11]. With the rapid development of Convolutional Neural Networks (CNNs) [12], state-of-the-art results in object detection have been achieved. The success of CNNs in image understanding is explained as their deeper architecture for complex feature learning, instead of depending on hand-engineered features as per the traditional approaches. Besides, CNNs are also beneficial from large visual databases such as ImageNet [13], the employment of GPUs for parallel computing, as well as the development of augmentation techniques [14] for incorporation of potential invariances.

4.1.2   Potential applications to social distancing to fight pandemics

Since the first outbreak of COVID-19 in November 2019, guidance from the World Health Organisation (WHO) have been released and updated regularly to cope with the situation. According to the advice from WHO in [15], which is last updated in October 2021, the following preventative measures are crucial: vaccination, physical or social distancing, usage of face masks or coverings, hand sanitising, cough and sneeze covering, and self-isolation upon positive test or symptom development. Among the above measures, many vision-based approaches have been developed to monitor physical or social distancing as well as face mask-wearing. Taking into account current advances in pedestrian [16] and face detection [17], a further processing step is developed to estimate the physical distance between detected pedestrians or to verify whether a mask is properly worn on the detected face. Applications are either implemented on static systems [18–20] or unmanned mobile systems, such as mobile robots [21,22] or unmanned aerial vehicles [23]. Compared to manual monitoring, these types of intelligent surveillance systems are more cost-effective. More importantly, the implementation of such systems does not violate the social distancing guidances since the monitoring and warning can be done remotely. Figure 4.1 illustrates the processing mechanism of a multi-camera intelligent surveillance system, where social distancing measures are detected from each view. The single view detection results are then fused together where a bird-eye map is generated showing objects labelled with their

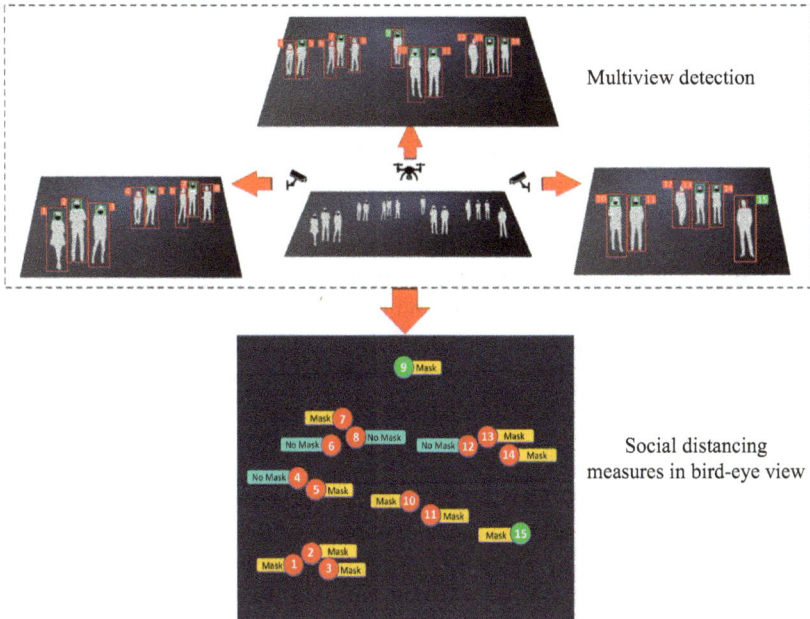

Figure 4.1   Illustration of a multi-camera intelligent surveillance system

engagement in social distancing, including following the physical distance guidelines (presented in green) or not (presented in red).

The rest of this chapter is organised as follows. First, vision-based applications for social distancing are introduced in Section 4.2. The core object detection approaches for those applications are reviewed and analysed in Section 4.3, including the classical and modern CNN-based ones. A simple masked face detection is built in Section 4.4 to analyse its effectiveness and limitation in an indoor environment, which will be extend to a discussion on open problems in Section 4.5.

4.2   Vision-based applications for social distancing

4.2.1   Intelligent surveillance for social distancing monitoring

To aid the social distancing monitoring using intelligent surveillance systems, many approaches have been proposed. In a typical system, static images or video streams of a monitored area are processed for pedestrian detection. Based on the detected objects, physical distances between pedestrians are estimated in a bird-view or a 3D map. In most systems, closed-circuit television (CCTV) cameras are employed due to their high availability and low cost. In [18,24,25], individuals with bounding boxes are detected from video stream frames via a pre-trained CNN. A perspective trans-formation is then applied to map image coordinates to real-world coordinates, where pairwise Euclidean distances between detections are calculated for social density evaluation. While a fixed threshold is applied in [18,25] to classify detections into low- and high-risk based on the calculated distance, a linear regression model is fitted to training data and then employed for an automatic real-time assessment of social distancing violations [24]. Encouraging outcomes have been achieved when verify-ing the approach on some public pedestrian crowd datasets, namely the Oxford Town Center [26], the Mall [27], and the Train Station Dataset [28]. Unlike the aforemen-tioned frame-to-frame-based detection and monitoring methods, a trajectory-based technique is proposed in [29], providing a more dynamic perspective on this matter. Here, Frechet distance [30] is additionally employed to measure pairwise distances between trajectories. This metric is then combined with the distance between detection pairs in a multi-scale social distancing evaluation scheme for a more comprehensive assessment of crowd gathering situations.

Since the performance of colour camera-based systems is impacted by light-ing conditions, other camera types have also been employed. Thermal cameras are employed to capture images and temperatures of the objects [19], which are auto-matically detected using the YOLO-v2 detector [31]. Euclidean distances between the centroid of identified object bounding boxes are calculated to evaluate whether physical distancing rules are followed. The approach is embedded on an on-board computer for real-time applications. In [21], human objects are first detected using Part Affinity Fields [32], the 3D position of which is estimated based on the mapping between colour and depth image inputs. The movements of the human objects are then tracked using DeepSort [33] for the generation of a human density heat map. The system is implemented on a mobile robot performing not only a social distance monitoring but also a cleaning service. The idea of taking object's trajectory into account is also employed in [22], where the motion of pedestrians is modelled via a velocity-obstacle-based algorithm, and matched between consecutive frames. The camera system, consists of a monocular camera and a LiDAR mounted on a mobile robot for gathered crowd detection. Both colour and LiDAR data are fused for not only navigation but also for evaluation of distances between members of the crowd.

Despite recent advances in human detection and pose estimation, occlusion still remains as a major challenge in this domain. Although the employment of mobile robots [21,22] or both CCTV cameras and robots [34] in the surveillance task can enhance the mobility and flexibility of the system, the problem has not been com-pletely solved due to the limited Field of View (FoV). The increasing popularity of Unmanned Aerial Vehicles (UAVs) in the last decade has led to a new and insightful approach for computer vision researchers to tackle the occlusion problem in crowd monitoring, due to their ability in wide-area coverage. The UAV-based social dis-tancing monitoring approach proposed in [23] detects human heads from the UAV's photography using PeleeNet [35] and multi-scale spatial attention. A meticulous calibration step is first required to transform the image to real-world coordinates under an assumption that pedestrians share the same height. The proposed detector is then pre-trained on some public human head datasets before being fine-tuned in the

UAV-collected human head dataset. UAV vision can be formulated as a small object detection, some major challenges of which are small objects and viewpoint changes. A four-stage architecture is then proposed in [36], where deformable convolution and interleaved cascade have been verified to be effective in dealing with those challenges.

4.2.2   Masked face detection

Machine learning-based or feature-based methods are still employed in some masked face detection applications, due to their advantage in computational effectiveness. This advantage has been reported in [37] where the average detection time of a compound technique formed by the Viola-Jones [38] and the Markov Clustering Networks [39] methods is slightly better than that of some well-known deep learning-based techniques, i.e. the You Only Look One (YOLO) [40] and the Single Shot MultiBox Detector (SDD) [41]. However, the detection accuracy of the techniques based on hand-crafted features is not promising [42,43]. In [43], two of the most popular feature-based approaches, the Viola-Jones [38] and the Histogram of Oriented Gradients [6] are evaluated with a variety of face coverings, including not only the disposal ones but also alternatives such as burka, helmet with goggles, costumes, etc. Experimental results reported in [43] have shown that the performance of these approaches is not comparable to that of deep CNN approaches.

In general, a CNN-based facemask detector finds faces from the input image before verifying whether a mask is worn. In [20], a hybrid approach is proposed, consisting of pre-detection and verification stages. Facial regions are first extracted in the former stage using Faster R-CNN [44] and Inception V2 [45], then verified in the latter by a broad learning system. In [46], a CNN-based detector is developed for detection and localisation of masked face(s) from input images. The architecture consists of ResNet-50 [47] for feature extraction and YOLO-v2 [31] for medical face mask detection. Unlike other approaches, the medical masked face is the focus in this research, hence the training data are mainly imported from the Kaggle's Medical Masks Dataset (MMD) [48]. The masked face detection in [49] is treated as an occluded face detection problem, where the mask is one of the possible objects covering the face, such as body parts or scarves. The proposed model looks for missing facial landmarks and combines with extracted facial features to identify whether a proposal is a real face using joint regression and classification.

In terms of real-time applications, light-weight models have been proposed for embedded devices such as NVIDIA Jetson Nano or Raspberry Pi. In [50], human faces are first extracted from the input frames using the SSD then fed into a Mobilenet-V2 architecture [51] for masked- and non-masked classification. Despite the promising accuracy, the processing time is currently limited at 15.71 frames per second (fps). A similar performance is reported in [52] when testing the masked face detection on a CPU. Experiment results on a low-power embedded system (Raspberry Pi 4) using a deep learning device (Intel's Neural Compute Stick (NCS) 2) show that a combination of distributed computing and multiple deep neural network accelerators can boost the performance on an edge device close to that on a CPU. The proposed framework consists of a face detection and a mask identification stage, where Inception modules are employed in the former for multi-scale detection and Mobilenet-V2 is adopted in the latter for reduction of computation cost.

Due to the inadequacy of dataset in masked face detection at the very beginning, transfer learning was employed in the major approaches, taking advantages of available general image datasets, such as ImageNet [13]. The pre-trained model is then repurposed on a smaller masked face dataset. However, researchers have to deal with a trade-off in the adaptability of the pre-trained model when transferring to the target domain. To overcome this disadvantage, a semi-synthetic dataset is generated and employed as the source domain in [53] for model training. By exploiting the distribution similarity between the synthetic and real datasets, the feature specificity in the deep networks is then retrained in the target domain to optimise the outcome. Later on, more and more masked face datasets have been generated and published to aid with the fight against COVID-19 pandemic. For instance, a large dataset is presented in [54] by mapping different mask poses to each human face obtained from the Flickr-Faces-HQ (FFHQ) dataset [55]. With this dataset, incorrectly masked faces can be further classified into three different types based on the mask coverage. Another popular dataset in the domain is Masked Faces (MAFA) [49] containing different mask types, the images of which are taken under various face orientations or occlusion level. Current datasets are either imbalanced between classes, limited in size, or biased towards a specific characteristic (such as skin colour). A more comprehensive review on masked face datasets can be found in [17].

4.3   Technical approaches

The core techniques employed in the aforementioned social distancing and face masked detection approaches are face detection and pedestrian detection, which are two sub-domains of object detection in computer vision. Figure 4.2 demonstrates the

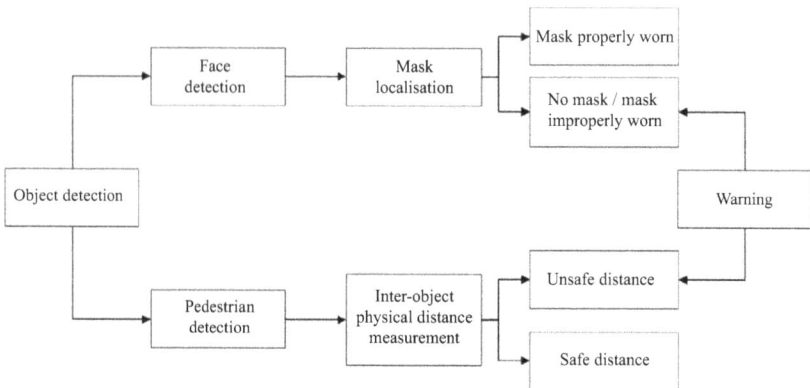

Figure 4.2   Processing pipeline of a general vision-based monitoring system

basic processing pipelines of the presented approaches and their relation to object detection. This section presents a brief introduction to classical and modern neural network-based techniques as well as sensor types and mobile systems that have been employed.

4.3.1   Classical approaches

Before the era of deep learning, traditional object detection methods are normally based on manually engineered low-level features. The extracted features are then trained using an appropriate classifier to better predict the interested objects. Among classical techniques using machine learning, the most two well-known ones are Viola-Jones [38] and Histogram of Oriented Gradients [6], due to their high accuracy and rapid detection rate. The robustness of these feature-based approaches, as pointed out in [4,16], is limited when dealing with the diversification of object appearance and surrounding environments.

4.3.1.1   Feature extraction

Haar Wavelet-based Detection Frameworks: An application of Haar-like features in computer vision is first proposed in [56], the capacity of which has been verified on both face and pedestrian detection tasks. For the former, it is observed in [5] that the significant face features, i.e. the eyes, mouth, and nose, are well captured by 'horizontal', 'vertical', and 'diagonal' wavelets. Although analysing the wavelet coefficients at different scales is required for a true scale-invariant system, it could also lead to a high computational cost. To overcome this challenge, integral images are proposed in [38], where each rectangular feature is represented by its sum of belonging pixels and can be calculated using a summed-area table. In the pedestrian detection task, although no specific pattern is observed as per the former one, the features is reported to be distributed along the object boundary [5].

Histogram of oriented gradients: Based on the idea that local objects can be well presented by oriented gradients, the Histogram of Oriented Gradients (HOGs) is first presented in [6] for pedestrian detection. In this approach, the input image is divided into cells, the pixels of which are employed for gradient calculation. The obtained information of the gradient magnitudes and directions is then incorporated to form a histogram representing each cell. To limit the effect of the lighting condition, histograms of gradients are normalised in blocks, the size of which is dependent on the size of the input image and the interested objects. The approach is then applied successfully to other domains, including face detection [57,58]. Figure 4.3 demonstrates the Haar-like and HOG features extracted from a face image from the FFHQ dataset [55].

4.3.1.2   Classification

Once the features have been extracted, their correlation with the interested object is learned via a shallow classifier (e.g. Support Vector Machine [SVM] [9], AdaBoost [10], or Deformable Part-based Model [DPM] [11]). While SVM stands out due to its low number of control parameters and capability to handle high dimensional data [59], AdaBoost is favourable due to its generalisation performance [60]. On the

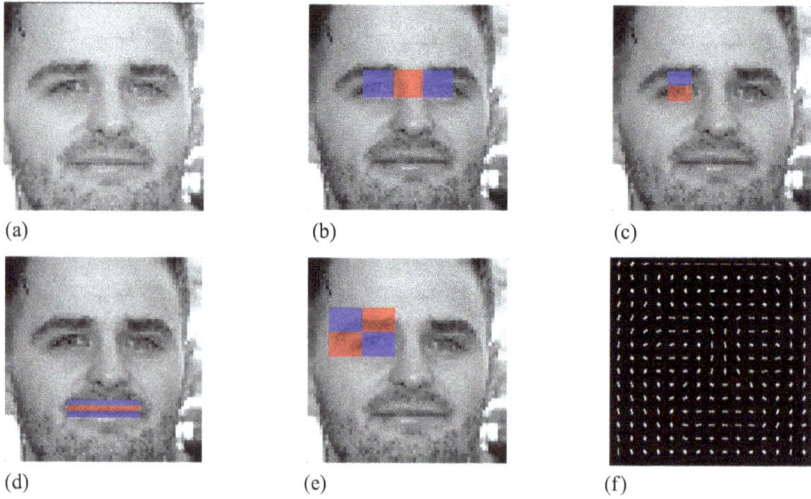

Figure 4.3   An example of Haar-like and HOG features extracted from a face image: (a) Original image, (b)–(e) horizontal, vertical and diagonal rectangle features, (f) HOG features

other hand, the DPM has shown its effectiveness in dealing with severe deformations by taking into account a root model and some deformation models.

4.3.2   Modern neural network-based methodologies

As reported in [4,16], the performance of classical approaches in the face and pedestrian detection tasks are outperformed by methods based on deep CNNs. The main reason for this matter is the variation in object poses, lighting conditions and background complexity in real-world applications. Since those challenges can be solved more effectively with deep features, most of the recent publications in this field are CNN-based.

In a typical CNN architecture, the input is sequentially convoluted through convolutional layers, the results of which is a feature map to feed to the next layer for further processing. Each convolutional layer can be followed by a pooling layer for reduction of feature map dimension by representing part of or a whole feature map with a particular value, such as its maximum or average. Fully connected layers are then applied to the pooled features to calculate the probability distribution for each class. Despite the dependence on the structure design [61], CNN-based techniques still outperform classical ones due to the following exponential advantages [12,62]: (i) the capability to generate hierarchical and discriminative feature representations that can be learned automatically from the provided data, (ii) the learning capacity that is beneficial from the growth of large-scale visual databases such as ImageNet [13]. Depending on the structure of the detection process, CNN-based object detection techniques can be categorised into two groups: two- and one-stage detectors. A brief introduction to these detectors is presented in the next subsection. The detectors are normally equipped with a backbone for feature extraction such as AlexNet [63], VGGNet [64], or GoogleNet [65], which is designed for the ImageNet classification. For embedded applications, light architectures such as ResNet-50 [47], MobileNet-V1 [66], MobileNet-V2 [51] have been developed, taking into account the restricted resources of on-board computers. Since the direct utilisation of classification backbones might be sub-optimal for the object detection task, some recent studies have considered searching backbones in object detectors as an alternative [67].

4.3.2.1 Two-stage detectors

The detection process in this detector class consists of two steps: the region proposal step where Regions of Interest (RoI) are extracted from the input image, and the classification step where found candidates are assigned to the corresponding classes. The most representative detectors in this category are: Region-based CNN (R-CNN) [68], Fast R-CNN [69], and Faster R-CNN [44].

R-CNN: Instead of using sliding windows as per other CNNs, region proposals are employed in R-CNN for potential object candidate identification. A CNN is then utilised for feature extraction, the score of which is calculated by an SVM. The proposed approach also pointed out the advantage of supervised pretraining on abundant data and fine-tuning when adapting the trained CNN to a specific domain. Despite its detection accuracy, R-CNN is constrained by the pipeline complexity where multi-stage fine-tuning is required, the inefficiency in training time and feature storage, and especially the slow detection speed, i.e., close to 50 s per image at test time [69]. The drawback in terms of detection speed and training time of R-CNN is then overcome in [70] using Spatial Pyramid Pooling (SPP). However, the pipeline complexity still persists in SPP, hence the development of other R-CNN variants, such as the Fast R-CNN and Faster R-CNN.

Fast R-CNN: Based on the input image and object proposals, a convolutional feature map is first generated, where each RoI is downsampled before being mapped into a feature vector. Here, features of the RoI are represented by a small fixed-scale feature map using max pooling. The output of the network is fed into a multitask loss for classifier training and bounding-box regression. Compared to the previous work, i.e. the R-CNN and the SPP, hierarchical sampling and single-tuning stage are utilised in Fast R-CNN, leading to its advantages in detection quality as well as training and inference efficiency.

Faster R-CNN: Another approach to improve the computational efficiency of region-based techniques is to change the proposal computation itself [44]. The proposals in Faster R-CNN are computed with Region Proposal Networks (RPNs) where convolutional feature maps contribute to both the region proposal generation and the detection network. The idea of sharing convolutional layers is implemented by first training the RPN and Fast R-CNN separately, where both models are initialised by an ImageNet pre-trained model. The region proposal training is then initialised by the detector network, where only unique layers are fine-tuned. Eventually, Fast R-CNN's fully connected layers are fine-tuned, while the shared ones are fixed. The performance of Faster R-CNN with VGG-16 backbone [71] on a GPU is reported to achieve 5 fps with state-of-the-art accuracy on both PASCAL VOC 2007 and 2012 datasets.

4.3.2.2   One-stage detectors

Different from the aforementioned group, the prediction process of the detectors in this group is conducted directly, bypassing the region proposal step. As a result, these detectors are well-known for their computational efficiency, which is more suitable for real-time applications. On the contrary, a trade-off for system accuracy should also be considered in the implementation of such detectors. In this category, YOLO [40], SSD [41], and RetinaNet [72] are the most popular due to their promising performance [73].

YOLO: Unlike other techniques, the object detection problem is approached as a regression problem in YOLO. Besides, the bounding box prediction and class probability calculation is unified in a single network and processed in a parallel manner. The input image is divided into a grid of $S \times S$ cells, where each cell is responsible for the prediction of up to two objects. The real-time performance of YOLO on PASCAL VOC 2007 [74] is reported to be at 45 fps while its fast version is recorded at 155 fps [40]. Its accuracy is also close to that of Fast R-CNN and Faster R-CNN in the same data set. The limitation of YOLO is exposed on images with small objects such as that from the PASCAL VOC 2012 dataset [75]. In addition, YOLO is also vulnerable to scale and aspect ratio variance due to its single-scale learning mechanism. Many models in the YOLO family have been developed, such as YOLOv2 [31], YOLOv3 [76], and YOLOv4 [77], the improvements of which include not only the multi-scale prediction but also the accuracy and detection speed.

SSD: To overcome the disadvantage of YOLO's single-scale approach, multi-scale feature maps are proposed in SSD using additional convolutional feature layers with different sizes. Here, multiple default bounding boxes at different feature maps are evaluated and compared with the ground truth box of each object for class prediction. Experiment results on the PASCAL VOC 2007 dataset at test time have shown the promising performance of SSD on both 300×300 and 512×512 input images in terms of accuracy and detection speed. The accuracy of SSD on the smaller input is higher than that of YOLO while the detection speed on the larger input is close to real-time [41]. The performance of SDD on small and context-specific objects can be further improved by combining with the Residual-101 [47] classifier and augmenting the proposed approach with deconvolution layers [78].

RetinaNet: In an attempt to improve the accuracy of one-stage detectors while maintaining their computational efficiency, the class imbalance has been pointed out in [72] as the main challenge impacting the performance of those detectors. To deal with this issue, the focal loss is proposed in RetinaNet by using a modified cross-entropy to bring the hard examples into learning focus. Anchor boxes are then utilised for multi-class detection based on some pre-defined Intersection-over-Union (IoU) thresholds. Experiment results on the COCO dataset [79] at test time have verified the comparable performance of RetinaNet to other two-stage detectors such as Faster R-CNN, especially on larger input images. As discussed in [80], the dependence on pre-defined anchor boxes leads to the detectors' performance sensitivity, generalisation ability, class imbalance in training as well as complexity in the calculation.

A number of approaches have been developed based on the aforementioned state-of-the-art two- and one-stage detectors to deal with the masked face detection [20,46,50,52,53] and social distancing monitoring problems [18,22–25,29,36]. However, as pointed out in [81,82], the results are still limited due to the adaptability of the proposed methods in scenario variations. For instance, one of the major reasons for the constraints in the performance of current masked face detection techniques is the lack of quality datasets with sufficient variations in terms of mask types, lighting conditions, occlusion levels, and so on. Nevertheless, the generation of such datasets required a tedious, meticulous, and sometimes expensive process of data collection and annotation, which can be partially solved using semisupervised learning methods [81]. On the other hand, the wearing of masks in the Covid-19 era also leads to the occlusion of discriminative face features and has created some concerns on current face recognition systems [83], especially to their operation and security [84].

4.3.3 Sensor types and systems

Despite the progressive improvement of object detection techniques over the past decade, the employment of only colour images as input could be sensitive in most scenarios where a good lighting condition is not guaranteed [85]. For improvement of the detection systems' illumination invariance, several add-ons have been considered, among which the most popular ones are stereo vision, Time-Of-Flight (ToF), and thermal cameras. Besides the lighting variance challenge, surveillance applications, especially for pedestrian detection, also struggle with occlusions. While many advances in occluded pedestrian detection have been proposed [86–88], the effectiveness of which is still constrained at slightly occluded objects. As discussed in [88], further improvement is required for heavily occluded scenarios. Another research direction in this field is multicamera tracking, where targets are tracked by individual cameras, and tracks are then re-identified if they belong to the same object [89,90]. Compared to the usage of a single camera only, the considerable impact of multimodal fusion on social distancing monitoring has been verified in [82] where the number of people in a scene can be identified more accurately, leading to enhanced performance in inter-person distance estimation. Besides, it has been discussed in [91] that feet location information can also be taken into account in addition to head location information [23], to increase the accuracy in assessing a crowd's social distancing compliance. Since extra annotations are required, the camera arrangement must be designed appropriately so that both head and feet images are well captured. Thus, when dealing with a multicamera system, an accurate calibration is key to take advantage of the overlapping field of view and enhance the collaborative performance of the system [92].

On the other hand, static systems, such as CCTV ones, are limited in the area coverage and might face challenges for social distancing tasks where objects are in motion [22]. Recently, unmanned mobile systems, such as a ground robots or aerial vehicles, have been increasingly utilised in computer vision-based research for

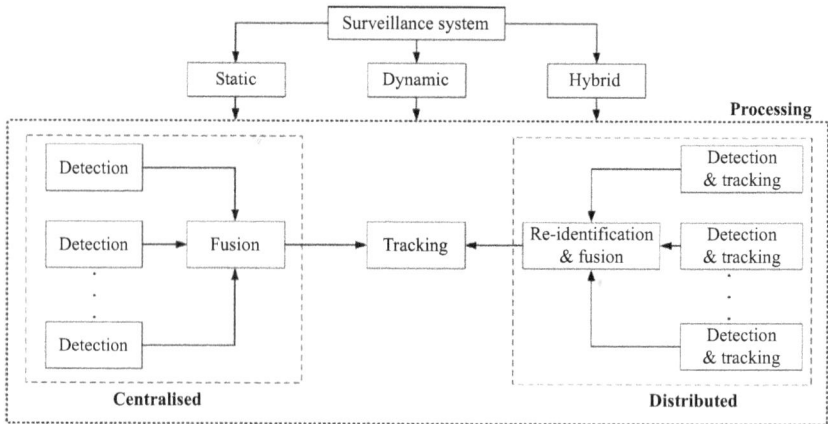

Figure 4.4 Classification of multi-camera surveillance systems

inspection [93], monitoring [94], and surveillance tasks [23,95,96], due to their autonomy, mobility, and accessibility. As suggested in [96], a network of camera-equipped
UAVs will be effective in mass surveillance for assessment of preventive measures in
COVID-like pandemics including mask detection, social distancing monitoring, and
suspected case alert.

Depending on the arrangement, surveillance camera systems can be categorised
into static, mobile and hybrid systems. In the first group, camera is fixed on infrastructures [97,98] while they are mounted on autonomous mobile devices [99,100]
in the second group. The last group consists of systems that utilise the setting of
both categories [34,101]. The detection and tracking tasks are processed either in
a centralised or a distributed manner, depend on the task requirement. The former
utilises the fused detection results from individual cameras to perform the tracking
tasks and is more suitable for arrangements with overlapped FoV. The latter looks for
the correlation between individual tracking results to fuse them into the final trajectory. The distributed processing is more favourable in tasks that lack the overlapping
FoV between cameras [92]. Figure 4.4 summarises the classification of multi-camera
surveillance systems.

The following subsections give a brief overview of the aforementioned sensor
types and unmanned mobile platforms that have been employed for surveillance
systems with applications in social distancing.

4.3.3.1 Sensor types

3D cameras: Stereo configurations have been applied in surveillance systems [102,
103], due to their ability to overcome the limitations of monocular imagery in terms of
reliable depth estimation. In general, a traditional stereo vision system consists of two
cameras being mounted next to each other in order to replicate the binocular vision
of human eyes. Depth information from the obtained stereo images is computed via

calculating the pair-wise distance between matched features from the rectified images. Stereo vision allows researchers to utilise the rich information obtained from digital images while also taking the advantage of depth information. Since the technology is based on the calculation of matching points, the generated disparity map normally includes a high level of noise when some matches can not be found [104].

ToF sensors calculate the distance from the object to the source by measuring the time required for a wave to travel to and reflect from an object. This process can be repeated on high-end sensors at the rate of up to million times per second, generating a real-time 3D map of the sensor's surrounding environment. With the development of affordable ToF-based RGB-D cameras, these types of sensors are becoming increasingly favourable in computer vision applications, including video surveillance systems [105,106]. ToF sensors are better than single- or stereo camera systems in terms of providing accurate depth measurements [107] and the fusion of RGB and depth data could provide a source of information that is more robust to illumination and colour variations [108]. More importantly, the effectiveness of the complementary between depth and colour information has been verified in [108] where a higher correlation between the detection results and the ground truth of crowd number has been obtained. This achievement is crucial in monitoring social distancing since a correct count is required to better estimate the inter-object physical distance. Besides these advantages, researchers have been studying to overcome the current limitations of the ToF-based RGB-D cameras, which are the alignment between the depth and RGB images as well as the erroneous measurements from the ToF sensors [109].

Thermal cameras: generate images of an interested object by converting the thermal energy of the object and its surrounding environment into temperature ranges. Thermal cameras are well known for their robustness in different light and weather conditions, especially the ability to operate in darkness. Besides the obvious application in non-contact temperature screening [110], thermal cameras have been applied widely in intelligent surveillance systems [111–113], mostly in a combination with colour cameras. The effectiveness of thermal cameras in the fight against COVID-19 has been verified in [111] where the body-induced thermal signatures are employed for not only physical distancing estimation but also temperature screening. As discussed in [114], the complementarity between thermal and colour cameras is emphasised under daylight conditions, where the single performance of the latter is better than that of the former. Under poor lighting conditions, the detection rate of thermal cameras alone outperforms not only that of colour cameras, but also any other fusion strategies [114].

4.3.3.2 Unmanned mobile systems

Compared to static surveillance systems, mobile platforms equipped with mounted cameras are more favorable due to their deployability and patrol ability [22], the application of which has been no longer limited to military but also extended to civilian problems [115]. Mobile platforms that have been utilised for pedestrian detection and tracking tasks, which can be further developed for social distancing applications, can be categorised into unmanned ground vehicles (UGVs) and unmanned aerial vehicles

(UAVs). Besides existing challenges in object detection, other major problems for UGV-based surveillance systems are localisation, navigation, and obstacle avoidance. When performing the task in environments with dynamic obstacles, these problems are more severe due to the landmark occlusion as well as the difficulty in trajectory prediction. UAVs, on the other hand, are more flexible and able to take photos or record videos of a wider concerned area at different altitudes and attitudes. However, current versions of UAVs are limited by the payload capability and weather vulnerability. While performing at different height levels, UAV-based surveillance systems have to deal with the changes in object scales and background complexity, which can be effectively solved with spatial attention [23] or cascade network [36]. Besides, the implementation of UAVs in the COVID-19 is also impacted by unclear regulatory policies in some countries as well as the unsafe operation beyond the visual line of sight [116].

4.4 Feasibility study and comparisons

In this section, a simple on-board masked face detection system is developed to verify its feasibility in an indoor environment and analyse the limitations for future studies. Although a detailed guidance on the correct use of masks have been provided by the WHO [117], there is still a percentage of users wearing their mask in a way that their noses and faces are not surely covered [118,119]. Hence, the system is designed to not only identify whether human objects are wearing a mask but also to verify if the mask is properly worn.

4.4.1 Model training and evaluation

4.4.1.1 Dataset generation
First, a new masked face dataset is generated by combining three existing ones. The first one is the Mask Dataset [120], consisting of 853 images of face images, categorised into three classes: with properly worn mask (proper), with improper worn mask (improper), and without mask (no mask). Bounding box annotations of each image are also provided in this dataset. Since there is a limited number of improper samples in this dataset, the second dataset [121] is taken into account, which is obtained from Kaggle with 2079 images. While the samples in the second dataset are also categorised into three classes as per the first one, the mask of the correctly worn samples are machine generated. Besides, they are all medical white masks, and appear with a high contrast to the surrounding environment in all samples. Figure 4.5(a) shows an example of a machine-generated mask on a face image. It is significant to see that also the original image was taken in a poor lighting condition, the mask appears at a much higher intensity level and does not fit to the background. Such impractical samples are then manually removed in this experiment. The third dataset [122] added to the combination is the one with 208 images of proper and no-mask samples. Notably, proper samples of this dataset contain not only medical face masks, but also other type of coverings, as shown in Figure 4.5(b). The final

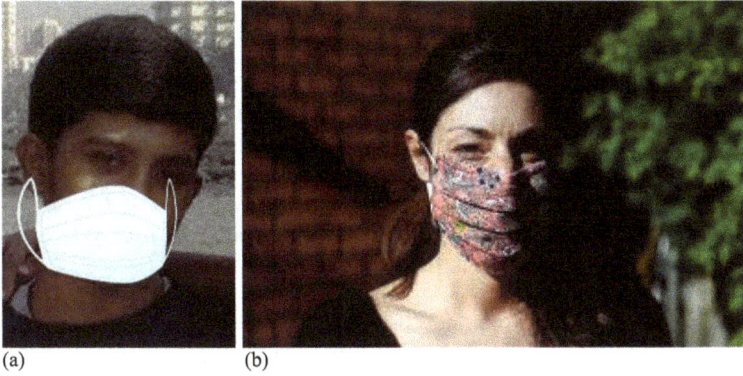

Figure 4.5   An example of (a) a virtual masked face and (b) a real masked face

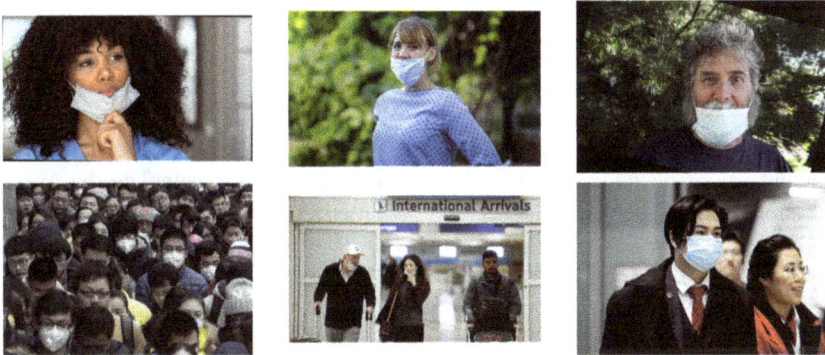

Figure 4.6   Different scenarios in masked face detection

dataset consists of 2241 images with bounding box annotations, which are split to training and testing set using the 80:20 ratio. Figure 4.6 presents some images of the employed dataset with all scenarios: proper, improper, and no mask.

4.4.1.2   Evaluation metrics

Intersection over Union (*IoU*) metric is widely used in object detection [123] to evaluate the performance of detectors. *IoU* employs the region properties, i.e. the width, the height, and the location, and compares the similarity between objects via their overlapping areas. Let O be the area of overlap between the prediction and the ground truth and U is their area of union, the evaluation metric is calculated as

$$IoU = \frac{O}{U}. \tag{4.1}$$

IoU can have any values between 0 and 1. The higher value the metric achieves, the better accuracy the detector performs. A pre-defined threshold t is then required for comparison with the calculated IoU and evaluation of predictions. A detection is considered correct if its IoU value is higher than the given threshold and vice versa. According to each established value of the threshold, predicted bounding boxes are classified into True Positive (tp), False Positive (fp) and False Negative (fn), which are defined below:

- tp: a correctly detected ground truth bounding box, i.e., $IoU \geq t$,
- fp: an incorrectly detected ground truth bounding box, i.e., $IoU \leq t$,
- fn: an undetected ground truth bounding box.

The obtained number of tp, fp, fn are then employed for calculation of other performance metrics, the precision p and the recall r. The former is the ratio of correct detections over the total detections while the latter is the ratio of correct detections over the total ground truth. These metrics are calculated as follows:

$$p = \frac{tp}{tp + fp}, \tag{4.2}$$

$$r = \frac{tp}{tp + fn}. \tag{4.3}$$

Since changing the threshold could lead to a different set of detections with different values of the precision and recall, it is desirable to calculate the Average Precision (AP), which takes into account both metrics [124] and defined as

$$AP = \int_0^1 p(r)dr. \tag{4.4}$$

Here, AP at different IoU values and AP across scales are employed for model evaluation [125]. The former group consists of AP calculated for 10 IoUs in the range from [50%, 95%], denoted as $AP50{:}5{:}95$, and AP evaluated at two single values 50% and 75%, denoted respectively as $AP50$ and $AP75$. The latter includes AP determined at different object sizes in number of pixels: smaller than 32^2 (denoted as APs), within the range $[32^2, 96^2]$ (denoted as APm), and larger than 96^2 (denoted as APl).

To select an appropriate model for the on-board detection system, some lightweight models that have been pre-trained on the COCO datasets [126] are selected to adapt to the masked face detection problem using transfer learning, i.e., SSD MobileNet-V1, SSD-MobileNet-V2, SSD ResNet50-V1, Faster R-CNN ResNet50-V1. Table 1.1 presents the performance of the participated models on the test set. Notably, SSD MobileNet-V1 outperforms other detectors on all evaluated metrics. Among the remaining detectors, SSD MobileNet-V2 performs well on medium and large images while Faster R-CNN ResNet50-V1 are more effective in terms of the metrics $AP50{:}5{:}95$ and $AP50$. On another note, the processing times of the detectors SSD MobileNetV1, SSD MobileNet-V2, SSD ResNet50-V1, Faster R-CNN ResNet50-V1 on a 2.3 GHz Intel Core i7 with 8 GB DDR3 are respectively 7, 9, 0.5, 0.2 fps. Considering the trade-off between the accuracy and processing time, SSD-MobileNet-V2 is selected to embed in our onboard detection system.

Table 4.1   Performance of the participated detectors on the test set

Accuracy	SSD MobileNet-V1	SSD MobileNet-V2	SSD ResNet50-V1	Faster R-CNN ResNet50-V1
*AP*50:5:95	0.769	0.730	0.713	0.738
*AP*50	0.968	0.935	0.940	0.964
*AP*75	0.893	0.842	0.846	0.837
APs	0.494	0.398	0.434	0.441
APm	0.731	0.647	0.580	0.630
APl	0.865	0.862	0.818	0.856

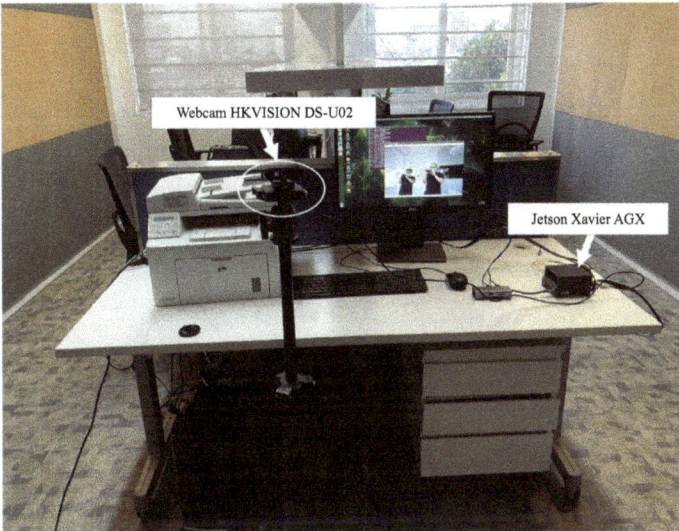

Figure 4.7   Setup of a simple masked face detection system

4.4.2   Case studies

In this subsection, the effectiveness and robustness of an on-board masked face detection system in an indoor environment is evaluated. The system consists of a Webcam HKVISION HS-Y02 and a Jetson Xavier AGX using the SSD MobileNet-V2 detector. The system setup is presented in Figure 4.7. The detector processes input frames at 14 fps, the performance of which has been evaluated in some scenarios, such as good and poor lighting conditions, variations of object scales, mask types, face orientations and occlusion levels.

4.4.2.1   Lighting condition variations

As can be seen in Figure 4.7, the test environment is a workspace with both natural and LED lighting. The test was taken from mid- to late-afternoon of a sunny day, the lighting conditions of which are changed according to the time and the adjustable

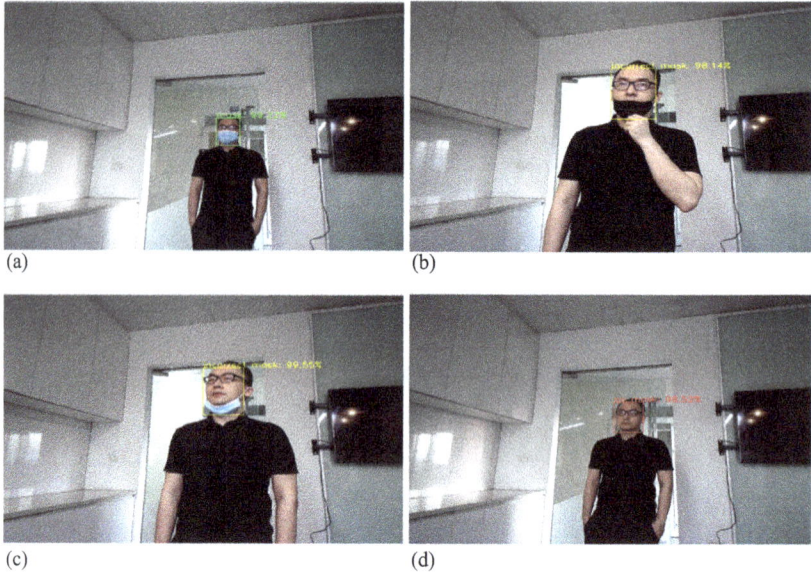

Figure 4.8   *Performance under good lighting condition: Correct detection of (a)*
proper, (b)–(c) improper and (d) no mask

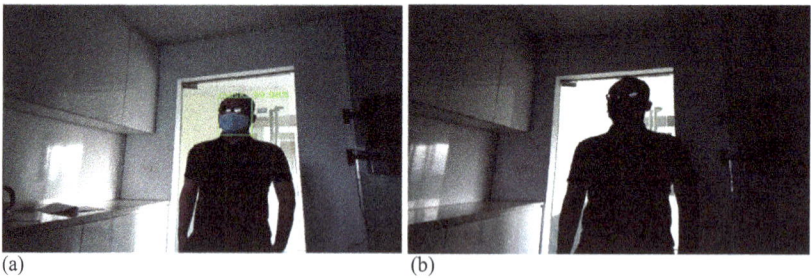

Figure 4.9   *Performance under poor lighting condition: (a) correct detection and*
(b) false detection

status of the LED lights. Under good lighting conditions with both LED and natural lights, the system can detect correctly all cases where a mask is properly, improperly, or not worn on the object as can be seen in Figure 4.8. Under a decent lighting condition with natural light only, the system is still capable to detect a masked face. However, in the dim lights, the face detector is no longer function (Figure 4.9).

4.4.2.2   Other variations

Besides the lighting condition that majorly impacts a masked face detection system using colour cameras, other factors affecting the system performance such as face orientations, scale variation, and occlusion are also demonstrated in this experiment.

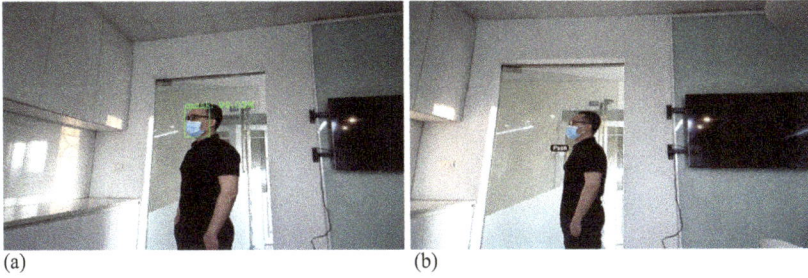

Figure 4.10 Performance under pose variation (a) correct detection and (b) false detection

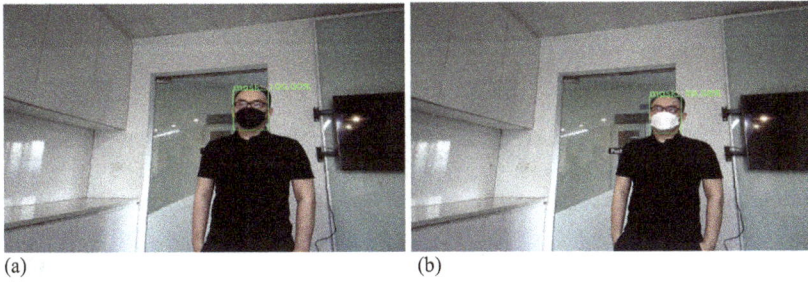

Figure 4.11 Performance on different masks: (a) black 3D mask and (b) white 3D mask

Since most of the training samples are front face images, the system is vulnerable to the side face input due to the lack of facial landmarks. Figure 4.10 shows the detection results at two different orientations where only one is successful. On another note, to test the sensitivity of the system on the scale variation of the input image, the subjects were asked to stay at 1, 2, and 3 m far from the camera. While the detector works well with subjects at a close distance, the subjects face is no longer detected from 3 m onwards.

The performance of the system is also validated on different types and colours of face masks. The subject was asked to wear a 2D blue, a 3D black, and a 3D white face mask to verify the detector's accuracy. In this test, all cases are successfully detected as shown in Figures 4.8 and 4.11.

Eventually, the occlusion impact on the detector's performance is tested by arranging two subjects standing after each other as demonstrated in Figure 4.12. Similar to the face orientation case, both subjects faces are detected in the partial occlusion scenario where ample facial landmarks of the rear subject are observable. In the heavy occlusion case where one-third of the rear subject's face is covered by the front subject, only the face of the front object is detected.

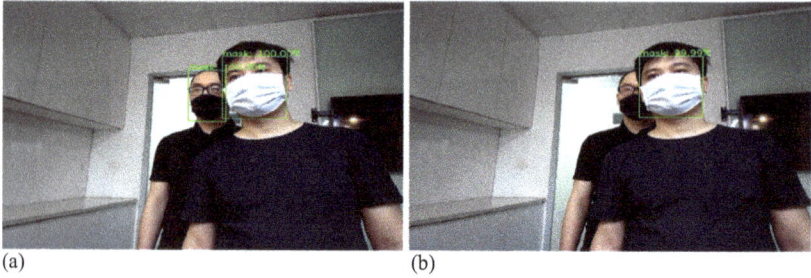

(a)   (b)

Figure 4.12   Performance under occlusion: (a) partial occlusion and (b) heavy
occlusion

4.4.3   Lesson learned and remarks

As pointed out in the literature and verified in the above experiment, face detection systems using colour cameras are heavily impacted by the lighting conditions. Although this issue can be easily controlled in in-door environments, the problem is more severe outdoors, especially during night times. Besides existing hardware solutions such as the employment of thermal cameras [127], image enhancement can be considered to improve the contrast and brightness of the input [128]. On the other hand, the system's invariance to object scales and face directions can be further improved by taking into account multi-scale or scale-adaptive detectors as well as providing more training samples of face images at various orientations, scales and image resolutions. In addition, although the test system described in this section is able to correctly detect faces wearing different masks, its robustness on many variations of cloths, disposable masks as well as respirators has not been tested in the experiment. As suggested in [17], more mask variations should be added into existing public datasets to increase the diversity of the training samples. Last but not least, the occlusion problem can be addressed in different manners, depending on the purpose and scope of the application, from changing the camera height and angle to applying advance techniques such as object reconstruction [129].

4.5   Open problems and discussions

Despite the timely development of vision-based social distancing approaches in the last two years, there are still open problems that require further research. Unlike other applications that can be processed off-line, surveillance systems for social distancing should be able to run accurately in real-time to provide timely warnings to violations. To enhance the performance of such systems, the following problems should be addressed.

4.5.1   Occlusion

As discussed above, surveillance systems, especially the ones that deal with crowd monitoring, are heavily impacted by occlusion where pedestrians are occluded by other pedestrians or by non-pedestrian objects. This problem is commonly approached with detectors trained on full-body annotations or based on visible body region information where guided attention mechanisms have shown promising results [86,87]. Alternatively, it has been pointed out in [130] that the occlusion issue can be effectively solved when presented in bird-view map, which can be generated via Cycle GAN [131], obtained directly from a UAV [23], or via multi-view.

4.5.2   Scale and pose variance

Objects normally appear in different scales and poses in surveillance video frames, generating difficulties for detectors in adapting to a specific domain. On one hand, detectors should be trained with high-quality samples of objects at different scales and poses. On the other hand, different architectures have been proposed to extract scale-invariant features using feature pyramid [132] or to reduce the difference between small and large objects using generative adversarial network [133,134]. Recently, with the widespread popularity of UAVs, small object detection [23,36] has become a critical problem in the scale variance domain and is suggested to be treated as a separate issue [16].

4.5.3   Illumination variance

When dealing with input taken in poor lighting conditions, the performance of surveillance systems using colour cameras is normally degraded due to the lack of sufficient brightness and contrast in the input frame for feature extraction. To overcome this obstacle, many approaches have considered the employment of an extra source of information, such as thermal [112], infrared [102], or laser cameras [22]. The adoption of multiple kinds of sensors on a system has led to the requirement of effective solutions for information fusion, especially for the heterogeneous data obtained from colour and depth cameras.

4.5.4   Interperson physical distance estimation

Besides general obstacles in object and pedestrian detection, i.e., occlusion, scale, pose, and illumination variations, the estimation of the physical distance between pedestrians is also a serious challenge in the COVID-era. Since the recommended physical distance to others is at least 1 m [15], an accuracy in distance estimation is crucial. For colour camera-based systems [18,23,25], either a prior knowledge or an assumption about the monitored environment is required for an effective calibration. The employment of depth or laser cameras can increase the estimation accuracy [21, 22,108] via an alignment between the depth and colour map. However, a detailed report on the impact of the alignment on the accuracy tolerance was not provided in the aforementioned studies.

4.6 Conclusion

Due to the recent advance in CNN-based object detection frameworks, many vision-based approaches have been successfully developed in the last two years to aid with the fight against COVID-19 via intelligent surveillance systems. This chapter presents some recent approaches dealing with social distancing monitoring and masked face detection. A brief review of core computer vision techniques for the aforementioned approaches is also provided, along with an introduction to related sensor types and possible systems. A simple onboard masked face detection system for indoor applications is then developed and tested against typical challenges of the domain, such as lighting conditions, occlusion, scale and pose variations. Finally, the extent of these problems in intelligent surveillance systems is discussed and some potential solution directions are suggested.

Acknowledgement

This work was supported in part by the Joint Technology and Innovation Research Centre – a partnership between the University of Technology Sydney and the University of Engineering and Technology, Vietnam National University, Hanoi. The authors would like to thank Cong Hoang Quach, Quang Manh Doan, and Cong Hieu Le for their help in the experiments.

References

[1] Szeliski R. *Computer Vision: Algorithms and Applications*. Springer Science & Business Media; 2010.

[2] Livingstone M, Hubel DH. *Vision and Art: The Biology of Seeing*. vol. 2. Harry N. Abrams New York; 2002.

[3] Andrew AM. Multiple view geometry in computer vision. *Kybernetes*. 2001;30(9/10):1333–1341.

[4] Zhao ZQ, Zheng P, Xu St, *et al.* Object detection with deep learning: A review. *IEEE Transactions on Neural Networks and Learning Systems*. 2019;30(11):3212–32.

[5] Papageorgiou CP, Oren M, Poggio T. A general framework for object detection. In: *Sixth International Conference on Computer Vision (IEEE Cat. No.98CH36271)*; 1998. pp. 555–62.

[6] Dalal N, Triggs B. Histograms of oriented gradients for human detection. In: *2005 IEEE Computer Society Conference on Computer Vision and Pattern Recognition (CVPR'05)*. vol. 1. IEEE; 2005. pp. 886–93.

[7] Lowe DG. Distinctive image features from scale-invariant keypoints. *International Journal of Computer Vision*. 2004;60(2):91–110.

[8] Bay H, Tuytelaars T, Van Gool L. Surf: Speeded up robust features. In: *European Conference on Computer Vision*. Springer; 2006. pp. 404–17.

[9]   Vapnik V. *The Nature of Statistical Learning Theory*. Springer Science & Business Media; 2013.

[10] Freund Y, Schapire RE. A decision-theoretic generalization of on-line learning and an application to boosting. *Journal of computer and system sciences*. 1997;55(1):119–39.

[11] Felzenszwalb PF, Girshick RB, McAllester D, *et al.* Object detection with discriminatively trained part-based models. *IEEE Transactions on Pattern Analysis and Machine Intelligence*. 2009;32(9):1627–45.

[12] LeCun Y, Bengio Y, Hinton G. Deep learning. *Nature*. 2015;521(7553): 436–44.

[13] Deng J, Dong W, Socher R, *et al.* Imagenet: a large-scale hierarchical image database. In: *2009 IEEE Conference on Computer Vision and Pattern Recognition*. IEEE; 2009. pp. 248–55.

[14] Cubuk ED, Zoph B, Mane D, *et al.* AutoAugment: learning augmentation strategies from data. In: *Proceedings of the IEEE/CVF Conference on Computer Vision and Pattern Recognition*; 2019. pp. 113–23.

[15] Advice for the public: Coronavirus disease (COVID-19). The content is last updated on 1 October 2021. https://www.who.int/emergencies/diseases/novel-coronavirus-2019/advice-for-public.

[16] Cao J, Pang Y, Xie J, *et al.* From handcrafted to deep features for pedestrian detection: a survey. *IEEE Transactions on Pattern Analysis and Machine Intelligence*. 2021;1. doi: 10.1109/TPAMI.2021.3076733.

[17] Nowrin A, Afroz S, Rahman MS, Mahmud I, Cho Y-Z. Comprehensive review on facemask detection techniques in the context of Covid-19. *IEEE Access*. 2021;9:106839–864.

[18] Shorfuzzaman M, Hossain MS, Alhamid MF. Towards the sustainable development of smart cities through mass video surveillance: a response to the COVID-19 pandemic. *Sustainable Cities and Society*. 2021;64:102582.

[19] Saponara S, Elhanashi A, Gagliardi A. Implementing a real-time, AI-based, people detection and social distancing measuring system for Covid-19. *Journal of Real-Time Image Processing*. 2021;18:1937–1947.

[20] Wang B, Zhao Y, Chen CLP. Hybrid transfer learning and broad learning system for wearing mask detection in the COVID-19 era. *IEEE Transactions on Instrumentation and Measurement*. 2021;70:1–12.

[21] Le AV, Ramalingam B, Gómez BF, *et al.* Social density monitoring toward selective cleaning by human support robot with 3D based perception system. *IEEE Access*. 2021;9:41407–16.

[22] Chen Z, Fan T, Zhao X, *et al.* Autonomous social distancing in urban environments using a quadruped robot. *IEEE Access*. 2021;9:8392–403.

[23] Shao Z, Cheng G, Ma J, *et al.* Real-time and accurate UAV pedestrian detection for social distancing monitoring in COVID-19 pandemic. *IEEE Transactions on Multimedia*. 2021. doi: 10.1109/TMM.2021.3075566.

[24] Yang D, Yurtsever E, Renganathan V, *et al.* A vision-based social distancing and critical density detection system for COVID-19. *Sensors*. 2021;21(13):4608.

[25]   Ahmed I, Ahmad M, Rodrigues JJPC, *et al*. A deep learning-based social distance monitoring framework for COVID-19. *Sustainable Cities and Society*. 2021;65:102571.

[26]   Benfold B, Reid I. Stable multi-target tracking in real-time surveillance video. In: *Proceedings of the Conference on Computer Vision and Pattern Recognition (CVPR)*; 2011. pp. 3457–64.

[27]   Chen K, Loy CC, Gong S, *et al*. Feature mining for localised crowd counting. In: *Proceedings of the British Machine Vision Conference (BMVC)*. BMVA Press; 2012. pp. 21.1–11.

[28]   Zhou B, Wang X, Tang X. Understanding collective crowd behaviors: learning a mixture model of dynamic pedestrian-agents. In: *2012 IEEE Conference on Computer Vision and Pattern Recognition (CVPR)*; 2012. pp. 2871–8.

[29]   Su J, He X, Qing L, *et al*. A novel social distancing analysis in urban public space: A new online spatio-temporal trajectory approach. *Sustainable Cities and Society*. 2021;68:102765.

[30]   Eiter T, Mannila H. Computing discrete Fréchet distance. *Citeseer*; 1994.

[31]   Redmon J, Farhadi A. YOLO9000: better, faster, stronger. In: *Proceedings of the IEEE Conference on Computer Vision and Pattern Recognition*; 2017. pp. 7263–71.

[32]   Cao Z, Hidalgo G, Simon T, *et al*. OpenPose: realtime multi-person 2D pose estimation using Part affinity fields. *IEEE Transactions on Pattern Analysis and Machine Intelligence*. 2019;43(1):172–86.

[33]   Wojke N, Bewley A. Deep cosine metric learning for person re-identification. In: *2018 IEEE Winter Conference on Applications of Computer Vision (WACV)*. IEEE; 2018. pp. 748–56.

[34]   Miseikis J, Borges PVK. Joint human detection from static and mobile cameras. *IEEE Transactions on Intelligent Transportation Systems*. 2015;16(2): 1018–29.

[35]   Wang RJ, Li X, Ling CX. Pelee: A real-time object detection system on mobile devices. *Advances in Neural Information Processing Systems*. 2018;31: 1963–72.

[36]   Zhang X, Izquierdo E, Chandramouli K. Dense and small object detection in uav vision based on cascade network. In: *Proceedings of the IEEE/CVF International Conference on Computer Vision Workshops*; 2019.

[37]   Hendra T, Spolaor R, Chen Z. A compound technique for multiple objects detection based on Markov clustering networks and Viola-Jones algorithm. In: *2019 IEEE 2nd International Conference on Information Communication and Signal Processing (ICICSP)*; 2019. pp. 459–63.

[38]   Viola P, Jones MJ. Robust real-time face detection. *International Journal of Computer Vision*. 2004;57(2):137–54.

[39]   Liu Z, Lin G, Yang S, *et al*. Learning Markov clustering networks for scene text detection. In: *2018 IEEE/CVF Conference on Computer Vision and Pattern Recognition*; 2018. pp. 6936–44.

[40]   Redmon J, Divvala S, Girshick R, *et al*. You Only Look Once: Unified, Real-Time Object Detection. In: *Proceedings of the IEEE Conference on Computer Vision and Pattern Recognition (CVPR)*; 2016.

[41] Liu W, Anguelov D, Erhan D, *et al*. SSD: Single shot multibox detector. In: *European Conference on Computer Vision*. Springer; 2016. pp. 21–37.

[42] Podbucki K, Suder J, Marciniak T, *et al*. CCTV based system for detection of anti-virus masks. In: *2020 Signal Processing: Algorithms, Architectures, Arrangements, and Applications (SPA)*; 2020. pp. 87–91.

[43] Prinosil J, Maly O. Detecting faces with face masks. In: *2021 44th International Conference on Telecommunications and Signal Processing (TSP)*; 2021. pp. 259–62.

[44] Ren S, He K, Girshick R, *et al*. Faster R-CNN: Towards real-time object detection with region proposal networks. *Advances in Neural Information Processing Systems*. 2015;28:91–9.

[45] Szegedy C, Vanhoucke V, Ioffe S, *et al*. Rethinking the inception architecture for computer vision. In: *Proceedings of the IEEE Conference on Computer Vision and Pattern Recognition*; 2016. pp. 2818–26.

[46] Loey M, Manogaran G, Taha MHN, *et al*. Fighting against COVID-19: a novel deep learning model based on YOLO-v2 with ResNet-50 for medical face mask detection. *Sustainable Cities and Society*. 2021;65:102600.

[47] He K, Zhang X, Ren S, *et al*. Deep residual learning for image recognition. In: *Proceedings of the IEEE Conference on Computer Vision and Pattern Recognition*; 2016. pp. 770–8.

[48] Medical Mask Dataset. Available from: https://www.kaggle.com/shreyashw aghe/medical-mask-dataset.

[49] Ge S, Li J, Ye Q, *et al*. Detecting masked faces in the wild with lle-cnns. In: *Proceedings of the IEEE Conference on Computer Vision and Pattern Recognition*; 2017. pp. 2682–90.

[50] Nagrath P, Jain R, Madan A, *et al*. SSDMNV2: a real time DNN-based face mask detection system using single shot multibox detector and MobileNetV2. *Sustainable Cities and Society*. 2021;66:102692.

[51] Sandler M, Howard A, Zhu M, *et al*. MobileNetV2: Inverted residuals and linear bottlenecks. In: *Proceedings of the IEEE Conference on Computer Vision and Pattern Recognition (CVPR)*; 2018. pp. 4510–20.

[52] Kong X, Wang K, Wang S, *et al*. Real-time Mask Identification for COVID-19: An edge computing-based deep learning framework. *IEEE Internet of Things Journal*. 2021;8(21):15929–15938.

[53] Chen S, Liu W, Zhang G. efficient transfer learning combined skip-connected structure for masked face poses classification. *IEEE Access*. 2020;8: 209688–98.

[54] Cabani A, Hammoudi K, Benhabiles H, *et al*. MaskedFace-Net – A dataset of correctly/incorrectly masked face images in the context of COVID-19. *Smart Health*. 2021;19:100144.

[55] NVIDIA Flickr Faces HQ Dataset. Accessed: 2021-10-07. Available from: https://github. com/NVlabs/ffhq-dataset.

[56] Oren M, Papageorgiou C, Sinha P, *et al*. Pedestrian detection using wavelet templates. In: *Proceedings of IEEE Computer Society Conference on Computer Vision and Pattern Recognition*; 1997. pp. 193–9.

[57]   Déniz O, Bueno G, Salido J, *et al.* Face recognition using Histograms of Oriented Gradients. *Pattern Recognition Letters*. 2011;32(12):1598–603.

[58]   Zafeiriou S, Zhang C, Zhang Z. A survey on face detection in the wild: past, present and future. *Computer Vision and Image Understanding*. 2015;138: 1–24.

[59]   Osuna E, Freund R, Girosit F. Training support vector machines: an application to face detection. In: *Proceedings of IEEE computer society conference on computer vision and pattern recognition*. IEEE; 1997. pp. 130–6.

[60]   Bartlett P, Freund Y, Lee WS, *et al.* Boosting the margin: A new explanation for the effectiveness of voting methods. *The Annals of Statistics*. 1998;26(5):1651–86.

[61]   Sun Y, Xue B, Zhang M, *et al.* Completely automated CNN architecture design based on blocks. *IEEE Transactions on Neural Networks and Learning Systems*. 2019;31(4):1242–54.

[62]   Alzubaidi L, Zhang J, Humaidi AJ, *et al.* Review of deep learning: Concepts, CNN architectures, challenges, applications, future directions. *Journal of Big Data*. 2021;8(1):1–74.

[63]   Krizhevsky A, Sutskever I, Hinton GE. Imagenet classification with deep convolutional neural networks. *Advances in Neural Information Processing Systems*. 2012;25:1097–105.

[64]   Simonyan K, Zisserman A. Very deep convolutional networks for large-scale image recognition. *arXiv preprint arXiv:14091556*. 2014.

[65]   Szegedy C, Liu W, Jia Y, *et al.* Going deeper with convolutions. In: *Proceedings of the IEEE Conference on Computer Vision and Pattern Recognition*; 2015. pp. 1–9.

[66]   Howard AG, Zhu M, Chen B, *et al.* Mobilenets: Efficient convolutional neural networks for mobile vision applications. *arXiv preprint arXiv:170404861*. 2017.

[67]   Chen Y, Yang T, Zhang X, *et al.* Detnas: Backbone search for object detection. *Advances in Neural Information Processing Systems*. 2019;32:6642–52.

[68]   Girshick R, Donahue J, Darrell T, *et al.* Rich feature hierarchies for accurate object detection and semantic segmentation. In: *Proceedings of the IEEE Conference on Computer Vision and Pattern Recognition*; 2014. pp. 580–7.

[69]   Girshick R. Fast R-CNN. In: *Proceedings of the IEEE International Conference on Computer Vision (ICCV)*; 2015.

[70]   He K, Zhang X, Ren S, *et al.* Spatial pyramid pooling in deep convolutional networks for visual recognition. *IEEE Transactions on Pattern Analysis and Machine Intelligence*. 2015;37(9):1904–16.

[71]   Simonyan K, Zisserman A. Very deep convolutional networks for large-scale image recognition. *arXiv preprint arXiv:14091556*. 2014.

[72]   Lin TY, Goyal P, Girshick R, *et al.* Focal loss for dense object detection. In: *Proceedings of the IEEE International Conference on Computer Vision*; 2017. pp. 2980–8.

[73]   Wu X, Sahoo D, Hoi SC. Recent advances in deep learning for object detection. *Neurocomputing*. 2020;396:39–64.

[74] Everingham M, Van Gool L, Williams CKI, *et al. The PASCAL Visual Object Classes Challenge 2007 (VOC2007) Results.* Available from: http://www.pascal-network.org/challenges/VOC/voc2007/workshop/index.html.

[75] Everingham M, Van Gool L, Williams CKI, *et al. The PASCAL Visual Object Classes Challenge 2012 (VOC2012) Results.* Available from: http://www.pascal-network.org/challenges/VOC/voc2012/workshop/index.html.

[76] Redmon J, Farhadi A. Yolov3: An incremental improvement. *arXiv preprint arXiv:180402767.* 2018.

[77] Bochkovskiy A, Wang CY, Liao HYM. Yolov4: Optimal speed and accuracy of object detection. *arXiv preprint arXiv:200410934.* 2020.

[78] Fu CY, Liu W, Ranga A, *et al.* DSSD: Deconvolutional single shot detector. *arXiv preprint arXiv:170106659.* 2017.

[79] Lin TY, Maire M, Belongie S, *et al.* Microsoft COCO: Common objects in context. In: *European Conference on Computer Vision.* Springer; 2014. pp. 740–55.

[80] Tian Z, Shen C, Chen H, *et al.* Fcos: Fully convolutional one-stage object detection. In: *Proceedings of the IEEE/CVF International Conference on Computer Vision;* 2019. pp. 9627–36.

[81] Prasad S, Li Y, Lin D, *et al.* maskedFaceNet: A Progressive Semi-Supervised Masked Face Detector. In: *Proceedings of the IEEE/CVF Winter Conference on Applications of Computer Vision;* 2021. pp. 3389–98.

[82] Varghese EB, Thampi SM. A multimodal deep fusion graph framework to detect social distancing violations and FCGs in pandemic surveillance. *Engineering Applications of Artificial Intelligence.* 2021;103:104305.

[83] Jeevan G, Zacharias GC, Nair MS, *et al.* An empirical study of the impact of masks on face recognition. *Pattern Recognition.* 2022;122:108308.

[84] Fang M, Damer N, Kirchbuchner F, *et al.* Real masks and spoof faces: On the masked face presentation attack detection. *Pattern Recognition.* 2022;123:108398.

[85] Chen Z, Huang X. Pedestrian Detection for Autonomous Vehicle Using Multi-Spectral Cameras. *IEEE Transactions on Intelligent Vehicles.* 2019;4(2):211–9.

[86] Zhang S, Yang J, Schiele B. Occluded pedestrian detection through guided attention in CNNS. In: *Proceedings of the IEEE Conference on Computer Vision and Pattern Recognition;* 2018. pp. 6995–7003.

[87] Xie J, Pang Y, Khan MH, *et al.* Mask-guided attention network and occlusion-sensitive hard example mining for occluded pedestrian detection. *IEEE Transactions on Image Processing.* 2021;30:3872–84.

[88] He Y, Zhu C, Yin XC. Occluded pedestrian detection via distribution-based mutual-supervised feature learning. *IEEE Transactions on Intelligent Transportation Systems.* 2021;pp. 1–16.

[89] Specker A, Stadler D, Florin L, *et al.* An occlusion-aware multi-target multi-camera tracking system. In: *Proceedings of the IEEE/CVF Conference on Computer Vision and Pattern Recognition;* 2021. pp. 4173–82.

[90]   Stadler D, Beyerer J. Improving multiple pedestrian tracking by track management and occlusion handling. In: *Proceedings of the IEEE/CVF Conference on Computer Vision and Pattern Recognition*; 2021. pp. 10958–67.

[91]   Dai Z, Jiang Y, Li Y, *et al*. BEV-Net: Assessing social distancing compliance by joint people localization and geometric reasoning. In: *Proceedings of the IEEE/CVF International Conference on Computer Vision*; 2021. pp. 5401–11.

[92]   Olagoke AS, Ibrahim H, Teoh SS. Literature survey on multi-camera system and its application. *IEEE Access*. 2020;8:172892–922.

[93]   Hoang VT, Phung MD, Dinh TH, *et al*. System architecture for real-time surface inspection using multiple UAVs. *IEEE Systems Journal*. 2020;14(2):2925–36.

[94]   La HM, Dinh TH, Pham NH, *et al*. Automated robotic monitoring and inspection of steel structures and bridges. *Robotica*. 2019;37(5):947–67.

[95]   Du B, Zhang C, Shen J, *et al*. A dynamic sensitivity model for unidirectional pedestrian flow with overtaking behaviour and its application on social distancing's impact during COVID-19. *IEEE Transactions on Intelligent Transportation Systems*. 2021.

[96]   Hossain MS, Muhammad G, Guizani N. Explainable AI and mass surveillance system-based healthcare framework to combat COVID-I9 like pandemics. *IEEE Network*. 2020;34(4):126–32.

[97]   Chavdarova T, Baqué P, Bouquet S, *et al*. Wildtrack: A multi-camera HD dataset for dense unscripted pedestrian detection. In: *Proceedings of the IEEE Conference on Computer Vision and Pattern Recognition*; 2018. pp. 5030–9.

[98]   Lima JP, Roberto R, Figueiredo L, *et al*. Generalizable multi-camera 3D pedestrian detection. In: *Proceedings of the IEEE/CVF Conference on Computer Vision and Pattern Recognition*; 2021. pp. 1232–40.

[99]   Saska M, Krajnik T, Pfeucil L. Cooperative μUAV-UGV autonomous indoor surveillance. In: *International Multi-Conference on Systems, Signals & Devices*. IEEE; 2012. p. 1–6.

[100] Štěpán P, Krajník T, Petrlík M, *et al*. Vision techniques for on-board detection, following, and mapping of moving targets. *Journal of Field Robotics*. 2019;36(1):252–69.

[101] Schranz M, Andre T. Towards resource-aware hybrid camera systems. In: *Proceedings of the 12th International Conference on Distributed Smart Cameras*; 2018. pp. 1–7.

[102] Krotosky SJ, Trivedi MM. On color-, infrared-, and multimodal-stereo approaches to pedestrian detection. *IEEE Transactions on Intelligent Transportation Systems*. 2007;8(4):619–29.

[103] Luo H, Liu J, Fang W, *et al*. Real-time smart video surveillance to manage safety: A case study of a transport mega-project. *Advanced Engineering Informatics*. 2020;45:101100.

[104] Kadambi A, Bhandari A, Raskar R. 3D depth cameras in vision: Benefits and limitations of the hardware. In: *Computer vision and machine learning with RGB-D sensors*. Springer; 2014. pp. 3–26.

[105] Layne R, Hannuna S, Camplani M, *et al*. A dataset for persistent multi-target multi-camera tracking in RGB-D. In: *Proceedings of the IEEE Conference on Computer Vision and Pattern Recognition (CVPR) Workshops*; 2017.

[106] Sun CC, Wang YH, Sheu MH. Fast motion object detection algorithm using complementary depth image on an RGB-D camera. *IEEE Sensors Journal*. 2017;17(17):5728–34.

[107] Hansard M, Lee S, Choi O, *et al*. Time-of-flight cameras: principles, methods and applications. *Springer Science & Business Media*; 2012.

[108] A cross-modal fusion based approach with scale-aware deep representation for RGB-D crowd counting and density estimation. *Expert Systems with Applications*. 2021;180:115071.

[109] Qiu D, Pang J, Sun W, *et al*. Deep end-to-end alignment and refinement for time-of-flight RGB-D module. In: *Proceedings of the IEEE/CVF International Conference on Computer Vision*; 2019. pp. 9994–10003.

[110] Dell'Isola GB, Cosentini E, Canale L, *et al*. Noncontact body temperature measurement: Uncertainty evaluation and screening decision rule to prevent the spread of COVID-19. *Sensors*. 2021;21(2):346.

[111] Savazzi S, Rampa V, Costa L, *et al*. Processing of body-induced thermal signatures for physical distancing and temperature screening. *IEEE Sensors Journal*. 2021;21(13):14168–79.

[112] Shahbaz A, Jo KH. Deep atrous spatial features-based supervised foreground detection algorithm for industrial surveillance systems. *IEEE Transactions on Industrial Informatics*. 2021;17(7):4818–26.

[113] Kang JK, Hoang TM, Park KR. Person re-identification between visible and thermal camera images based on deep residual CNN using single input. *IEEE Access*. 2019;7:57972–84.

[114] Li C, Song D, Tong R, *et al*. Illumination-aware faster R-CNN for robust multispectral pedestrian detection. *Pattern Recognition*. 2019;85: 161–71.

[115] Huang H, Savkin AV. An algorithm of reactive collision free 3-D deployment of networked unmanned aerial vehicles for surveillance and monitoring. *IEEE Transactions on Industrial Informatics*. 2019;16(1):132–40.

[116] Chamola V, Hassija V, Gupta V, *et al*. A comprehensive review of the COVID-19 pandemic and the role of IoT, drones, AI, blockchain, and 5G in managing its impact. *IEEE Access*. 2020;8:90225–65.

[117] World Health Organization. Mask use in the context of COVID-19: Interim guidance, 1 December 2020. World Health Organization; 2020. https://apps.who.int/iris/handle/10665/337199.

[118] Machida M, Nakamura I, Saito R, *et al*. Incorrect use of face masks during the current COVID-19 pandemic among the general public in Japan. *International Journal of Environmental Research and Public Health*. 2020;17(18):6484.

[119]   Szepietowska M, Krajewski PK, Matusiak Ł, *et al.* Do University Students Adhere to WHO Guidelines on Proper Use of Face Masks during the COVID-19 Pandemic?—Analysis and Comparison of Medical and Non-Medical Students. *Applied Sciences*. 2021;11(10):4536.

[120]   Mask Dataset. Available from: https://makeml.app/datasets/mask.

[121]   Three-Class Faced Mask Dataset. Available from: https://www.kaggle.com/spandanpatnaik09/face-mask-detectormask-not-mask-incorrect-mask.

[122]   Two-Class Faced Mask Dataset. Available from: https://www.kaggle.com/sumansid/facemask-dataset?select=No+Mask.

[123]   Rezatofighi H, Tsoi N, Gwak J, *et al.* Generalized intersection over union: a metric and a loss for bounding box regression. In: *Proceedings of the IEEE/CVF Conference on Computer Vision and Pattern Recognition*; 2019. pp. 658–66.

[124]   Zhu M. Recall, precision and average precision. *Department of Statistics and Actuarial Science, University of Waterloo, Waterloo*. 2004;2(30):6.

[125]   Padilla R, Netto SL, da Silva EA. A survey on performance metrics for object-detection algorithms. In: *2020 International Conference on Systems, Signals and Image Processing (IWSSIP)*. IEEE; 2020. pp. 237–42.

[126]   Huang J, Rathod V, Sun C, *et al.* Speed/accuracy trade-offs for modern convolutional object detectors. In: *Proceedings of the IEEE Conference on Computer Vision and Pattern Recognition*; 2017. pp. 7310–11.

[127]   Knapik M, Cyganek B. Driver's fatigue recognition based on yawn detection in thermal images. *Neurocomputing*. 2019;338:274–92.

[128]   Shen J, Li G, Yan W, *et al.* Nighttime driving safety improvement via image enhancement for driver face detection. *IEEE Access*. 2018;6:45625–34.

[129]   Cai J, Han H, Cui J, *et al.* Semi-Supervised Natural Face De-Occlusion. *IEEE Transactions on Information Forensics and Security*. 2021;16:1044–1057.

[130]   Luo Y, Zhang C, Zhao M, *et al.* Where, What, Whether: Multi-modal learning meets pedestrian detection. In: *Proceedings of the IEEE/CVF Conference on Computer Vision and Pattern Recognition*; 2020. pp. 14065–73.

[131]   Zhu JY, Park T, Isola P, *et al.* Unpaired image-to-image translation using cycle-consistent adversarial networks. In: *Proceedings of the IEEE International Conference on Computer Vision*; 2017. pp. 2223–32.

[132]   Wang X, Zhang S, Yu Z, *et al.* Scale-equalizing pyramid convolution for object detection. In: *Proceedings of the IEEE/CVF Conference on Computer Vision and Pattern Recognition*; 2020. pp. 13359–68.

[133]   Goodfellow I, Pouget-Abadie J, Mirza M, *et al.* Generative adversarial nets. *Advances in Neural Information Processing Systems*. 2014;27.

[134]   Yin R. Multi-resolution generative adversarial networks for tiny-scale pedestrian detection. In: *2019 IEEE International Conference on Image Processing (ICIP)*. IEEE; 2019. pp. 1665–9.

Chapter 5
Artificial intelligence and big data for COVID-19 and social distancing

Quoc-Viet Pham[1], Dinh C. Nguyen[2], Thien Huynh-The[3],
Pubudu N. Pathirana[4] and Won-Joo Hwang[5]

Since the first confirmed case in December 2019, the coronavirus disease-19 (COVID-19) pandemic has affected every aspect of our lives and caused severe difficulties to healthcare systems in the world. Many approaches have been investigated to mitigate this pandemic, such as lockdown, social distancing, mask wearing, and working from home. At the same time, a great deal of COVID-related data (e.g., X-ray images, transportation histories, and confirmed case statistics) has been generated, which can be effectively used for various purposes. In the battle against the pandemic, artificial intelligence (AI) and big data have found various applications thanks to their distinctive capabilities in analyzing the data and finding notable features from the massive and heterogeneous data. In this chapter, we focus on highlighting the applications of AI and big data for the COVID-19 outbreak. In particular, we review state-of-the-art solutions using AI and big data to fight against COVID-19, such as detection, diagnosis, tracking, and social distancing. We also present a set of challenges and recommendations, which may provide new insights and drive novel research solutions to stop the COVID-19 pandemic.

5.1 Introduction

5.1.1 Preliminary

Since late 2019, coronavirus disease-19 (COVID-19) has changed the world significantly, from healthcare and education to transportation and politics. Infected

[1] Korean Southeast Center for the 4th Industrial Revolution Leader Education, Pusan National University, Busan, Republic of Korea
[2] School of Engineering, Deakin University, Waurn Ponds, VIC, Australia
[3] ICT Convergence Research Center, Kumoh National Institute of Technology, Gyeongsangbuk-do, Republic of Korea
[4] School of Engineering, Deakin University, Waurn Ponds, VIC, Australia
[5] Department of Biomedical Convergence Engineering, Pusan National University, Yangsan, Republic of Korea

COVID-19 patients typically develop respiratory disease and need to be treated by proper treatment methods. Since the COVID-19 virus is efficient in human-to-human transmissions and common symptoms do not appear immediately, this virus is more dangerous than previous coronavirus families. From a few thousand affected cases, the COVID-19 virus has spread over almost all countries and areas globally and caused millions of deaths [1]. Although many policies have been executed and a large number of people have been vaccinated, no obvious sign shows that the COVID-19 pandemic is under control and can stop in the near future. New virus variants, such as Alpha, Beta, Gamma, and Delta, have raised many challenges to the existing solutions and healthcare systems in combating the pandemic. Moreover, new variants of the COVID-19 virus are much more contagious than the previous ones. For example, the Delta variant, which originated in India, has caused a spike in many countries world-wide, where the healthcare systems become overwhelmed with many new infected cases and deaths.

As a result of the global impact of the COVID-19 pandemic, a great deal of effort has been made to find effective methods to combat the outbreak. Playing a key role, the governments mostly focus on preventing the pandemic, such as locking down the country to limit the virus spreading, issuing a set of adaptive policies according to the pandemic situation, improving the capability of the healthcare systems, and investigating more effective solutions and clinic approaches. In the meanwhile, people are requested to follow a set of guidelines and recommendations from local authorities and health organizations, such as the World Health Organization (WHO) and the Centers for Disease Control and Prevention. For example, individuals should wear a mask in public places, wash their hands frequently, adhere to the social distancing (also known as physical distancing) rules, and report any latest symptoms to the local healthcare centers*. Furthermore, research and development related to COVID-19 is now a priority, and many stakeholders, ranging from governments, medicine companies, big techs, and academics, have shown a keen interest in it. As an effort to contribute to COVID-19 research, IBM, Amazon, Google, and Microsoft collaborated with the White House to build a supercomputing system [2]. In addition, computer scientists have made preliminary steps to combat the virus aside from the worldwide endeavor to create viable solutions for the COVID-19 coronavirus. The two key technologies, AI and big data, have shown their important role and applications in the global fight against COVID-19.

5.1.2 *Motivation and contributions*

Driven by the remarkable success and applications of AI in many disciplines, from computer vision and natural language processing to fully autonomous driving and smart healthcare, AI provides a powerful tool to combat the COVID-19 pandemic [3]. For example, the work in [4] built a deep learning network to identify the currently commercial medicines for drug-repurposing (i.e., drug repositioning). This work is helpful to find a fast treatment strategy based on the existing drugs, especially

---

*https://www.who.int/health-topics/coronavirus

during 2020, when the commercial vaccine was still under research and not clinically approved. In an attempt to stop the outbreak, a biomedicine company in Hongkong, namely Insilico Medicine, proposed a deep model for drug discovery[†]. In particular, the deep model generates several COVID-19 protease structures used in computer simulations to find new molecular entity compounds against the COVID-19 virus. Application of deep learning for fast COVID-19 diagnoses can be found in [5], where a deep learning model, namely DRENet, was developed. The DRENet model was then evaluated on a collected dataset of computed tomography (CT) images from 88 confirmed COVID-19 patients, 86 healthy people, and 100 patients infected with bacteria pneumonia. It was shown in [5] that the DRENet could achieve competitive results, such as the sensitivity of 0.96 and precision of 0.79. Another application of AI is in the real-time prediction of the dynamics of the COVID-19, such as the number of infected patients, end time, and virus trajectory [6]. A competitive accuracy of the AI technique developed in [6] was useful for tracking the COVID-19 outbreak and enhancing the organizational strategies in response to the COVID-19 pandemic. Besides continuous efforts from the research communities, various giant techs and healthcare companies have also been involved in the fight against the pandemic. As an example, Moderna Inc. collaborated with IBM Corporation to develop a technical solution to overcome the issue of vaccine distribution disruption[‡], which would help the governments and healthcare service providers accelerate the vaccine injection rate.

Similar to AI, big data has been leveraged to develop effective solutions to infectious diseases like the COVID-19 pandemic. In this regard, big data has been found in a large number of solutions and approaches used to combat the COVID-19 pandemic. In the context of the COVID-19 pandemic, big data can be generated by many sources, such as patient historical data, physician notes, X-ray reports, CT images, tracking, and statistical information. Big data allows us to understand various aspects of the pandemic, such as outbreak tracking, viral structure, illness treatment, and vaccine manufacture. When big data is combined with clever AI-based technologies, we can conduct complicated simulation models based on coronavirus data streams for pandemic prediction through via a comprehensive utilization of big data and AI. Further, big data can be used to estimate human mobility during the pandemic and the effects of social distancing policies [7]. Big data can process a vast volume of data for automatic recognition and predict the future of the COVID-19 pandemic owing to its data aggregation capacity. Moreover, big data analytics from a number of real-world sources, such as images of infected patients, social media, and statistical data, can aid in the implementation of large-scale COVID-19 research and the development of high-reliability treatment methods [8]. Besides, big data helps the researchers and healthcare providers understand the evolution of the COVID-19 virus and its variants, thus better responding to the outbreak and medical treatment approaches.

Moreover, social distancing is a great non-pharmaceutical method to present the spread of the outbreak, especially when vaccines are not available to the majority of the population and COVID-19 specifics are still under research. AI and big data have

[†]https://insilico.com/ncov-sprint/
[‡]www.reuters.com/article/us-health-coronavirus-moderna-ibm-idUSKBN2AW2DT

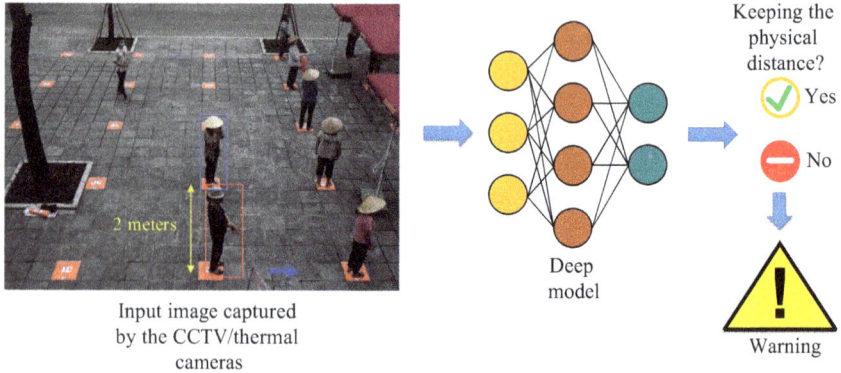

Figure 5.1   Illustration of an AI-enabled framework for social distancing

shown their potential for social distancing via numerous applications and use cases in this context. For example, an AI-enabled video and image professing framework can be implemented at an edge computing cloud of shopping malls to check the existence of people in the prohibited areas. Such a framework can also help check whether or not a person does not strictly keep the minimum distance from others. In this context, existing cameras, traffic cameras, CCTV systems, and individual cameras are potential sources of data to train the deep network. For example, the work in [9] developed a deep neural network using YOLOv2 for three basic tasks: people tracking, social distancing, and temperature checking, as illustrated in Figure 5.1. In particular, the images were extracted from video streams of thermal cameras, and the proposed deep network can achieve high accuracy of 95.6% while it can be effectively ported in a low-cost embedded device. While both AI and big data have important applications for the pandemic, we emphasize that big data analytics refers to the data collection and data analysis processes over a large volume of data. In the meanwhile, AI aims at applying logic and human intelligence for related tasks, such as virus detection, diagnosis, tracking, and social distancing.

Obviously, AI and big data have shown their effectiveness in fighting against the COVID-19 pandemic. To better understand the role of AI and big data in preventing COVID-19, in this chapter, we review the state-of-the-art solutions from the literature based on these two key thrusts: AI and big data. In particular, they have been used to develop innovative approaches of detection, diagnostic, medicine, tracking, and social distancing. In a nutshell, the features and contributions offered by this chapter are as follows.

- **Applications of AI**: First, we present an in-depth review of the role of AI in combating the COVID-19 pandemic via important applications, including virus detection, outbreak tracking, infodemiology, biomedical engineering, and social distancing. From the reviewed literature, these AI-enabled applications can help to mitigate the impacts of the COVID-19 pandemic, thus reducing the risk of close contact and better implementing social distancing policies.

- **Applications of big data**: Second, we also focus on the process of data collection and processing and the applications of big data for the COVID-19 pandemic. In particular, we review five main applications of big data, such as outbreak estimation, virus spreading, clinical treatment, drug discovery, and social distancing. It is reviewed that big data with AI techniques can be used to design useful solutions to fight COVID-19 and implement social distancing policies in practice.
- **Recommendations and Challenges**: Finally, based on the extensive review AI and big data for COVID-19 and social distancing, we present important challenges, lessons, and recommendations for the communities, governments, and societies and further identify a set of key challenges that need to be considered in future research actions.

5.1.3   The chapter organization

The organization of this chapter is as follows. In Sections 5.2 and 5.3, we present various applications of AI and big data in combating the pandemic, such as infected case identification, outbreak tracking, drug research, medical treatment, and social distancing. In Section 5.4, we present a set of important recommendations and challenges that need to be taken into consideration to mitigate the pandemic better. We draw a conclusion in Section 5.5.

5.2   AI for COVID-19 and social distancing

Motivated by the success of AI in different engineering disciplines, such as computer vision, natural language processing, financial services, transportation, military, and mobile communications, AI has been used to investigate effective solutions to the COVID-19 pandemic. Generally, machine learning is an important branch of AI that enables the extraction of relevant information from data, and deep learning can improve the performance of intelligent systems by learning from simple features in a deep manner. In the meanwhile, federated learning, invented by Google in 2015, is a new AI concept to enable collaborative AI solutions for the COVID-19 pandemic without the need to share COVID-19 data with the learning server, thus improving the data privacy of health users. In need, various AI techniques (e.g., deep learning, machine learning, and federated learning) can be used to learn from the COVID-19 data. In this section, we review prospective applications of AI/machine learning available in the literature to combat the COVID-19 pandemic, including detection and diagnosis, prediction and monitoring, infodemiology and infoveillance, biomedicine, and social distancing.

5.2.1   AI for COVID-19 detection and diagnosis

Accurate detection and early treatment are the key solutions to prevent the COVID-19 pandemic effectively. At present, the reverse transcription-polymerase chain reaction (RT-PCR) is realized as the standard method to identify respiratory viruses. Several intensive efforts have been made to improve the accuracy of RT-PCR and to dis-cover other alternatives. However, besides being costly and time-consuming, these techniques usually have a low true positive rate and consequently require some spe-cific materials, equipment, and instruments to achieve a high rate. Importantly, many countries have lacked testing kits due to budget limitations, supply deficiency, and difficulties with logistic services. In this context, the standard method may not be suitable to satisfy the very strict requirement of fast-accurate detection and tracking in the period of the COVID-19 pandemic. A cost-efficient COVID-19 detection solu-tion is exploiting the convenient use of smart devices and the power of AI technology, which is widely known as mobile health or mHealth in the literature. Remarkably, the development of edge and cloud computing can effectively handle the essential draw-backs of smart devices, including low battery, small storage, and limited computing capability.

Another COVID-19 detection approach is applying AI/machine learning algo-rithms for medical image processing, which has been found in many research works on coronavirus. Most of the existing medical image analysis methods use X-ray images and CT scans as the input of deep learning models to automatically detect coronavirus infection. The authors in [10] developed a deep learning model with convolutional neural network (CNN) architectures to identify COVID-19 cases based on the abnor-malities in the chest radiography images of COVID-19 patients. This three-class (normal, COVID-19 infection, and non-COVID-19 infection) classification model supports the medical staff in deciding which cases, between normal and COVID-19 infection, should be tested using standard methods, and which treatment strategies should be followed for COVID-19 and non-COVID-19 infection cases. Based on the simulations for performance evaluation, the proposed CNN can reach the accu-racy of approximately 93.3% on a free-access dataset having 13,975 images of 13,870 patients. Because X-ray images are usually more cost-efficient than CT scans, numer-ous research works of COVID-19 detection were performed on available datasets of X-ray images. In [11], a deep CNN model, namely Decompose-Transfer-Compose (DeTraC), was developed to learn the pattern recognition of COVID-19 from chest X-ray images. In this model, the decomposition layer aims to reduce feature maps and generate more sub-regions for independent analysis, whereas the composition layer aims to combine sub-regions from the decomposition layer to conduct the final classification result. Moreover, a transfer layer interconnects the composition and decomposition layers to accelerate the training speed, reduce the computational cost, and facilitate the training on small datasets. A general framework of deep learning-based COVID-19 detection and diagnosis is presented in Figure 5.2, where an output of binary classification with the normal class (i.e., negative) and COVID-19 infected class (i.e., positive) is illustrated.

Developing an effective diagnosis and treatment method is important to miti-gate the impact of the COVID-19 coronavirus. Some methods utilized deep learning and deep reinforcement learning to evaluate abnormalities in the COVID-19 dis-ease in terms of quantitative metrics. In these kinds of learning models, the input is non-contrasted chest CT images, and the output is severity scores, which include opacity percentage, high opacity percentage, and lung high opacity score. Based on

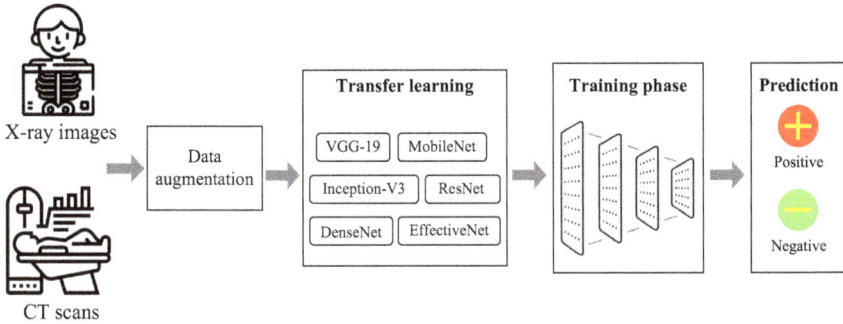

Figure 5.2 Deep learning-based framework with transfer learning to detect and diagnose COVID-19 from X-ray image and CT scans

the performance evaluation using a small dataset of CT images, the DRL model can achieve over 96% for all severity scores and outperform many conventional methods with image processing algorithms and machine learning classification models. As an example, a random forest classification model was adopted in [12] to learn COVID-19 diagnosis patterns with 63 extracted features, such as the infection ratio of the lung and ground truth class opacity lesions. Although the numerical results with different metrics, including sensitivity, selectivity, accuracy, and area under the curve, were incomparable to deep learning models, the approach argued that the descriptive features extracted from the right lung are more prominent than those of the left lung via severity level-based assessment. Some works focus on a scalable and cost-efficient COVID-19 diagnosis solution. For example, an AI-based diagnosis framework, namely AI4COVID-19, was introduced in [13], in which the expert medical knowledge was combined with AI models to analyze the cough/sound signals of mobile users.

5.2.2 AI for outbreak prediction and monitoring

Susceptible-Infected-Removed (SIR) is a traditional compartment model to predict infectious diseases based on the transmission rate and the recovering rate. However, this statistical model may not be suitable for the COVID-19 pandemic because of the following two assumptions.

1. The cases that have been recovered will not be infected.
2. The model does not consider the time-varying characteristics of the transmission rate and the recovering rate.

Some studies have been presented to properly adapt to the COVID-19 pandemic to tackle the above-mentioned issues of the SIR model. For example, an advanced version of the SIR model with a time-dependent constraint was proposed in [14], which remained adaptable to the changes in infectious disease control and preventive legislation (e.g., traffic halt and city lockdown). Furthermore, the improved model considered two types of infected patients: detectable and undetectable, in which the transmission rate of the former is lower than that of the latter. Some numerical results indicated that the time-dependent model could estimate the confirmed cases in China and the time duration of outbreaks in certain countries (e.g., Italy and Iran) accurately. Lately, the SIR model was adapted with different economic conditions and outbreak scenarios of different countries in the world, where space-time dependence and additional compartments can be considered to improve estimation precision. These models can predict the coronavirus outbreak and send warning messages back to the governments and medical centers, and accordingly, some responses should be made forcefully before the outbreak.

Some research works on the SIR model were also introduced by exploiting various machine learning and deep learning models to estimate the outbreak dimension of the COVID-19 pandemic. For example, the authors in [15] argued that the traditional SIR model was incapable of mining the effects of quarantine policy and social distancing. Instead, the quarantine policy can be formulated as a strength function, which is then learned by an artificial neural network to predict the outbreak size in Wuhan, China. Based on the public dataset provided by the Chinese National Health Commission, the experimental results demonstrated the importance of the right quarantine policy in controlling the COVID-19 outbreak, and the number of infected cases will increase exponentially because of some mistakes in quarantine policies. This learning model also obtained the same result when being used to estimate the global outbreak dimension. Lately, several statistical models, such as the logistic model, Bayesian nonlinear model, and autoregressive integrated moving average model, were applied to the size of COVID-19 outbreak in time. Besides, data-driven modeling was utilized to learn the phenomenology of the COVID-19 pandemic via asymptomatic information.

To conduct an analysis of the global transmission dynamics of COVID-19, a modified autoencoder (MAE) model [16] was proposed for mining the correlations between different concerning attributes. Relying on the simulations using the data collected from WHO reports, The MAE model has an average error of less than 2.5

5.2.3   AI for infodemiology and infoveillance

Most trustworthy information about the COVID-19 pandemic is circulated via the official websites of the ministry of health and welfare and health organizations like the WHO. However, social media platforms (e.g., Facebook, Twitter, and Instagram) have shown great potential in broadcasting useful pandemic information timely. The information from these platforms is highly accessible and attractive; hence some comprehensive analyses can be carried out if the data is acquired and processed properly. Being a powerful tool to process big data characterized by volume, velocity, and variety, AI/machine learning algorithms have been utilized to analyze the social network dynamics, which in turn helps to understand the COVID-19 situation and then fight against the pandemic.

Some AI-based real-life applications have been considered in the past during the pandemic period: using Twitter data for public behavior surveillance, examining the health-searching behavior of the Ebola breakout, and public response to the

Chikungunya outbreak. Similarly, the sudden emergency of COVID-19 was attracted several studies about infodemiology and infoveillance. In [17], the authors collected the data from three platforms, including the Baidu search engine, Sina Weibo, and Ali e-commerce marketplace, to investigate public concerns, evaluate risk perception, and trace public behaviors. Notably, the data of public emotion was evaluated by a computerized text analysis program, namely Linguistic Inquiry and Word Count (LIWC), whereas the data of public attention and awareness were assessed via a Weibo daily index indicating the number of posts and comments related to the COVID-19 pandemic. Furthermore, the daily indices of two platforms, Baidu and Ali, were used to examine the behaviors complying with the recommended/mandatory protection measures and also the rumors about ineffective treatments in the breakout period. AI-enabled speech and audio analysis (so-called computer audition) was considered for infodemiology and infoveillance. Several advanced AI/machine learning algorithms can be leveraged to analyze spoken conversations acquired from news, advertising videos, and other social media. Some probable use-cases were taken into accounts, such as risk prediction and assessment, social distancing effect analysis and monitoring, treatment, and recovery tracing. Besides the potentials of computer audition in the combat against COVID-19, some practical challenges should be tackled completely, such as how to collect the privacy data of COVID-19 patients, how to process raw audio and speech data effective with the presence of noise and outlier, and how to explain the results generated by AI black-box models.

AI/machine learning algorithms have been exploited to process the COVID-19 data acquired from heterogeneous sources, including mobility data, demographic data, social media data, and public health and medical data, at multiple levels, from individuals to official organizations. Some conditional generative adversarial models can be deployed to augment the limited data, and heterogeneous graph auto-encoder models can be studied to estimate some risks in a hierarchical structure. Some specialized models allowed us to pay high attention to a specific region having the pandemic risk, which in turn responds to appropriate actions to minimize the effect of the COVID-19 pandemic. Recently, an effective method, namely Augmented ARGONet, was introduced in [18] to predict the number of confirmed COVID-19 cases two days ahead. Accordingly, a novel statistical learning model was developed to effectively handle the multimodality data collected from multiple sources, such as health and medical reports from the Chinese Center for Disease Control and Prevention (China CDC), COVID-19 Internet search activities from Baidu, media flows from Media Cloud, and daily forecast. For the purpose of dataset enrichment, each data point was augmented by adding white noise before it was fed into a Lasso regression model to estimate the number of confirmed COVID-19 cases. Based on the experimental results obtained from 32 provinces in China, the Augmented ARGONet method can perform more accurately than many baseline models in some testing scenarios.

5.2.4 AI for biomedicine and pharmacotherapy

The world has been being in the fierce race of finding highly-effective vaccines and treatments against COVID-19 virus. That demands huge efforts from not only health and medical science but also computer science, in which AI plays a vital role in accelerating the research's progress. A huge amount of healthcare and biomedical data has activated the development of AI in the biomedicine and pharmaceutical fields. Recently, deep learning with recurrent neural network (RNN) and CNN architectures is able to process unstructured and high-dimensional data for learning the underlying features at multi-scale representations, which in turn improves the learning efficiency. Indeed, the high capacity of deep learning in terms of high-level feature extraction allows analyzing, understanding, and interpreting heterogeneous medical data like DNA microarray data. In the last decade, AI/machine learning algorithms and deep learning architectures have been applied for a wide range of biomedical applications: medical image classification, object detection and recognition, genomic sequence analysis, structural biology analytic, and drug delivery. Due to the harshness of the COVID-19 pandemic, AI has intensively found medicine applications, such as drug delivery and repositioning, and protein structure analytics, to restrain the coronavirus from spreading.

Using AI to invent new medicines and to advance existing drugs for immediate usage to treat the coronavirus is beneficial from both the economic and scientific perspectives, especially in the conditions in which clinically verified vaccines and medicines are unavailable. Some studies applied deep learning models to COVID-19 virus-specific datasets to examine the availability of commercial drugs and find some potential high-affinity inhibitors. Meanwhile, few works leveraged data-driven models to repurpose drugs through the combination of machine learning algorithms and statistical analysis mechanisms. In [19], the authors selectively listed around 6,225 candidate drugs for examination, which were then refined using multiple consecutive phases with a network-based knowledge extraction mechanism, a deep learning-based relation extraction method, and a graph-based analysis approach. After the in silico[§] and in vitro[‖] studies, including the verification of safety trials on monkeys and rats, an innovative inhibitor, namely CLV218, was recommended as a potential drug candidate for the COVID-19 virus treatment. Deep learning was further exploited to identify the COVID-19 protein structure in comparison with Ebola and HIV-1 viruses. Notably, some deep learning models were adapted to deal with small-scale datasets and facilitate biological data. Lately, multitask learning models with deep neural network architectures were designed to estimate the toxicity of protein-bound molecules, which in turn improves the in silico examination process and increases the success rate of drug candidates.

5.2.5 AI for social distancing

AI plays as the core technology of social distancing applications in the COVID-19 pandemic. Most of the existing methods were developed for social distancing monitoring systems, wherein the presence of humans with interactive activities and movements was detected, recognized, and tracked automatically using different sensors, such

---

[§]Experiments conducted via computer simulation.
[‖]Experiments conducted outside a living organism.

as camera, thermal, and LiDAR. The comprehensive survey in [21] discussed that intelligent techniques, e.g., computer vision, machine learning, and deep learning, can play an important role for different tasks in social distancing, such as contact tracing, traffic density prediction, infected trajectory prediction, sickness trend, and self-quarantined and self-protection location prediction. Some studies on AI for social distancing are discussed below.

In [20], an intelligent visual-based social distancing surveillance system was developed by combining AI with other modern technologies, including internet-of-things and unmanned aerial vehicles. The image sequence acquired by drones with a camera was transmitted and processed at the cloud to detect people in images using a widely used high-performance CNN model, namely YOLO-v3. The model was improved to identify the violation of social distancing policy by measuring the normalized Euclidean distance, after dimension scaling and coordinate calibration, between two or more people in a group. Drones with the ability of localization and navigation can be suitable for different indoor and outdoor scenarios. Compared with conventional systems using robots taking the front view and side view, drones performed object detection and distance measurement more accurately with the top view (or bird view) besides monitoring a wider area. By equipping thermal sensors, the drone-based social distancing monitoring systems can identify COVID-19 inspection. The framework of drone-based surveillance system is illustrated in Figure 5.3, which has three primary modules, including (1) localization (e.g., GPS), (2) navigation, and (3) visual-based social distancing detection using AI and deep learning approaches.

In another work [22], a drone-based active surveillance system was designed for multiple tasks, including social distancing detection, monitoring, analysis, and control. Interestingly, the system can send an advisory inflow notice to prevent over-crowding if the social density parameter is larger than a pre-defined warning threshold. If the pedestrian density parameter is handled to be smaller than the threshold, the social distancing violation is controlled at a low level. Furthermore, the regulator is able to manage offline aggregate statistics and online status of social distancing simultaneously. The effectiveness of the proposed system was demonstrated by measuring social distancing and critical density statistics of crowded indoor locations,

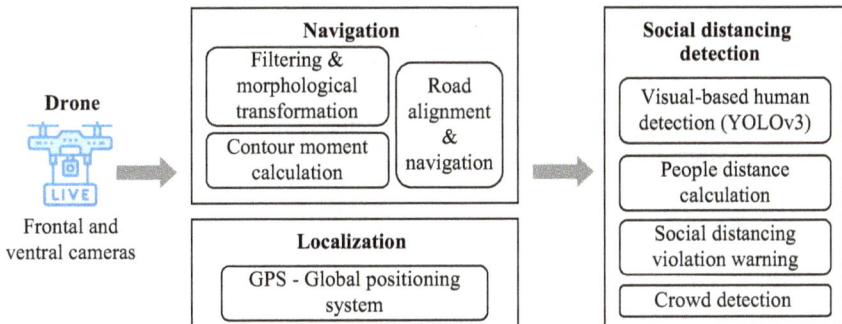

Figure 5.3 A framework of drone-based social distancing surveillance system [20]

such as New York Central Station. Some quantitative results were conducted on the Oxford Town Center dataset, in which the system reported some incorrect detections in some areas having extremely high pedestrian density with occlusion. Notably, some people in a single family (or having very close relationship) may not be required for social distancing, which needs to be considered and examined automatically by more intelligent surveillance systems in the future.

5.2.6   Summary

In this section, we summarize the role of AI for important applications in the COVID-19 pandemic, including detection and diagnosis, epidemic monitoring and identification, infodemiology and infoveillance, biomedicine and pharmacology, and social distancing. We believe that the AI-based framework is well-suited to mitigating the increasing impact of the COVID-19 pandemic since a large amount of COVID-19 data can be handled effectively by advanced AI techniques, such as machine learning and deep learning. Although AI-based approaches cannot be deployed as substitutes for the pandemic, they are useful since they can assist medical staff and policymakers in making better decisions and policies. However, there are still big challenges, such as the quality of the COVID-19 data, accuracy of machine learning/deep models, and large-scale deployment. Solving these challenges necessitates more efforts and novel studies from the scientific community as well as full involvement from other parties, such as authorities and healthcare providers, who usually have a massive amount of high-quality data and share them for scientific purposes.

5.3   Big data for COVID-19 and social distancing

Big data has been used in a variety of medical applications, such as mobile health and electronic health records, where big data analytics are used to facilitate the provision of health services. Big data is characterized by a set of features of large volume, velocity, and variety. More specifically, volume refers to an enormous amount of COVID-19 data generated by a very large number of smart and Internet-of-Things (IoT) devices, and variety refers to the generation rate of different data sources, such as healthcare applications, agriculture, and smart industries. Meanwhile, variety implies that data can be generated and stored in different formats, including structured data (e.g., text, images, and streaming videos) and unstructured data (e.g., emails, voice calls, and audio recordings).

The extraordinary increase in the amount of COVID-19 data and the success of AI for COVID-19 applications have led to the rapid emergence in the global battle against the COVID-19 pandemic. In the context of the COVID-19 pandemic, big data has been employed for various applications, including outbreak prediction, virus spread tracking, diagnosis and treatment, vaccine and drug discovery, and social distancing, as summarized in Figure 5.4 and discussed in this section.

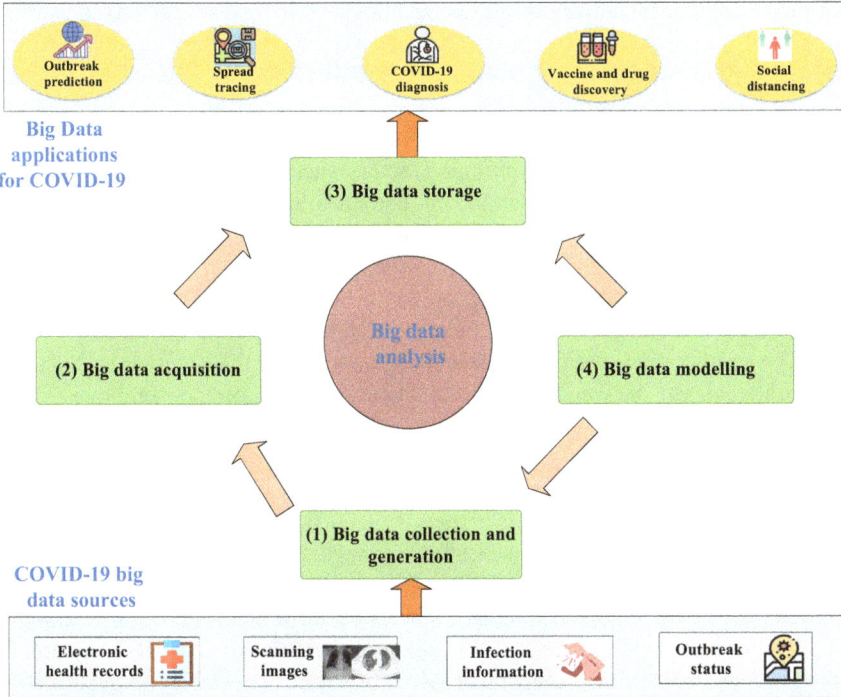

Figure 5.4 Applications of big data for COVID-19

5.3.1 Big data for outbreak prediction

Big data provides strong solutions to assist the prediction of the outbreak based on its ability to handle large-scale datasets. Indeed, the analysis of the massive data collected from distributed sources is useful to extract the most important and meaningful features of COVID-19, such as outbreak trends and localization of new outbreak areas [23]. In this work, a case study was conducted in Italy to predict the trend of outbreaks in different areas by using a set of complex models that can accurately characterize the dynamics of the pandemic. This is enabled by the aggregation of large data sets from Italian Civil Protection sources. This aims to overcome the limitations of the prediction ability of simple training models at a single hospital institution, which simple and deterministic models enable via human transmission modeling. This allows for using real datasets collected from the COVID-19 pandemic in Italy to reveal the outbreak trends in high-risk areas, for helping the local authorities in planning effective disease control strategies. Public datasets are also potential data sources for big data analytics that would be helpful to visualize the geographical areas with possible outbreaks [24]. In this work, a study was conducted in Wuhan, the people migration from and to Wuhan city, which would help health agencies to determine how many people are infected with COVID-19 for applying quarantine.

Big data also helps healthcare professionals to simulate the data models to estimate the tendency of the outbreak. For instance, it recognizes regions with high risks and identifies where people are infected with the COVID-19 virus.

Big data is also a strong tool for the outbreak prediction on the global scale, by enhancing the data learning accuracy via comprehensive investigations and data aggregation from a number of contributed factors, such as infected cases, population, living conditions, environments, and so on. The study in [25] used a large dataset from different countries, such as the Republic of Korea and China, to predict the pandemic based on a logistic model that can enforce the reliability of the predictions. Google Trends was used as a data engineering tool in [26] to collect COVID-19-related information from different countries, including China, the Republic of Korea, Italy, and Iran. The combination of different data sources facilitates the accurate visualization of the outbreak tendency and thus supports to estimate the possible outbreak.

5.3.2 Big data for virus spread tracing

Another potential role of big data is to trace the COVID-19 spread, which is necessary for healthcare professionals to successfully manage the pandemic. For example, big data analytics were introduced in [27], where a high quantity of social media search indexes (SMSI) was adopted to detect new cases for monitoring the spread of the COVID-19 virus in public areas. All information relevant to COVID-19 symptoms such as cough, pneumonia, fever and chest distress were collected by analyzing SMSI that also consists of human movement data that assists the modeling of infection trends. The research in [28] considered a big data-based approach for monitoring the COVID-19 spread. A large dataset was used and trained by a multiple-linear model to calculate the variance of infected cases in China cities. The analytic results demonstrate a high detection rate of positive cases against the population. The authors in [29] studied a joint approach using both big data and Geographic Information Systems (GIS) that are able to detect locations of people in the quarantine areas. Big data algorithms based on AI are integrated to build a correlation model between COVID-19 spreading patterns and user movement patterns.

Another study in [30] collected datasets from multiple countries in Europe to construct an analytic data model for the task of spread tracking. A macroscopic growth law was then developed that allowed estimating the maximum number of infectious patients in the quarantine areas. This is essential for the evaluation of COVID-19 prevention and tracing the potential spread of COVID-19 diseases. Different from [30], the research in [31] studied the COVID-19 pandemic from the temperature perspective, aiming to evaluate the impacts of temperature on the spread of the virus. The simulation outcomes verify that the growth rate in northern hemisphere countries significantly decreases because of the warmer weather and lockdown policies, while this growth rate in southern hemisphere countries increases. To address the problem of accuracy limitation in previous works, the authors in [32] suggested an unsupervised data learning model for controlling the COVID-19 disease by using a transfer learning method for mapping the infected cases with different regions at different infection rates.

5.3.3   Big data for COVID-19 diagnosis and treatment

Big data is of importance to COVID-19 diagnosis and treatment. The solutions with big data are highly important for analyzing virus transmission procedures, infection statistics, and disease characteristics for developing vaccine candidates, which is a critical stage to contain the spread and mitigate the death rate. Indeed, the potential for using big data to diagnose COVID-19-like diseases has been demonstrated via recent trials on early diagnosis [33], treatment, and predictive surgery. For example, a study in [34] used a robust, sensitive, and quantitative solution enabled by multiplex polymerase chain reactions to diagnose the SARS-CoV-2. The model in [34] is made up of 172 pairs of particular primers linked to the SARS-CoV-2 genome, which are available at the Chinese National Center for Biological Information (https://bigd.big.ac.cn/ncov). Multiplex PCR was demonstrated to be an effective and low-cost method for infection diagnosis, showing competitive performance of mean coverage 99% and specificity 99.8%. Another study in [35] designed a new diagnostic framework for genomic evaluations. A number of Australian people returning from overseas are studied based on genome data in the data analytic process. This work is helpful to overcome the limitation of genome data and helps to diagnose COVID-19.

The work in [36] relied on proteomics cells infected with the COVID-19 virus for diagnosis. In particular, a large dataset including 6381 proteins in human cells is used in the experiment. Thus, cooperative and network analyses were carried out to evaluate the data obtained from the Kyoto Genes storage. The work in [37] focused on the non-neural expression of SARS-CoV-2 entry genes. In this context, all types of cells are gathered from bulk and single-cell RNA-Seq datasets for diagnosis and/or prognosis in COVID-19. More experimental data was reproduced in [38] to create new epitopes to deal with the limitation of datasets that are useful to produce antibodies in the later stages.

5.3.4   Big data for vaccine and drug discovery

Another promising application of big data is vaccine development which is paramount for preventing the COVID-19 epidemic. It is since the development of a new vaccine is critical for combating the increasing worldwide impact of the COVID-19 pandemic. In this regard, a number of studies have been investigated to develop feasible vaccines against COVID-19 utilizing big data. For example, a database called GISAID was built in [39] to extract the amino acid residues. The key purpose of this work is to find potent targets for developing future vaccines against the COVID-19 pandemic. In [40], an investigation was conducted on the spike proteins of SARS CoV, MERS CoV, and SARS-CoV-2 with respect to out-breaking human coronavirus strains. This study is important to evaluate the spike sequence and structure from SARS CoV-2 necessary for developing a suitable vaccine. To support vaccine development, the study on reverse vaccinology and immune informatics is also necessary [41]. By investigating the entries of an online database, the strain of the SARS-CoV-2 can be detected and collected.

Big data also aids in the development of drug manufacturing methods, thus helping to fight the pandemic. For example, in [42], more than 2500 small molecules were validated through a molecular docking procedure, where fifteen out of twenty-five drugs are used for the evaluation of significant inhibitory potencies. Another study in [19] considered a big data-driven drug repositioning method, where AI techniques were integrated to combine both graph and literature knowledge, aiming to serve the COVID-19 vaccine development. Moreover, big data has great potential to help understand drug therapies by mining datasets collected from different factors such as disease severity, infection rates in human body, and enzyme inhibitors. It is necessary to evaluate the influence of drugs used for the treatment of COVID-19. Big data sources from a wide range of candidate drugs with different factors can be useful for finding effective drugs for COVID-19 by using a set of highly related information, such as distinct classes of drugs, chemical names in drugs, component ingredients in drugs, the levels of drug exposure during and after infection. For example, the medical big data organization, called the British National Formulary [43], conducted a comprehensive study to collect drug information and descriptive characteristics of people infected with COVID-19 to serve the drug manufacturing process. The work in [44] exploited the data learning feature of big data analytic tools to obtain the most important data patterns, aiming to study new adverse drug reactions during the trial process of drug development. In this regard, the identification of drug interactions in the human body, drug safety signals, effectiveness levels of drug usage, and the patterns of drug reactions is crucial to determine which drug is appropriate to patients based on their gender, health conditions, and infection severity. machine learning approaches can be used to support big data processing for such drug examination and to classify and predict the drug reactions before and during the use of drugs, which helps mitigate the risks of drug developments.

5.3.5   Big data for social distancing

Big data can be useful for social distancing applications in the COVID-19 pandemic. Indeed, big data analytics with AI techniques provide interesting solutions for building social distancing policies by data mining and data learning. For example, big data models are able to detect the movement of people in a certain area, aiming to alert people to maintain the social distance [45]. In such a case, data collected from public sources, e.g., CCTV videos, shopping mall cameras, and personal photographs can be used to build big data models to detect at-risk and infected people in public areas. When any infected individual is found in public areas, the authorities can take appropriate actions to prevent the virus spreading [21]. The locations of users equipped with mobile devices, such as smartphones, can be useful datasets for learning the infected movement patterns, keeping physical distance, social behaviors, and trajectories on a daily basis to enforce social distancing rules, e.g., in public areas. Moreover, big data can also be integrated with other technologies, such as IoT and wearable devices equipped with cameras for assessing the social distance between different people in working areas, such as children and teachers in the class. This helps to reduce the possibility of interaction between infected patients and uninfected people for preventing the COVID-19 spread [46].

The work in [47] carried out a holistic study to model and track the spread of COVID-19 by employing a big data-based social distancing method. Specifically, a large dataset about underlying personal relationships, health center locations, human movement trajectories was used to model the spreading patterns of virus and the possibility of outbreaks with respect to the dynamics of populations and the growth of reported cases. Particularly, an epidemiological model was employed to estimate the variance of susceptibility, infection, deaths-at-peak, and recovery as well as the expected changes in human movement behaviors. These can be achieved by using a short-term Social Distance Indicator (SDI) metric that measures the possibility of social distancing enforcement at a population. Accordingly, a social graph is built to simulate the distance calculation where the highly infected areas are denoted as primary nodes along with an edge infection weight. The proposed model is thus able to estimate the human mobility (via analysis of cluster dynamics) and predict appropriately the distance of human movement as well as physical cluster expansion and population density for tracing the disease spread [48].

Recently, Australia used drones to capture big image data to monitor the social distancing behaviors of residents in public areas such as recreational parks, beaches, and bus stations [49]. The movement patterns of the public can be studied via statistical data modeling that can help health authorities to enforce necessary lockdown rules at areas with high infection and high population without complying with distancing policies. Furthermore, this drone-based big data solution can combine with IoT, where sensors can be used to measure human temperatures. Accordingly, by considering human motion behaviors via images and possible infection statistics from sensor measurement, timely social distancing enforcement can be introduced, aiming to prevent the virus spread and protect safe areas against infection risks. Another possible application of big data analytics can be the social distancing support in smart parking areas, in which data modeling is useful to determine how many vehicles should be allowed in a particular parking spot [50]. Moreover, traffic flow, especially in big cities with a high density of vehicles, can be adjusted to ensure distances among people on the roads and parking sites, especially to keep distancing rules to elderly people who are vulnerable to the COVID-19 virus.

5.3.6   Summary

In this section, we review the significance of big data in preventing the COVID-19 pandemic via potential applications: outbreak prediction, spread tracking, diagnosis and treatment, vaccine and drug discovery, and social distancing. Utilizing big data analytics on massive COVID-19 data help anticipate the outbreak on a worldwide scale. Furthermore, combined with AI techniques, e.g., machine learning and deep learning, big data can be used to investigate creative approaches of COVID-19 spread tracking, which helps authorities and governments predict the outbreak in the future. Further, big data shows great potential to support COVID-19 diagnosis, develop effective vaccines and potential drugs against COVID-19, and maintain the implementation of social distancing rules.

5.4   Challenges, lessons, and recommendations

Many studies and projects have shown that that AI and big data are powerful tools to design effective solutions for the COVID-19 pandemic. However, there are still great challenges that need to be considered in future research in order to control the COVID-19 pandemic. In this section, we discuss these challenges along with a set of potential solutions, important lessons, and recommendations for the research communities and authorities.

5.4.1   Challenges and solutions

5.4.1.1   Regulation

Since the pandemic is still not under control in most countries and there are many new variants of the virus, different measures and policies have been used, such as locking down, social distancing, large-scale testing, and vaccination. In this context, regulatory authorities play a vital role in executing the policies, issuing new approaches in response to the outbreak, stimulating the involvement of multiple stakeholders, and ensuring the harmonization of clinical regulations.

From the first confirmed COVID-19 in China in December 2019 until the current state, numerous attempts have been made to mitigate this issue. The quarantine policy in the Republic of Korea, effective from April 1, 2020, is an example. All people entering the Republic of Korea must be quarantined for 14 days at either their registered residence or an approved facility. Further, all people must perform daily self-diagnosis twice a day and submit the results using self-diagnosis applications on their phones. In addition, the Seoul Metropolitan Government has created a "AI monitoring calling system" to automatically examine the health status of persons who have not uploaded self-diagnosis results to the application [51]. According to [52], the Zhejiang Provincial Government teamed up with the Alibaba DAMO Academy to design a new COVID-19 detection system. The developed system can reach an accuracy of 96%, and thus, it was implemented in more than 100 hospitals in China. Besides, the detection system can help the doctors reduce the detection time and clinical diagnosis significantly, from around 15 min in the conventional approach to less than 20 s.

5.4.1.2   Lack of standard datasets

The absence of standard datasets is a key difficulty in making AI and big data platforms and apps reliable for combating the COVID-19 pandemic. From the extensive review in the previous sections, AI and big data have been leveraged in many approaches to the COVID-19 pandemic, but they have normally been evaluated using different datasets. Indeed, when two datasets with differing amounts of samples are utilized to evaluate two AI-based virus detection approaches, we cannot conclude the superior approach for virus detection. It is since a detection approach may perform well on a dataset but may have poor performance when tested in another dataset. Moreover, the majority of datasets used in the literature were created by individual efforts, such as CT images from the hospitals and available data from the Internet. As a result, creating standard datasets that can be made available on the Internet and sharing them among researchers and scientists is a great challenge. It is noted that the COVID-19 data is available in different types and sources, such as text, images, biomedical sequence data, and individual Twitter posts.

A great deal of effort from numerous organizations has been devoted to addressing this issue. For example, a website-based COVID-19 content portal was designed by Johns Hopkins University and is available at https://systems.jhu.edu/research/public-health/ncov/. This portal can provide a large number of data and statistical information, such as the number of daily and weekly infected cases, fatality ratio, vaccinations. In addition, the CORD-19 dataset was created by the Georgetown Center for Security¶ and its collaborators, such as Chan Zuckerberg Initiative and Microsoft Research. This dataset contains valuable data related to COVID-19, SARS-CoV-2, and related viruses from a massive number of research articles. An example of the use of the CORD-19 dataset was investigated in [53], where a recommendation system, namely KnowCOVID-19, was designed to recommend high-quality publications from a large pool of scientific papers. Another effort to create a COVID-19 dataset was made by Alibaba DAMO Academy and a number of Chinese hospitals**. In particular, they developed an AI system for identifying COVID-19 infected cases, and the input to the AI model is more than 5000 CT scans of confirmed cases. Notably, the AI system was deployed by more than 20 hospitals and achieved outstanding performance, such as accuracy of 96% and the detection time is 20 s.

5.4.1.3 Privacy and security challenges

In addition to keeping people healthy and controlling the COVID-19 pandemic, data security and privacy are also important issues and should be explored further. The scandal surrounding the Zoom videoconferencing application's security and privacy problems is one example of this dilemma. To control the COVID-19 pandemic, authorities and healthcare providers may ask people to provide personal information, such as real-time locations, CT scans, diagnostic reports, facial images, historical trajectory, and daily activities. The COVID-19 data is essential to help authorities and healthcare providers make immediate decisions and control the situation; however, personal data is normally sensitive and people are usually not willing to share their data. Moreover, people may be more inclined to share their data with the authorities and healthcare providers if they are incentivized enough and sensitive data is inaccessible to unauthorized parties.

Various technical solutions have been proposed based on emerging technologies, such as blockchain, federated learning, and unmanned aerial vehicle (UAV) to overcome the privacy and security related to the COVID-19 data. For example, the work in [54] leveraged blockchain and federated learning to build a global learning model for virus detection and privacy-preserving, as shown in Figure 5.5. In particular, federated learning enables different hospitals to train different models using their local CT images, and thus hospitals do not need to share their CT images with

---

¶www.kaggle.com/allen-institute-for-ai/CORD-19-research-challenge
**www.alizila.com/how-damo-academys-ai-system-detects-coronavirus-cases

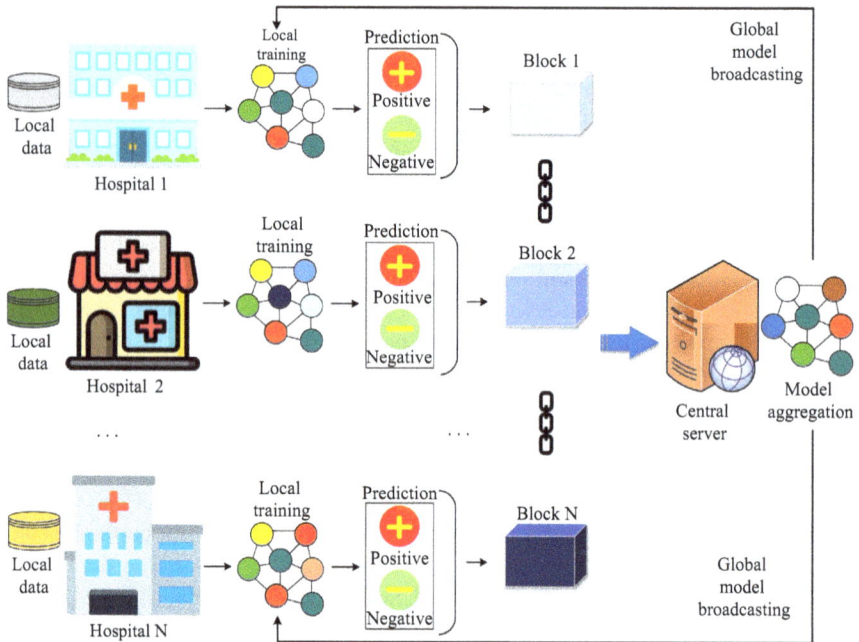

Figure 5.5 *A blockchain-enabled federated learning framework for COVID-19 detection, where blockchain is used to verify the local model updates from different hospitals and verified models are aggregated at an edge server*

other hospitals, and the data privacy could be improved. Then, blockchain is used to verify the data owner whenever new CT images are available from the hospitals. Experimental results showed that the deep model proposed in [54] could achieve competitive performance, such as the sensitivity of 98% and accuracy of 98.68%, while it was suggested that the performance could further increase if more and more hospitals are willing to share their local models with the central entity. Furthermore, the use of UAV is also promising for social distancing owing to distinctive features of UAV, e.g., high mobility, fast deployment, and easy implementation in hard-to-reach areas. Motivated by these advantages, the work in [55] investigated a deep learning model and deployed it in UAVs to detect pedestrians in real-time and calculate the distance among pedestrians. By enhancing multi-scale features of small objects (i.e., pedestrian heads), the deep model in [55] showed that it could accurately detect pedestrians with an average precision of 90% and provide real-time detection with a frames-per-second of 75.

5.4.2 Lessons and recommendations

From the open literature, we found that AI and big data are used to develop many novel approaches to the COVID-19 pandemic through a range of compelling applications, spanning from outbreak tracking and virus detection and to clinical diagnosis and social distancing. AI has the potential to deliver realistic COVID-19 pandemic remedies in a variety of ways. By utilizing AI approaches like machine learning, deep learning, and federated learning, AI has proven highly effective for aiding pandemic prediction, coronavirus detection, drug manufacturing, and social distancing. AI employs analytic techniques to forecast effective vaccines and clinical drugs against COVID-19 using existing information, which would be obtained from a large number of available datasets. Big data has also demonstrated great potential to combat the COVID-19 pandemic. Big data and AI aid our understanding of the COVID-19 in terms of viral structure and illness progression. Big data can help healthcare providers in a variety of medical activities, including early diagnosis, illness analysis, and treatment result prediction. With their great applications, AI and big data are expected to help devise more novel approaches in the future and further contribute to the battle against the COVID-19 pandemic.

Based on the extensive review, we have some recommendations that should be considered in future research and actions to better combat the pandemic. First, most AI and big data-based systems are used to aid doctors and medical staff for better COVID-19 diagnosis and treatment due to limitations of poor performance, low reliability, and inadequate explainability of AI and big data-based algorithms. Second, AI and big data can be integrated with other promising techniques to investigate new COVID-19-fighting solutions. For example, Oracle cloud computing data analysis capabilities were used to develop vaccination, a novel vaccine candidate against the COVID-19 virus[††]. In addition, recent studies have shown the potential of wireless technologies (e.g., UAV, IoT, localization) in response to the pandemic through a variety of applications, including delivery of testing samples, goods transportation, social distancing, and movement monitoring. Finally, vaccination plays a key role in controlling the pandemic, but vaccination distribution is still a challenge due to many reasons. Hence, AI and big data can be leveraged to devise new solutions to accelerate the vaccination rate and support the vaccine distribution supply chain.

5.5   Conclusion

The COVID-19 pandemic has caused many difficulties to our lives and healthcare systems in almost all countries in the world. In response to the pandemic, computer scientists have made considerable efforts to investigate technical solutions via the use of different AI approaches and big data analytics. In this chapter, we have reviewed the use of AI and big data for various use cases in response to the COVID-19 pandemic. In particular, we have presented that AI and big data can be used for detection, diagnosis, virus spread tracking, biomedicine, infodemiology and infoveillance, and social distancing. Moreover, we have presented two illustrative examples to show the practical use of and big data-enabled solutions during the COVID-19 pandemic. Finally, based

---

[††]www.drugtargetreview.com/news/59650/ai-and-cloud-computing-used-to-develop-covid-19-vaccine

on this extensive review, we have also discussed a number of significant challenges, lessons, and recommendations for the authorities and research communities.

References

[1]   Pham QV, Nguyen DC, Huynh-The T, *et al*. Artificial intelligence (AI) and big data for coronavirus (COVID-19) pandemic: a survey on the state-of-the-arts. *IEEE Access*. 2020;**8**:130820–39.

[2]   White House Announces New Partnership to Unleash U.S. Supercomputing Resources to Fight COVID-19; 2020. Accessed date: March 23, 2020. Available from: https://www.whitehouse.gov/briefings-statements/white-house-announces-new-partnership-unleash-u-s-supercomputing-resources-fight-covid-19/.

[3]   Bragazzi NL, Dai H, Damiani G, *et al*. How big data and artificial intelligence can help better manage the COVID-19 pandemic. *International Journal of Environmental Research and Public Health*. 2020;**17**(9):3176.

[4]   Beck BR, Shin B, Choi Y, *et al*. Predicting commercially available antiviral drugs that may act on the novel coronavirus (SARS-CoV-2) through a drug–target interaction deep learning model. *Computational and Structural Biotechnology Journal*. 2020;**18**:784–90.

[5]   Song Y, Zheng S, Li L, *et al*. Deep learning enables accurate diagnosis of novel coronavirus (COVID-19) with CT images. *IEEE/ACM Transactions on Computational Biology and Bioinformatics*. 2021;**18**(6):2775–80.

[6]   Katris C. A time series-based statistical approach for outbreak spread forecasting: Application of COVID-19 in Greece. *Expert Systems with Applications*. 2021;**166**:114077.

[7]   Hu S, Xiong C, Yang M, *et al*. A big-data driven approach to analyzing and modeling human mobility trend under non-pharmaceutical interventions during COVID-19 pandemic. *Transportation Research Part C: Emerging Technologies*. 2021;**124**:102955.

[8]   Zhang S, Diao M, Yu W, *et al*. Estimation of the reproductive number of novel coronavirus (COVID-19) and the probable outbreak size on the Diamond Princess cruise ship: A data-driven analysis. *International journal of infectious diseases*. 2020;**93**:201–4.

[9]   Saponara S, Elhanashi A, Gagliardi A. Implementing a real-time, AI-based, people detection and social distancing measuring system for COVID-19. *Journal of Real-Time Image Processing*. 2021;**18**(6):1937–47.

[10] Wang L, Wong A. COVID-Net: a tailored deep convolutional neural network design for detection of COVID-19 cases from chest X-ray images. *Scientific Reports*. 2020;**10**:1–12.

[11] Abbas A, Abdelsamea MM, Gaber MM. Classification of COVID-19 in chest X-ray images using DeTraC deep convolutional neural network. *Applied Intelligence*. 2021;**51**:854–64.

[12] Tang Z, Zhao W, Xie X, *et al.* Severity assessment of coronavirus disease 2019 (COVID-19) Using quantitative features from chest CT images. *arXiv preprint arXiv:200311988*. 2020.

[13] Imran A, Posokhova I, Qureshi HN, *et al.* AI4COVID-19: AI Enabled preliminary diagnosis for COVID-19 from cough samples via an app. *Informatics in Medicine Unlocked*. 2020;**20**:100378.

[14] Chen YC, Lu PE, Chang CS. A time-dependent SIR model for COVID-19. *arXiv preprint arXiv:200300122*. 2020.

[15] Dandekar R, *et al.* Neural Network aided quarantine control model estimation of global COVID-19 spread. *arXiv preprint arXiv:200402752*. 2020.

[16] Hu Z, Ge Q, Li S, *et al.* Forecasting and evaluating intervention of COVID-19 in the World. *arXiv preprint arXiv:200309800*. 2020.

[17] Hou Z, Du F, Jiang H, *et al.* Assessment of public attention, risk perception, emotional and behavioural responses to the COVID-19 outbreak: social media surveillance in China. *Risk Perception, Emotional and Behavioural Responses to the COVID-19 Outbreak: Social Media Surveillance in China (3/6/2020)*. 2020.

[18] Liu D, Clemente L, Poirier C, *et al.* A machine learning methodology for real-time forecasting of the 2019-2020 COVID-19 outbreak using Internet searches, news alerts, and estimates from mechanistic models. *arXiv preprint arXiv:200404019*. 2020.

[19] Ge Y, Tian T, Huang S, *et al.* A data-driven drug repositioning framework discovered a potential therapeutic agent targeting COVID-19. *bioRxiv*. 2020.

[20] Somaldo P, Ferdiansyah FA, Jati G, *et al.* Developing smart COVID-19 social distancing surveillance drone using YOLO implemented in robot operating system simulation environment. In: *2020 IEEE 8th R10 Humanitarian Technology Conference (R10-HTC)*; 2020. pp. 1–6.

[21] Nguyen CT, Saputra YM, Van Huynh N, *et al.* A comprehensive survey of enabling and emerging technologies for social distancing – Part II: Emerging technologies and open issues. *IEEE Access*. 2020;**8**:154209–36.

[22] Yang D, Yurtsever E, Renganathan V, *et al.* A vision-based social distancing and critical density detection system for COVID-19. *Sensors*. 2021;**21**(13):4608.

[23] Giordano G, Blanchini F, Bruno R, *et al.* Modelling the COVID-19 epidemic and implementation of population-wide interventions in Italy. *Nature Medicine*. 2020;**26**(6):855–60.

[24] Chen B, Shi M, Ni X, *et al.* Visual data analysis and simulation prediction for COVID-19. *arXiv preprint arXiv:200207096*. 2020.

[25] Tátrai D, Várallyay Z. COVID-19 epidemic outcome predictions based on logistic fitting and estimation of its reliability. *arXiv preprint arXiv:200314160*. 2020.

[26] Strzelecki A. The second worldwide wave of interest in coronavirus since the COVID-19 outbreaks in South Korea, Italy and Iran: a google trends study. *Brain, Behavior, and Immunity*. 2020;**88**:950–1.

[27]   Qin L, Sun Q, Wang Y, *et al*. Prediction of number of cases of 2019 novel coronavirus (COVID-19) using social media search index. *International Journal of Environmental Research and Public Health*. 2020;**17**(7):2365.

[28]   Zhao X, Liu X, Li X. Tracking the spread of novel coronavirus (2019-nCoV) based on big data. *medRxiv*. 2020. doi: https://doi.org/10.1101/2020.02.07.20021196.

[29]   Zhou C, Su F, Pei T, *et al*. COVID-19: Challenges to GIS with Big Data. *Geography and Sustainability*. 2020;**1**(1):77–87.

[30]   Castorina P, Iorio A, Lanteri D. Data analysis on Coronavirus spreading by macroscopic growth laws. *International Journal of Modern Physics C*. 2020;**31**(07):2050103.

[31]   Notari A. Temperature dependence of COVID-19 transmission. *Science of the Total Environment*. 2021;**763**:144390.

[32]   Lampos V, Majumder MS, Yom-Tov E, *et al*. Tracking COVID-19 using online search. *NPJ digital medicine*. 2021;**4**(1):1–11.

[33]   Garattini C, Raffle J, Aisyah DN, *et al*. Big data analytics, infectious diseases and associated ethical impacts. *Philosophy and Technology*. 2019;**32**(1): 69–85.

[34]   Li C, Debruyne DN, Spencer J, *et al*. High sensitivity detection of coronavirus SARS-CoV-2 using multiplex PCR and a multiplex-PCR-based metagenomic method. *bioRxiv*. 2020.

[35]   Eden JS, Rockett R, Carter I, *et al*. An emergent clade of SARS-CoV-2 linked to returned travellers from Iran. *Virus Evolution*. 2020;**6**(1):veaa027.

[36]   Bock JO, Ortea I. Re-analysis of SARS-CoV-2 infected host cell proteomics time-course data by impact pathway analysis and network analysis. A potential link with inflammatory response. *Aging (Albany NY)*. 2020;**12**(12):11277.

[37]   Brann DH, Tsukahara T, Weinreb C, *et al*. Non-neural expression of SARS-CoV-2 entry genes in the olfactory epithelium suggests mechanisms underlying anosmia in COVID-19 patients. *Science Advances*. 2020;**6**(31):eabc5801.

[38]   Lon JR, Bai Y, Zhong B, *et al*. Prediction and Evolution of B cell epitopes of surface protein in SARS-CoV-2. *Virology Journal*. 2020;**17**(1):1–9.

[39]   Ahmed SF, Quadeer AA, McKay MR. Preliminary identification of potential vaccine targets for the COVID-19 Coronavirus (SARS-CoV-2) based on SARS-CoV immunological studies. *Viruses*. 2020;**12**(3):254.

[40]   Banerjee A, Santra D, Maiti S. Energetics and IC50 based epitope screening in SARS CoV-2 (COVID 19) spike protein by immunoinformatic analysis implicating for a suitable vaccine development. *Journal of Translational Medicine*. 2020;**18**(1):1–14.

[41]   Sarkar B, Ullah MA, Johora FT, *et al*. The essential facts of Wuhan novel coronavirus outbreak in China and epitope-based vaccine designing against 2019-nCoV. *bioRxiv*. 2020.

[42]   Li Z, Li X, Huang YY, *et al*. FEP-based screening prompts drug repositioning against COVID-19. *bioRxiv*. 2020.

[43]   Dambha-Miller H, Griffin SJ, Young D, *et al*. The use of primary care big data in understanding the pharmacoepidemiology of COVID-19: a consensus

statement from the COVID-19 primary care database consortium. *The Annals of Family Medicine*. 2021;**19**(2):135–40.

[44] Hussain R. Big data, medicines safety and pharmacovigilance. *Journal of Pharmaceutical Policy and Practice*. 2021;**14**(1):1–3.

[45] Hartley DM, Reisinger HS, Perencevich EN. When infection prevention enters the temple: Intergenerational social distancing and COVID-19. *Infection Control and Hospital Epidemiology*. 2020;**41**(7):868–9.

[46] Ye J. Pediatric mental and behavioral health in the period of quarantine and social distancing with COVID-19. *JMIR Pediatrics and Parenting*. 2020;**3**(2):e19867.

[47] Sadowski A, Galar Z, Walasek R, *et al*. Big data insight on global mobility during the COVID-19 pandemic lockdown. *Journal of Big Data*. 2021;**8**(1): 1–33.

[48] Villanustre F, Chala A, Dev R, *et al*. Modeling and tracking Covid-19 cases using Big Data analytics on HPCC system platform. *Journal of Big Data*. 2021;**8**(1):1–24.

[49] Gupta M, Abdelsalam M, Mittal S. Enabling and enforcing social distancing measures using smart city and its infrastructures: a COVID-19 use case. *arXiv preprint arXiv:200409246*. 2020.

[50] Ferguson K, Bradford J. A parking-lot injection clinic: an adaptation to the COVID-19 pandemic. *Canadian Family Physician*. 2021. Available from: https://www.cfp.ca/news/2020/07/14/07-14.

[51] Seoul Introduces the COVID-19 AI monitoring call system; 2020. Available from: http://english.seoul.go.kr/seoul-introduces-the-covid-19-%E3%80%8Cai-monitoring-call-system%E3%80%8D/.

[52] How next-generation information technologies tackled COVID-19 in China; 2020. Available from: www.weforum.org/agenda/2020/04/how-next-generation-information-technologies-tackled-covid-19-in-china/.

[53] Oruche R, Gundlapalli V, Biswal AP, *et al*. Evidence-based recommender system for a COVID-19 publication analytics service. *IEEE Access*. 2021;**9**:79400–15.

[54] Kumar R, Khan AA, Kumar J, *et al*. Blockchain-federated-learning and deep learning models for COVID-19 detection using CT imaging. *IEEE Sensors Journal*. 2021;**21**(14):16301–14.

[55] Shao Z, Cheng G, Ma J, *et al*. Real-time and accurate UAV pedestrian detection for social distancing monitoring in COVID-19 pandemic. *IEEE Transactions on Multimedia*. 2022;**24**:2069–83.

Chapter 6

Advanced sensing and automation technologies

Yuris Mulya Saputra[1], Nur Rohman Rosyid[1], Dinh Thai Hoang[2] and Diep N. Nguyen[2]

In the context of social distancing, various sensing and automation technologies can bring key roles in providing intelligent social distancing scenarios. In this chapter, we present overviews and state-of-the-art applications of emerging sensing and automation technologies including ultrasound, inertial sensor, visible light, and thermal. These four technologies can provide small-to-large coverage, sufficiently low deployment and operational costs, and high accuracy as well as privacy for indoor and/or outdoor environments. Specifically, ultrasound technology that leverages periodical ultrasonic beacons (UBs) can be utilized to keep distance among people, real-time monitoring in public buildings, and mobile robot navigation in indoor environments. Inertial sensor technology, which contains a gyroscope and accelerometer, is useful to recognize positions of pedestrians for keeping distance and perform automation using autonomous vehicles, e.g., medical robots and unmanned aerial vehicles (UAVs). The use of visible light coming from the light-emitting diodes (LEDs) can also provide low-cost crowd monitoring system as well as a navigation assistance system in a large-scale indoor area, and smart traffic control among vehicles on the roads. In a low-light condition, thermal-based system using infrared and imaging camera can be used to control distance among people, physical contact tracing, and real-time traffic monitoring over long distances. To this end, each emerging sensing/automation technology has unique features that are expected to be the potential solution for specific social distancing scenarios. In the following sections, we discuss the aforementioned sensing/automation technologies and their scenarios for social distancing applications in more detail.

6.1 Ultrasound

Ultrasound communication, an especially ultrasonic positioning system (UPS), has been commonly used for indoor positioning applications thanks to its short coverage and high accuracy of centimeters [1]. Different from the frequency range of normal

[1]Department of Electrical Engineering and Informatics, Universitas Gadjah Mada, Yogyakarta, Indonesia
[2]Faculty of Engineering and Information Technology, University of Technology Sydney, Sydney, Australia

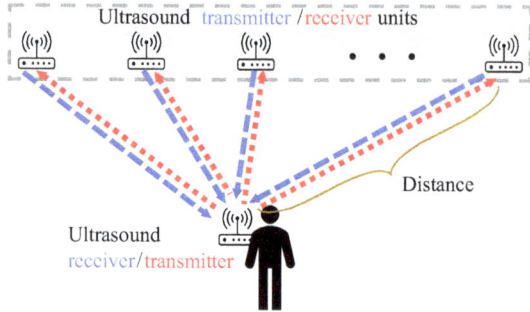

Figure 6.1   A general ultrasound system for positioning application

human hearing, i.e., between 20 Hz and 20 kHz, ultrasound operates with a frequency higher than 20 kHz, e.g., UPS with 40 kHz. Specifically, as shown in Figure 6.1, the UPS may contain ultrasound beacons (UBs) which work as tags or nodes to transmit ultrasound signals of the target users (red dotted lines) or the array of sensing nodes on the walls or ceiling (blue dotted lines) for further processing. These UBs will broadcast the analog ultrasound and radio frequency (RF) signals periodically and simultaneously using their information, e.g., unique IDs and current timestamps. Leveraging the ultrasound and RF signals, the position of target users can be calculated using positioning techniques, e.g., *trilateration* or *triangulation* mechanism [2], upon passing through an analog-to-digital converter and a digital signal processor.

Compared with other RF-based positioning methods, ultrasound has some benefits. In particular, ultrasound does not need a line-of-sight between the ultrasound transmitter and ultrasound receiver. Additionally, it does not interfere with electromagnetic waves when the ultrasound signals are transmitted. However, due to its limited ultrasound signal propagation, ultrasound is only effective to be used for social distancing scenarios in indoor environments. In the following, we provide several social distancing scenarios including distance among people, real-time crowd monitoring, and automation, that can be implemented using ultrasound devices.

6.1.1   Distance among people

In this scenario, UPS can be utilized to notify people when their positions are too close to each other using Active Bat and Cricket systems.

Active Bat-based UPS
An Active Bat system (AB) [3], which is one of the first popular UPS, can be applied to keep distance among people. This AB relies on the time-of-flight (TOF), i.e., a technique to measure the distance between a sensing device and an object, of the ultrasonic signal. Specifically, as illustrated in Figure 6.2(a), an AB system contains an ultrasound transmitter, i.e., ultrasound tag, embedded on the target people whose

positions are required to be located. This transmitter of each target can broadcast ultrasound signals within its coverage in an indoor environment. To receive the ultrasound signals from the target people, the system deploys an ultrasonic receiver matrix on the ceiling or the walls, aiming at forwarding the sensing information to a centralized computation system using wired or wireless connections. This centralized computer will calculate the positions of target people based on the ultrasound time-of-arrival (TOA), i.e., the constant time when the transmitted ultrasound signal is received by the ultrasound receiver. Based on the generated positions of the target people, the system can raise alarm when two people are too close from each other according to a pre-defined minimum distance between two people in an indoor environment. Due to the existence of multiple sensors at the ultrasound receiver matrix, the AB system can achieve a high positioning accuracy, i.e., 14 cm. Nonetheless, this system still suffers from several drawbacks. First, the system may face a high complexity problem, especially when the ultrasound transmitters at the target people are required to transmit signals to a large number of ultrasound receiver sensors. Second, the exact locations of the target people may be disclosed as their positions are computed at the centralized conputer, thereby leading to a high privacy concern.

Cricket-based UPS
To address the limitations of using AB system, the Cricket (CK) system is developed in [4]. Instead of using the centralized computer to calculate the positions of the target people, this calculation is performed by the target people themselves who act as the receivers (as illustrated in Figure 6.2(b)). In particular, the UBs as the ultrasound transmitters can be first deployed on the ceiling or walls. Then, the ultrasound signals

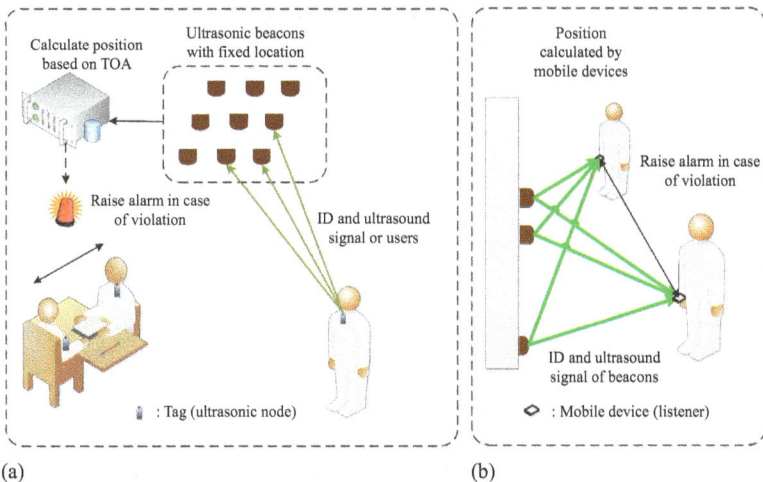

Figure 6.2   Ultrasound application to keep distance among people using (a) active bat method and (b) cricket method [5]

along with RF signals can be transmitted to the target people. To this end, the ultrasound receiver embedded at each target people will passively receive the transmitted signals. Using the received ultrasound signals, the ultrasound receiver can compute each target people's position by itself according to the IDs and coordinates of UBs. As the ultrasound receiver only receives the signals without transmitting any ultrasound signals, the target people's privacy can be securely preserved and the complexity of the system can be reduced. In this case, the target people can send their privacy-protected positions with each other to determine whether they are too close or not and activate the alarm when they violate the pre-defined minimum distance among them.

6.1.2 Real-time crowd monitoring

In addition to the UPS system using TOF, UPS system can be developed based on the signal strength of ultrasound signals [6]. Thanks to the limited propagation and *confinement* feature (i.e., signal is restricted within the same room as the UBs), of the ultrasound signals, we can use UPS as an efficient solution to detect the crowd using a real-time monitoring in public buildings, i.e., whether the target people are within the same building or not. This confinement feature is also adopted by infrared-based positioning system. However, infrared signal is not robust to the interference from heat sources including sunlight and other thermal sources, and it faces line-of-sight loss issue [6]. Therefore, ultrasound is still the most efficient technology for crowd monitoring using a signal strength-based binary decision method (i.e., whether the target people is in the same room as the UBs or not). For example, assume there exist some adjacent small rooms/buildings with a certain number of target people. One target people X has a full ultrasound signal strength coverage within its room/building. However, another target people Y has part of ultrasound signal strength coverage outside the room/building. In this case, using the binary decision approach, the system can decide whether both X and Y are in the same room/building or not without knowing the exact locations of the target people. If a certain number of people are currently in the same room/building, the system can activate the alarm to warn them that they are in a close contact, and thus some of them can move out from the room/building. Using this method, the implementation costs can be remarkably reduced as the system only require few UBs to decide the binary positions of the target people.

6.1.3 Automation

For the automation scenario, ultrasound can be utilized together with mobile robot(s) or UAV(s) in indoor environments.

Mobile robot-based UPS
A mobile robot, e.g., a medical robot, may have a significant role in help minimizing physical contact levels between healthcare workers, e.g., doctors and nurses, and patients/visitors within a hospital. Using such an approach, they can keep the level of social distancing to contain some contagious viruses. For that, the use of ultrasound can take part to enhance the navigation system of the robots. For example, a robot

navigation system leveraging the combination of Wi-Fi signal strength and ultrasound for indoor environments is introduced in [7]. This robot can deal with the uncertainties due to the noisy wireless channel and RF signal's multiple path effects (which is very general in crowded indoor places such as hospitals) through using an autonomous algorithm, i.e., partially observable Markov decision process (POMDP). In this case, the POMDP learning model can be used to generate an environment map with a more accurate localization, planning, and learning process for the mobile robot by observing the Wi-Fi signal strength and ultrasound. The observation of Wi-Fi signal strength is performed to obtain the global localization estimation, while the monitoring of ultrasound is conducted to preserve a good local environment estimation. The experimental results show that by combining Wi-Fi and ultrasound, the mobile robot can achieve a high estimation rate by more than 95% of true locations and much faster convergence compared with those of using independent Wi-Fi or ultrasound observation only. This performance can be further improved when there are many people walking in the indoor environment.

UAV-based UPS
In addition to mobile robots, UAVs can also be applied to minimize physical contact of people in indoor environments. For example, autonomous UAVs can be used for goods delivery within a big building or inventory management inside the warehouse without the presence of many workers in the same building/warehouse. Nonetheless, existing works mostly study on UAV navigation for outdoor environments using global navigation satellite (GNSS), which is only efficient and accurate to be used for UAV positioning in outdoor settings. For indoor settings, the accuracy of GNSS will reduce gradually, thereby applying such the technique inside the building cannot be used directly since it may degrade the performance of UAV navigation. To address the aforementioned problem, an ultra-wideband (UWB)-based indoor localization system using autonomous UAV systems is proposed in [8]. Particularly, in addition to GNSS, the UAV systems contain ultrasound, accelerometer, magnetometer, cameras, and pressure sensor. Here, the ultrasound takes control of UAVs' height when the UAVs are in a close proximity to the ground. Experimental results reveal that by following a pre-defined trajectory 10 times, the UAV systems can achieve an accurate localization with alignment error less than 10 cm in an indoor setting, and thus can provide an efficient autonomous goods delivery and management to support social distancing.

Important points
Ultrasound can be utilized for social distancing scenarios including keeping distance among people, crowd monitoring, and automation. For distance among people, UPS including AB and CK systems with ultrasound transmitters and receivers are available to be used immediately to localize and inform people to safely control distance with each other. For crowd monitoring, the existence of ultrasound's confinement feature can provide efficient signal strength-based binary positioning method to detect and notify the crowd or the number of people within a building. Finally, for automation, ultrasound can help orchestrating the navigation of mobile robots and UAVs for indoor environment purposes.

6.2   Inertial sensor

Inertial sensor, referred to as inertial navigation system (INS), is one type of sensors that can calculate the position, orientation, and movement speed of a mobile object with the absence of external sources. Particularly, the INS includes two subtypes of sensors, i.e., three-axis accelerometers (which can measure how fast the mobile object moves linearly with respect to mobile object direction in x, y, and z coordinates) and three-axis gyroscopes (which can calculate the angular velocity in regards to maintaining the orientation of the mobile object in *yaw*, *pitch*, and *roll*). Leveraging both accelerometers and gyroscopes, the rotation and acceleration information can be obtained to determine the orientation and position deviation of the mobile object [9]. Furthermore, the INS utilizes a *dead reckoning* approach (i.e., current position, orientation, and movement speed of a mobile object can be calculated based on the previous positions, orientations, and heading directions for particular periods) such that the INS does not require to obtain information from external references. Using such an approach, the INS can produce an accurate positioning within a short time period. Nevertheless, the above approach suffers from accumulation error over time (referred to as *integration drift*) when the current position is always measured according to the previous positions. Consequently, to maintain high-accurate positioning, the INS is usually combined with other positioning systems, e.g., a global positioning system (GPS), to re-initialize the base position periodically based on the current observation of GPS [9]. As illustrated in Figure 6.3, the INS first captures position, velocity, and orientation parameters. Then, the GPS compares its captured position parameters with the captured parameters from the INS. Here, the comparison difference is applied as the input of the Kalman filter to estimate the positioning error. This estimation error is finally used to adjust the output positioning parameters from the INS.

In the following, we provide several social distancing scenarios that can be performed using INS devices. Particularly, the INS can be applied to keep distance and provide automation. For keeping distance, INS-based positioning applications inside smartphones can be deployed to notify people when they are in a close proximity to each other. Additionally, the INS can be embedded into mobile robots and UAVs

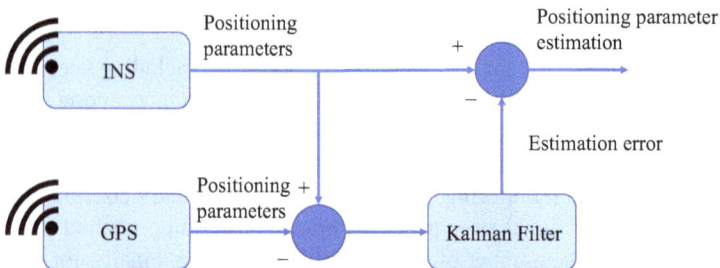

Figure 6.3   A general integrated INS and GPS system for positioning application

to control delivery and navigation applications, aiming at minimizing the physical close-contact among people.

6.2.1 Distance among people

The INS has been widely popular to be applied for navigation of marine, aviation, and ground vehicle applications. Due to the proliferation of smart devices (e.g., smartphones and smartwatches) which include the embedded INS, the use of INS can be recently extended to pedestrian navigation and positioning for social distancing application such as keeping distance among people in indoor/outdoor environments (using alarm to warn people when their positions are too close from each other) as illustrated in Figure 6.4(a). Leveraging the INS, we can obtain pedestrian positioning with a high-accuracy, especially when the INS is combined with other positioning systems for outdoor environments, e.g., GPS, Wi-Fi, and UWB.

Smartphone-based INS

A smartphone-based indoor positioning framework is proposed in [10]. This is to deal with the fact that attaching an inertial measurement unit (IMU) on human body is not cost-effective and inapplicable. Using the existing sensors in the smartphone including accelerometer, gyroscope, and magnetometer (i.e., a sensor that can measure magnetic fields direction and strength), the framework can estimate the smartphone's current position which reflects smartphone user movement in the indoor setting. Specifically, when a mobile user starts to make a move, the captured data from those three sensors can be used to estimate current position of the mobile user along with an estimation error. In this framework, those three sensors can be utilized to detect mobile user's step, estimate mobile user's step length, and heading direction. Once the mobile user's step is detected, magnetometer and gyroscope can be used to estimate heading direction between two consecutive steps. The combination of magnetometer and gyroscope can help improving the accuracy of heading orientation calculation. This can be executed by considering an improved heading estimation algorithm to obtain the weights of

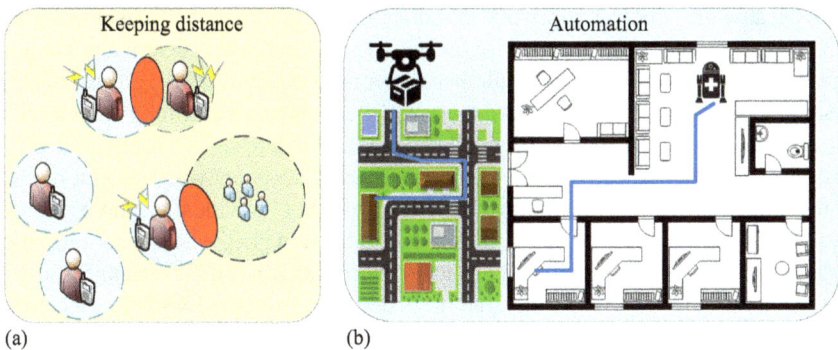

(a)   (b)

Figure 6.4   Inertial sensors-based applications for mobile user positioning to (a) keep distance and (b) automation using UAVs and medical robots [5]

magnetometer and gyroscope in the correlation function between them. The experimental results show that the proposed algorithm can enhance the positioning accuracy by 2.42 times compared with those of other conventional positioning methods.

Wi-Fi-aided INS

Extending from the work in [10], an indoor positioning system leveraging the complementary use of INS and Wi-Fi is investigated in [11]. In this case, the INS is applied to compensate the Wi-Fi's out of coverage area by providing real-time navigation with a high-accuracy using the dead-reckoning-based positioning estimation. Meanwhile, Wi-Fi positioning is utilized to deal with the common integration drift problem of the INS. To further optimally improve the positioning accuracy, a Kalman filter is used to filter duplicated INS information coming from the INS noise. Using such a system, the proposed scheme can reduce the mean positioning error by 1.53 m. Likewise, an indoor positioning using the INS and Wi-Fi fingerprint technologies is proposed in [12]. Particularly, Wi-Fi signals and magnetic fields in the positioning target map are first measured to generate fingerprint map. To reduce the computational complexity, the Wi-Fi fingerprints are modeled using Gaussian Mixture Models (GMMs) distribution. This Wi-Fi fingerprints can be then used to improve the accuracy of the pedestrian dead reckoning method performed by the INS. Specifically, due to the integration drift problem, the dead reckoning approach is required to update the position through monitoring a reference source. In this case, the generated fingerprint map can be utilized as the reference source to update the pedestrian's position. Based on the experiments, the proposed method can improve the indoor positioning with a mean error up to 8 m.

UWB-aided INS

In addition to Wi-Fi, the INS application can be applied together with UWB technology. For example, an indoor positioning system using the INS and UWB for pedestrian localization and tracking is discussed in [13]. Specifically, UWB, which is robust to multipath signal propagation, is used to estimate position by computing the distance between two transceivers using round-trip time (RTT) measurement of UWB signals. This UWB can also compensate the integration drift problem of the INS. Meanwhile, the INS can work to reduce high complexity and operational cost of UWB. Here, the INS can deal with the low dynamic range and external radio noise sensitiveness of the UWB. To implement the proposed framework, an information fusion method utilizing an extended Kalman filter is proposed, aiming at combining the captured data from both UWB and INS. Experimental results reveal that the UWB and INS combination can achieve better positioning accuracy with a deviation 5cm in the indoor environment compared with those of other individual technology-based positioning systems. The above work is then extended in [14], where the information fusion method issue between UWB and INS is optimized to deal with the uncertainties in the measurements. Specifically, the measurements from the INS with accelerometers and gyroscopes are combined with TOA measurements from UWB to estimate the 6-D pose of INS using multiple IMUs and UWB devices attached to a human body. Instead of using the extended Kalman filter, an optimization problem using a *maximum-a-posteriori* estimation and UWB calibration using a *maximum-likelihood* formulation are executed. From the experiments, the positioning accuracy can be significantly improved with a root mean square error 3 cm.

6.2.2 *Automation*

The use of INS can be extended for social distancing scenarios where autonomous devices are involved in, e.g., delivery services using UAVs and navigation using medical robots, as shown in Figure 6.4(b).

Mobile robot-based INS
For healthcare services, the INS has been widely used within the medical robots to assist surgeons and the movement of patients as well as deliver goods for patients with minimum physical contact between healthcare workers and patients. For example, an INS application leveraging a dead reckoning localization approach for a mobile robot is developed in [15]. Specifically, an odometry (i.e., a motion sensor to detect robot movement), which is embedded on the mobile robot's wheels, is used to predict the mobile robot's position. Nonetheless, the limitations of using odometry, i.e., big error due to wheels' slippage, system mismatches, inaccuracies of the measurements, and noise, trigger the use of INS with the dead reckoning approach without considering external sources in indoor settings. To enhance the accuracy of inertial measurements, an error model using Kalman filter is developed. This filter can also approximate the velocity and orientation of the mobile robot under the presence of noises from using the INS. The experimental results demonstrate that the proposed solution can achieve a higher accuracy compared with that of using individual odometry. In [16], a UWB-and-INS-based indoor robot positioning system using an adaptive Kalman filter is discussed. This aims to minimize the accumulation error generated by the INS. To derive error equations of the accelerometer and gyroscope, an *auto-regressive* algorithm is utilized. The simulation results show that the use of the proposed Kalman filter can improve the positioning accuracy compared with that of conventional Kalman filter (with an accuracy level less than 0.24 m, which is sufficient for practical settings).

UAV-based INS
For delivery services, UAVs can be used to deliver goods or other stuffs from one location to another location with minimum physical contact between the sender(s) and receiver(s). Different from mobile robots using the INS that are widely used for indoor settings, UAV-based INS is mostly utilized for outdoor settings. Additionally, the navigation of a UAV is also more complex than that of a mobile robot due to the altitude consideration in the UAV system. For example, the authors in [17] present a quad-rotor UAV-based navigation system using the INS and camera sensor which can perform multiple autonomous activities, e.g., take-off, positioning, and landing, in an outdoor environment. Specifically, the combination of INS and camera can help to obtain the position, velocity, and altitude of an UAV. In this case, the camera embedded on the UAV can first record images from a nearby outdoor environment and then send them to the centralized server. This server can process the images to generate the position of the UAV with respect to the captured environment. The generated

UAV's position can then be integrated with generated UAV's position from the INS through a Kalman filter, aiming at determining the final estimated position and velocity of the UAV. Through real-time experiments, the proposed framework can estimate 3-dimensional position and velocity of the UAV accurately with a frequency rate 13 Hz. In [18], the authors develop a UAV navigation and positioning system using a resilient fusion method between the INS and camera sensors. Particularly, two observers that utilize the INS and camera sensors perform two different activities. The first observer measures the UAV orientation according to the readings from the gyroscope and camera sensors. Meanwhile, the second observer calculates the UAV position and velocity based on the reading from the accelerometer and camera sensors. To reduce signal errors and estimate gyroscope as well as accelerometer biases efficiently, a nonlinear complimentary filter is used for each observer. The experimental results reveal that the measurement from the camera sensor can be applied to minimize the errors from using low-cost INS and shorten the moving time, and thus a high positioning accuracy can be achieved for the UAV.

Important points:
Inertial sensors (i.e., the INS) that are embedded in mobile and autonomous devices, bring opportunities for the development of social distancing scenarios including keeping distance among people and automation. For distance among people, mobile devices, e.g., smartphones and smartwatches, can be used to determine the position of mobile users, especially in indoor settings. In this case, the gyroscope of INS can measure the orientation of the mobile users while the accelerometer of INS can calculate the position and velocity of the mobile users. The fusion method using INS-GPS, INS-Wi-Fi, and INS-UWB can further improve the positioning accuracy and hide the limitations of using an individual INS. For automation scenario, the INS can be applied in indoor/outdoor settings using mobile robots and UAVs. The use of INS in this scenario, especially when it is integrated with camera sensors, can help to enhance the positioning and navigation efficiency through providing more accurate paths and short moving time.

6.3   Visible light

The emerging evolution of visible light using light-emitting diodes (LEDs) illumination for communication (referred to as visible light communication or VLC [20]) and positioning (referred to as visible light positioning or VLP [21]) applications has triggered various use of this technology to deal with social distancing scenarios. The key benefits of using visible light are its capability to provide high flexibility, accuracy, reliability, resiliency, energy efficiency, and security [20,22,23]. Additionally, the ability of LEDs to provide unnoticeably fast switching into different degrees of light intensity can be utilized as the potential efficient data communication method. For example, two different degrees of light intensity in a typical on-off-based modulation scheme can be represented as data bit "0" (low light intensity) and "1" (high light intensity). Visible light signals also do not drive any interference to RF signals.

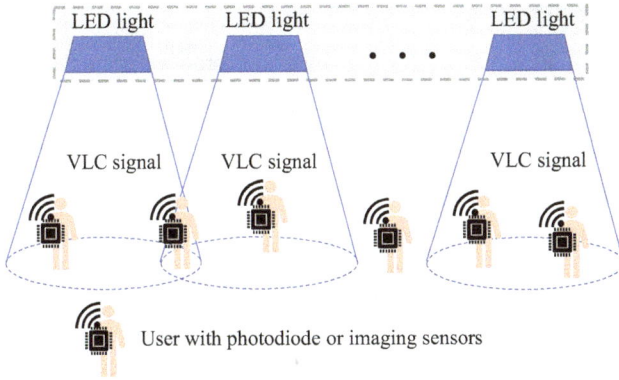

Figure 6.5   A general VLC system for positioning application

Hence, they can be deployed in various indoor settings efficiently such as schools, hospitals, schools, workplaces.

Generally, VLC systems incorporate two main components as illustrated in Figure 6.5. The first component, i.e., LEDs, works as the visible light transmitter to transmit important signals from users, e.g., positioning information. The second component, i.e., photodiodes or imaging sensors (e.g., cameras), operates as the visible light receiver to receive the transmitted signals [20]. In this case, white visible light has been widely utilized for common illumination for both indoor and outdoor environments due to similarity of an object's color when it is seen under white light and natural light. For communication purpose, VLC can be categorized into two modes including infrastructure-to-device communication and device-to-device (D2D) communication. For infrastructure-to-device communication mode, VLC via LEDs can illuminate an entire indoor setting, e.g., a room, such that the visible light can send user data to various devices within the room. Meanwhile, the LEDs attached at the street lamps or traffic lights can be applied to provide internet access to pedestrians or vehicular users. For D2D communication, pixels of LED display from a smartphone user can be utilized to exchange data to the camera of another smartphone user. It is worth noting that VLC is only commonly used for downlink transmission. Here, the uplink transmission from users' devices can be performed by using RF signals, Bluetooth [24,25], or infrared [26] to minimize observable distraction from using LEDs [20].

In the following, we present some social distancing scenarios that can be implemented using VLC due to the massive presence of LED lights in today's human daily life. Specifically, VLC can be applied for photodiode-based or camera-based real-time monitoring leveraging advanced positioning methods. Furthermore, VLC can be used to provide automation including information navigation system and autonomous robot. VLC can also be utilized to control traffic on the roads, especially when high road traffic condition may trigger a huge-packed people density in a particular area.

6.3.1   Real-time crowd monitoring

The utilization of VLC based on LED lights can produce accurate positioning and navigation applications for both indoor and outdoor settings. To keep up social distancing purpose, VLC can be executed to detect crowds in particular public places as illustrated in Figure 6.6(a). For that, there exist two types of receivers that can be applied to realize those scenarios, i.e., photodiode-based and camera-based VLC.

Photodiode-based VLC

The photodiode has been considered as one of the most well-known VLC receivers due to its inherent benefits including low cost and easy deployment. Here, a photodiode or a set of photodiodes can work as a moving tag that is embedded into mobile users or mobile objects, e.g., shopping carts, mobile robots, and vehicles. When each mobile user utilizes this moving tag, he/she can execute a self-positioning mechanism leveraging the triangulation method, aiming at keeping away from crowded places. Additionally, the collected positioning data from the sensing process of moving tag can be shared with the central server (which is owned by particular authorities/officials) to observe a large group of people in certain public places. Upon receiving and processing the positioning data, a particular service provider can notify or warn people in the considered public places by diversifying the color temperature of the lights in the crowded places. For example, when there exist high-density and low-density people in the considered areas, dark color and bright color can be applied, respectively. Note that the aforementioned approach will protect any sensitive information of the mobile users since the communication is only between the LED light fixtures as the VLC transmitters and photodiode-based moving tags as the VLC receivers.

To improve positioning accuracy for people wearing photodiode-based tags in indoor settings, an advanced method that uses a hybrid mechanism, i.e., angle-of-arrival (AOA) and received signal strength (RSS)-based VLC system, for 3-D localization is proposed in [27]. Specifically, the VLC receiver that uses an AOA-based VLC system can obtain its location by itself through observing the illumination direction of LEDs using a *least-square* estimator. To further improve the accuracy of the positioning based on AOA, RSS-based VLC system can be considered by utilizing a *weighted-least-square* estimator. From the simulation results using four visible light access points, the influence of LEDs-based VLC transmitters' orientations towards the estimator performance can be investigated with the positioning error of less than 1 meter. In [28], an AOA-based positioning leveraging multi-element LEDs-based VLC transmitters is investigated. In particular, each LED in multi-element transmitters can be directed into different targets. For example, in a circular deployment of LEDs, a VLC receiver can distinguish which LED fixture that the angle of visible light comes from through the AOA-based VLC detection. Upon collecting positioning data, a Kalman filter is used to boost the positioning precision. Based on the simulations with different topologies of VLC transmitters and receivers, the proposed system can achieve the positioning accuracy of less than 0.2 m in average. Nonetheless, the main drawback of photodiode-based VLC system is the mandatory requirement to attach the photodiode devices on the mobile users/objects to receive the LED illumination signals. This may incur a failure location detection when the mobile users/objects do not carry the photodiode devices. To address this issue, pure-LiFi company (https://purelifi.com) lately has developed a small optical device that can be embedded into smartphones to operate as photodiode-based VLC receiver, aiming at implementing localization services with high accuracy [29].

Camera-based VLC

The explosion of mobile traffic demand using smart devices, e.g., smartphones, has triggered the development of VLC-based applications, e.g., navigation and indoor positioning, on mobile users' smartphones, e.g., Lumicast [24], Atrius Personal Navigator [25], and Carrefour Retail Systems [30]). Instead of using photodiodes, front-facing cameras of the smartphones can receive VLC signals which may contain visible light beacons and localization information (i.e., LED fixture's ID and location), as similarly implemented in [31]. Specifically, using an on-off keying technique, the commercial LED lights can transmit their localization information to multiple mobile users' smartphones. By capturing one single image frame using AOA-based cameras, the smartphones can observe the light presence in the image, process the received LED fixture's ID and location, and generate the location and orientation of the smartphones with respect to the directions of LED lights. The image capturing process can also be frequently executed to update the positions of the smartphones. In this case, upon capturing photos from the cameras frequently, smartphones will send the photo information to the central server for centralized image processing. If visible light beacons exist, the important beacon information such as ID and coordinates can be extracted and transmitted back to smartphones. To this end, the AOA-based positioning algorithm is executed to accurately approximate the location and orientation of the smartphones. Recent works using camera-based VLC positioning systems are also investigated in [32–34].

From the practical implementations of camera-based VLC system in [24,25,30], we can develop an interesting social distancing scenario to help mobile users in the shopping malls or supermarkets finding their desired products quickly in a real-time manner. Furthermore, we can use the aforementioned implementations to help mobile users discover less crowded areas in a proactive way. Compared to the photodiode-based VLC, the camera-based VLC is more convenient to use since the mobile users can simply use its front-facing cameras in the active smartphones as the VLC receivers, and thus physical contacts among mobile users in the crowds can be detected and tracked immediately. Nevertheless, the camera-based VLC system also has a bottleneck, especially when image capturing is required to be performed frequently to obtain accurate positioning of the mobile users. Here, the smartphones of mobile users may consume high energy usage when tracking people in various locations.

6.3.2 Automation

In addition to real-time monitoring, VLC can be largely utilized to provide autonomous assistance for particular purposes (e.g., information or physical assistance for customers, elders, or disabled persons) when many people are located in crowd places such as shopping malls, supermarkets, city parks, hospitals, banks, and libraries [as illustrated in Figure 6.6(a) and (b)]. For example, autonomous robots can

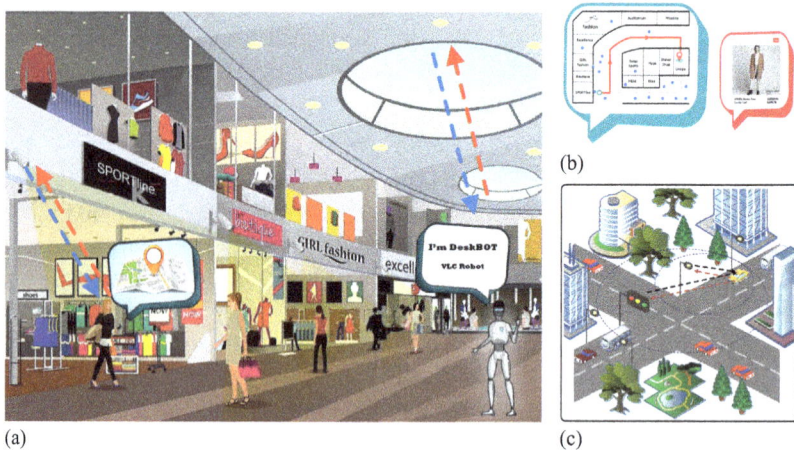

Figure 6.6   VLC-based applications for (a) real-time crowd monitoring and assistance systems (indoor), (b) information navigation systems (indoor), and (c) traffic control systems (outdoor) [5]

help customers, elders, or disabled persons to find and bring specific stuff. Additionally, autonomous information systems can assist customers to find locations of certain books, shops, or healthcare rooms. Such a system can help minimizing the physical contact between customers and assistant staff for the social distancing purpose.

Autonomous information-based VLC
As implemented in [24,25], VLC-based intelligent retail systems can also be applied to help customers in the shopping centres for autonomous information systems, e.g., sale information and product specification can be shown on the smartphone's screen when the smartphone is within the coverage of a particular LED illumination. In a museum or an exhibition hall, VLC can provide autonomous information for visitors to display certain augmented reality applications, e.g., running text or 3-D representation of paintings and portrayals [35]. Additionally, in [36], a low-budget surveillance and information system to automatically detect visitors within the museum and notify the visitors (notifications about current events or when they touch or take pictures of some prohibited exhibition objects) is discussed. This aims to help minimizing close physical contacts among visitors as well as between visitors and exhibition objects in these places. Specifically, white LEDs are used as the VLC transmitters in the dark or obscure environment with a 1-m distance to inform visitors about event notification or object description automatically. Meanwhile, light sensors based on photodiodes can be used as the VLC receivers to warn visitors when human contacts and photographing towards the exhibition objects are present. Here, the VLC receivers can replace the use of touch, infrared motion, and laser sensors due to its low-cost deployment and sensitivity to recognize bright light, e.g., when a visitor uses a smartphone's flashlight

to take photos of objects in a dark setting, without damaging any exhibition objects. For communication, two modulation schemes using on-off keying with Manchester coding, i.e., exclusive-OR operation, and variable pulse position modulation can be combined. Experimental results show that the proposed framework can provide better security and a faster data transfer rate for VLC systems.

Autonomous robot-based VLC

Similar to autonomous information systems that can minimize close contact among people, VLC-based autonomous robots can also be utilized to provide navigation and communication especially to help people on specific occasions, e.g., shopping-assistant, walking-assistant, and elderly-assistant robots [37,38]. In [37], an indoor positioning robot that can recognize LED lights based on the robot operating system is introduced. In particular, the positioning robot incorporates a VLP algorithm with three main processes including a dynamic tracking algorithm using LED and region-of-interest, LED and its ID detection using machine learning-based feature recognition, and a high-accurate cm-level positioning algorithm with two LEDs. Leveraging these processes, the robot can assist the aforementioned people automatically according to the movements of the assisted people. From the experiments, the proposed robot system can achieve indoor positioning accuracy of less than 1 cm and fast computing time within 0.08 s. Likewise, a low-cost and high-accurate indoor positioning mobile robot is presented in [38]. Here, the mobile robot leverages LED lights for the VLP system and ensemble Kalman filter to improve positioning accuracy. Specifically, a light sensor is first used to capture LED lights and extract the distance between the LED lights and robots. Then, a forward ensemble Kalman filter is used to estimate some positioning coordinates. This forward filter is followed by the backward ensemble Kalman filter to deal with the estimation error produced by the forward filter. Finally, the smoother is applied to provide a higher positioning accuracy. Using experimental settings, the proposed system can improve the positioning accuracy of less than 11.2 cm with the acceptable computational complexity for robot applications.

6.3.3 Traffic control

We can extend the applications of VLC to orchestrate road traffic volume on the roads. Typically, there exists a high demand road for traffic on peak-hour periods which can incur a massive density of people in certain road areas, e.g., roads between residential areas to workplace areas and vice versa. In the existing literature, smart traffic control systems, e.g., a fuzzy intelligent traffic signal leveraging a fuzzy logic approach with embedded software and hardware device [39] and a reinforcement learning-based intelligent traffic light system [40], have been discussed. Motivated from the aforementioned traffic light systems, we can also implement intelligent traffic control systems leveraging VLC systems to control traffic volume as illustrated in Figure 6.6(c). This aims to reduce the density of vehicles or pedestrians on crowded roads. Here, the VLC system can provide communications between vehicular users or pedestrians and light systems, e.g., traffic lights and street lamps. For example, current passing vehicles and pedestrians can send their IDs and positioning information to the light systems via VLC using the LED-based headlights and smartphones' LED display as VLC transmitters, respectively. In this case, the light systems can receive information from vehicles and pedestrians using VLC receivers, e.g., photodiodes or camera sensors. Then, upon processing and extracting meaningful information from the collected information, we can use LED-based street lamps or LED-based traffic lights to notify vehicular users and pedestrians on the roads regarding the current density of people on the considered road areas. The light systems on the roads can also orchestrate the vehicles and pedestrians by providing guides to let them avoid the crowded road areas and prohibit them to enter the areas. In this situation, vehicles and pedestrians can receive the information from the light systems using dash cameras or smartphones' cameras as the VLC receivers, respectively.

Important points

VLC system has a great potential to be widely used to support social distancing scenarios. In particular, the use of VLC can provide more advantageous performance, especially to achieve high-accurate indoor positioning and navigation, compared with the positioning and navigation applications using conventional RF technologies. Using the illumination of visible light, we can develop low-cost smart retail systems, real-time crowd monitoring, and automation assistance system using mobile robots on a large-scale setting in various crowded public indoor areas (including shopping malls/supermarkets, hospitals, and airports/train stations/bus stations), aiming at minimizing physical contacts among people in the considered places. Additionally, we can build an intelligent traffic control system for outdoor environments, e.g., on the roads, utilizing LED-based traffic lights and street lamps, to further avoid the crowds on peak-hours on certain roads. To this end, designing an autonomous framework that can instantly inform or warn people when they are within the crowds, e.g., by varying the colors of the lights in high-concentration areas, is of importance to provide an easy warning. Alternatively, people can move away from crowded areas when the lights are dimmed frequently. The use of VLC can also be combined with other RF technologies, e.g., Bluetooth and infrared, to guarantee that the positioning and navigation services are not disrupted when the smartphones camera or LED display is not being used by the mobile users. e.g., when the smartphones are in the pocket or bags. Nevertheless, VLC can only be used perfectly when there exist less ambient lights and sunlight [20,22]. The reason is that the ambient lights and sunlight may interfere with the communication of visible light, which then leads to the poor performance of RSS-based positioning systems and outdoor-based communications.

6.4   Thermal

Aside from ultrasound, an inertial sensor, and visible light, thermal technology which is referred to as a thermal-based positioning system (TPS) can be utilized to support social distancing scenarios, e.g., to detect people and objects, under severe conditions including dark settings as well as fog-filled, smoke-filled, or snow-filled environments. In thermal technology, objects or people can be recognized based on heat

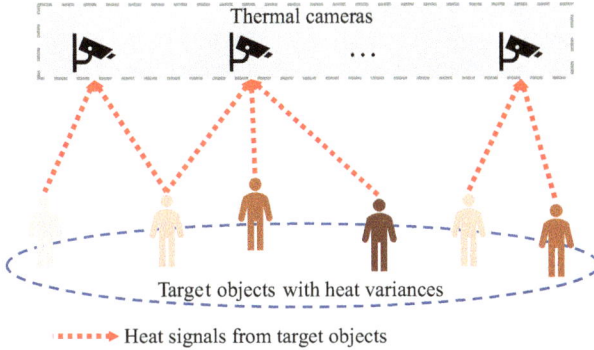

Figure 6.7 A general TPS system for positioning application

variances, instead of light intensity as illustrated in Figure 6.7. The hotter an object is, the more thermal energy it releases. Here, the heat variances can be displayed as shades of grey or various color palettes. For example, when two target objects have different thermal energy, it can be observed clearly based on their dissimilar colors regardless of the lighting conditions. Nonetheless, although TPS can monitor the objects through the aforementioned extreme environments, it is unlikely that the TPS can see a hidden object through thick walls (except that the hidden object triggers a sufficient temperature gap, the TPS can sense the object on the wall's surface).

Generally, TPS can be categorized into two major types: (1) infrared positioning (IRP) and (2) thermal imaging camera (TIC). For the IRP system, it usually requires a low-cost deployment with a short-range observation, i.e., up to few meters, to detect the position of target objects via infrared energy based on AOA or TOA techniques [41–43,45]. In this case, to localize a target object, the IRP receiver needs to be considerably close to the target object. Meanwhile, TIC uses a thermal camera to develop heat, i.e., thermal energy, images from the target objects with a longer range, i.e., up to few kilometers [46,47]. As such, to measure a body temperature via a human's surface skin, the TIC does not need to be physically close to the target people being examined. Compared to the camera-based VLC, TIC has a lower resolution (fewer pixels) sensor. This is because the thermal sensor is required to sense a larger-wavelength energy than the VLC receiver.

In the following, various social distancing scenarios considering IRP and THC are investigated. In particular, IRP can be applied to keep distance among people, physical contact tracing, and real-time monitoring. THC can be utilized to provide real-time monitoring including symptom and vulnerable group detection.

6.4.1 Distance among people

To keep distance among people, IRP systems including Active Badge [43], Firefly [44], and Optotrak [45] systems can be applied.

IRP-based active badge system

In the active badge system [43], a batch or tag that transmits a unique infrared beacon signal periodically, i.e., every 15 s, is attached to each target object, i.e., an office worker or staff. In this case, the unique signal can be received by fixed infrared sensors, and thus the distance between the target object and the fixed sensors can be generated to determine the location of the target object. This Active Badge system is applicable to be used for a short-range application up to 6 m without a capability to travel via walls. Consequently, this system can be used efficiently to check the close-distance between two persons and detect crowds in small indoor environments. Despite this system only requires low-cost and easy deployment, the use of periodic unique signals can only identify the location of target object periodically, i.e., 15-s time interval. Hence, it may reduce the accuracy of location when the target object moves frequently.

IRP-based Firefly and Optotrak system

To improve the localization accuracy especially when a target object moves quickly, Firefly [44] and Optotrak [45] systems can be executed. Particularly, for Firefly systems, an infrared camera, i.e., Firefly capturing camera, can calculate a real-time 3D position of a target object (i.e., a circle movement of an athlete) wearing an infrared transmitter, referred to as a light source marker. In this case, the infrared camera can identify the diameters and center positions of multiple circle shapes based on the light sources attached to the target object. Since a target object only wears one marker, the Firefly system can determine the 3D position of target object accurately. To further monitor the movement of a target object accurately, the Optotrak system can be implemented using multiple infrared cameras as the markers. The signals transmitted from these markers are measured at various depths of the sensor receivers in multiple conditions including motion, static tilted, and static vertical. The experimental results show that the Optotrak system can provide a very good precision for people's motion monitoring with an accuracy between 2 and 4 m. Nonetheless, the Firefly and Optotrak systems are not reliable to the interference from other light sources, e.g., sunlight and light bulbs. Based on the drawback, these IRP systems are mostly suitable to be implemented in a dark or less-light conditions within small rooms.

6.4.2   Physical contact tracing

In addition, to keep distance among people, Firefly and Optotrak systems [44,45] indirectly can be utilized to perform the contact tracing scenario since both systems can identify people's movement accurately. Specifically, multiple markers can be embedded into a body part, e.g., on the waist, fingers, or ears, of the target object, that is usually used to make a physical contact, e.g., hugs, handshakes, or cheek kissing. As observed in Figure 6.8, the IRP system can capture the movement of the body part and then send/store the motion information to a centralized server for contact tracing analysis, e.g., to determine the close contact between the target object and other people based on the pre-defined distance between the target object and other people through their IRP devices. This physical contact information can be

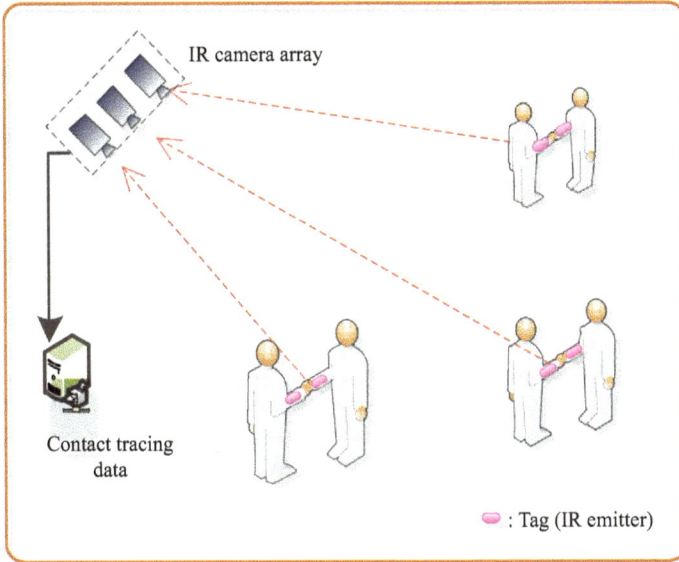

Figure 6.8   A physical contact monitoring using IRP system. If there exists a close contact between two persons, this activity can be stored for future use, e.g., contact tracing [5]

saved to efficiently trace and warn other people when the target object suffers from a contagious virus.

6.4.3   Real-time monitoring

TPS can also be utilized for real-time monitoring including road traffic monitoring, crowd monitoring, and vulnerable group monitoring.

IRP and TIC-based traffic monitoring

IRP and TIC systems can be implemented to support road traffic monitoring in less-light conditions, e.g., in the evening or at night, as illustrated in Figure 6.9(b). In [48], a robust vehicle monitoring using the IRP in diverse environments, e.g., thick fog and snow, to determine level and flow of the road traffic is proposed. Specifically, the reflected thermal energy from vehicles' tires can be used to detect the number of vehicles on the roads. In this case, the proposed system first performs a spatio-temporal image processing and vehicle pattern recognition. To improve accuracy of the vehicle detection, a misdetection correction method using a matching method is used. Then, to further determine each vehicle's position and categorize its movement speed, the combination between spatio-temporal image processing and vehicle pattern recognition can be utilized. From the experiments, the proposed system can accurately detect most of vehicles on the roads up to 92.8% with small false detection rate. Nevertheless, as IRP system can only work well for a short-range application, the

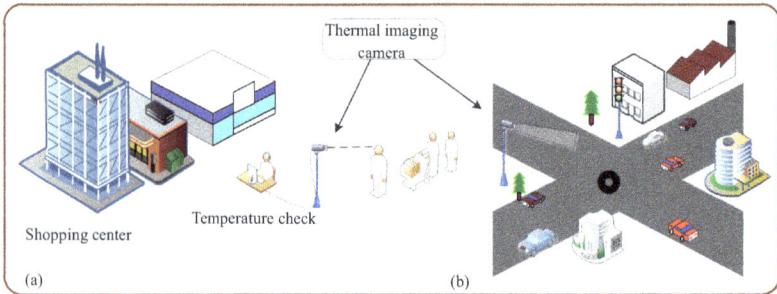

*Figure 6.9   TPS applications for (a) body temperature checking of a vulnerable
group using the thermal camera, i.e., TIC, and (b) real-time traffic
monitoring especially in a dark or less-light situation, e.g., in the
evening and at night, using IRP and TIC [5]*

use of TIC system can further improve the vehicle detection accuracy especially in a
larger area with high vehicle traffic volume.

TIC-based crowd monitoring

Leveraging the TIC's advantage to detect target objects in a large monitoring area up
to few kilometers, TIC can be useful to implement many real-time monitoring applications without high-accurate positioning requirement including crowd monitoring
in public buildings, closure violation monitoring, and non-essential trip monitoring.
For example, a joint framework which combines thermal-visible fusion sensors, i.e.,
infrared and visible cameras, for people tracking system using wide-range videos is
discussed in [49]. These thermal-visible fusion sensors are useful to provide more
accurate trajectories of a target object. Particularly, the thermal camera performs the
thermal video tracking while the visible camera implements the color video tracking.
The combination of both cameras then can be used to detect the target object's trajectories accurately. In [50], a UAV-aided TIC system to implement real-time target object
detection, categorization, and tracking in a large area is proposed. Specifically, a TIC
attached to a UAV is first utilized to capture a thermal video of the considered area.
Using an analog-to-digital converter, the analog thermal video can be converted into a
digital video stream. This digital video stream is then used to detect, classify, and track
target objects in the considered area. Experimental results demonstrate that the proposed system can classify the types of object target accurately by 93.3% with only 5%
false positives. In another work, the authors in [51] investigate human behavior-based
tracking system leveraging omnidirectional TIC system. This system can provide
straightforward heat-emitting object detection and uninterrupted long-term tracking.
Specifically, the TIC is attached to a mobile robot to track people in various environments and light conditions based on the people's body heat levels. To speed up
the people tracking estimation, a Kalman filter based on the movement behaviors of

people is used. The adoption of maximum-a-posteriori-based estimation can further increase the accuracy of the people's next position according to the people's previous movement behaviors. From the experiments, the proposed system can achieve a much lower prediction error up to 60.75% and faster estimation speed by 39%.

TIC-based temperature monitoring

The TIC system can also be utilized to detect vulnerable groups, e.g., elder people and sick people, accurately, aiming at monitoring their health conditions regularly. To this end, TIC system can be incorporated to check these people's body temperature instantly from a far distance since TIC can detect emitted heat from people from few kilometers [46,47,52]. Moreover, the capability of TIC to determine a slightly temperature differences, i.e., with a difference gap 0.01 degree [53] can be useful to determine the sickness and health condition trends of sick people. This temperature checking using TIC can be further applied in the public places, e.g., shopping malls and supermarkets [as shown in Figure 6.9(a)], to remotely determine the visitors' temperature. As such, we can detect infection symptoms via the high temperature of certain visitors earlier to notify other people in the same area avoiding the infected target visitors, aiming at minimizing the spread of contagious viruses.

Important points

Thermal-based positioning systems with short-range and long-range profiles are also useful to support social distancing scenarios especially in an environment when less-light conditions take place. Leveraging the short-range specification, IRP system is useful to provide accurate positioning to keep distance among people and contact tracing in poor-light small rooms. This is due to the fact that IRP system is easy to implement with the low-cost deployment. Meanwhile, TIC system is applicable to perform real-time monitoring for larger areas, e.g., real-time crowd monitoring and road traffic monitoring with poor-light condition, as it supports the long-range specification. Nonetheless, the high-cost of TIC system is required to be taken into account when deploying the system for commercial public areas in practice.

6.5 Summary

In this chapter, we have discussed various sensing and automation technologies to support social distancing applications. Typically, each emerging sensing and automation technology has unique features for specific social distancing scenarios. First, we have introduced the ultrasound technology, i.e., UPS, which is confined by walls and very applicable for indoor settings such as keeping distance among people, real-time crowd monitoring, and automation using mobile robots and UAVs. Second, we have presented the inertial sensor technology, i.e., INS, that can be widely integrated with smartphones or smart wearable devices to keep distance among people and automation for both indoor and outdoor environments. Third, we have investigated the visible light technology, i.e., VLC and VLP, especially using LED lights and photodiodes/camera sensors to provide autonomous assistance systems for crowded public areas and traffic control systems for outdoor settings. Finally, we have discussed the thermal-based

Table 6.1   Summary of Advanced Sensing and Automation Technologies

Technology	Ultrasound [3,6–8]	Inertial sensor [9–12,15,19]	Visible light [24,25,27,29 30,36–38]	Thermal [44,46,49 47,50]
Range	Short, restricted by walls	Not applicable	Short few kilometers (TIC)	few meters (IRP),
Cost	Low medium	Low	Low – medium	Medium – high
accuracy	Less than 14 cm [3]	Less than 1 m [9]	≤ 1 cm [37], ≤ 10 cm [25,27,38], ≤ 20 cm [28]	≤ 0.125 mm (IRP) [45], 0.9 m (TIC) [51]
Setting	Indoor	Indoor and outdoor	Indoor and outdoor	Indoor and outdoor
Privacy	Low [3] – high [4]	High	High	Low – high
Existing system integration	Active Bat and Cricket systems, mobile robots, UAVs	Smartphone, mobile robots, UAVs assistant robots	Smartphone, smart retail system,	Active Badge, Firefly, and Optotrak systems, UAVs

technology, i.e., IRP (for short-range systems) and TIC (for wide-range systems), that can be utilized to support keeping distance among people, contact tracing, and real-time monitoring scenarios especially when poor-light conditions exist. All the aforementioned sensing and automation technologies can provide positioning decision with high-accuracy especially when those technologies are integrated with other emerging technologies, e.g., Wi-Fi, GPS, UWB, RF signals, and Kalman filter. Furthermore, the integration among ultrasound, inertial sensor, visible light, and thermal technologies into one intelligent system with switching mode in the future may further improve accuracy and privacy for social distancing applications in both indoor and outdoor settings as well as light and dark environments. Nevertheless, the challenges in terms of software/hardware modification and deployment/operational costs are required to be further investigated. Table 6.1 provides the summary of advanced sensing and automation technologies for social distancing scenarios based on the explanation in Sections 6.1–6.4.

References

[1]   Holm S. 'Airborne ultrasound data communications: the core of an indoor positioning system'. *IEEE Ultrasonics Symposium*; Taipei, Taiwan, Oct 2005. pp. 1801–4

[2]   Zhang D., Xia F., Yang Z., Yao L., Zhao W. 'Localization technologies for indoor human tracking'. *Proc. 5th Int. Conf. Future Inf. Technol.*, Busan, South Korea, May 2010. pp. 1–6

[3] Ward A., Jones A., Hopper A. 'A new location technique for the active office'. *IEEE Personal Communications*. 1997;**4**(5):42–7

[4] Priyantha N.B. 'The cricket indoor location system'. Ph.D. dissertation. Department of Electrical Engineering and Computer Science, Massachusetts Institute of Technology, 2005

[5] Nguyen C.T., *et al.* 'A comprehensive survey of enabling and emerging technologies for social distancing—part II: emerging technologies and open issues'. *IEEE Access*. 2020;**8**:154209–36

[6] Holm S. 'Ultrasound positioning based on time-of-flight and signal strength'. *2012 International Conference on Indoor Positioning and Indoor Navigation (IPIN)*, Sydney, Australia, Nov 2012. pp. 1–6

[7] Ocana M., Bergasa L.M., Sotelo M.A., Flores R. 'Indoor robot navigation using a POMDP based on WiFi and ultrasound observations'. *2005 IEEE/RSJ International Conference on Intelligent Robots and Systems*, Alberta, Canada, Aug 2005. pp. 2592–7

[8] Tiemann J., Wietfeld C. 'Scalable and precise multi-UAV indoor navigation using TDOA-based UWB localization'. *2017 International Conference on Indoor Positioning and Indoor Navigation (IPIN)*, Sapporo, Japan, Sep 2017. pp. 1–7

[9] Grewal M.S., Weill L.R., Andrews A.P. *Global positioning systems, inertial navigation, and integration*. NJ: John Wiley & Sons; 2007

[10] Kang W., Nam S., Han Y., Lee S. 'Improved heading estimation for smartphone-based indoor positioning systems'. *2012 IEEE 23rd International Symposium on Personal, Indoor and Mobile Radio Communications*, Sydney, Australia, Sep 2012. pp. 2449–53

[11] Evennou F., Marx F. 'Advanced integration of WiFi and inertial navigation systems for indoor mobile positioning'. *EURASIP Journal on Advances in Signal Processing*. 2006;**2006**(1):1–11

[12] Ban R., Kaji K., Hiroi K. Kawaguchi N. 'Indoor positioning method integrating pedestrian dead reckoning with magnetic field and WiFi fingerprints'. *2015 Eighth International Conference on Mobile Computing and Ubiquitous Networking*, Hokkaido, Japan, Jan 2015, pp. 167–72

[13] Angelis A.D., Nilsson J., Skog I., Peter H., Carbone P. 'Indoor positioning by ultrawide band radio aided inertial navigation'. *Metrology and Measurement Systems*. 2010;**17**(3):447–60

[14] Kok M., Hol J.D., Schön T.B. 'Indoor positioning using ultrawideband and inertial measurements'. *IEEE Transactions on Vehicular Technology*. 2015;**64**(4):1293–303

[15] Cho B.S., Moon W.S., Seo W.J., Baek K.R. 'A dead reckoning localization system for mobile robots using inertial sensors and wheel revolution encoding'. *Journal of Mechanical Science and Technology*. 2011;**25**(11):2907–17

[16] Fan Q., Sun B., Sun Y., Wu Y., Zhuang X. 'Data fusion for indoor mobile robot positioning based on tightly coupled INS/UWB'. *The Journal of Navigation*. 2017;**70**(5):1079–97

[17]   Carrillo L.R., López A.E., Lozano R., Pégard C. 'Combining stereo vision and inertial navigation system for a quad-rotor UAV'. *Journal of Intelligent & Robotic Systems*. 2012;**65**(4):373–87

[18]   Cheviron T., Hamel T., Mahony R., Baldwin G. 'Robust nonlinear fusion of inertial and visual data for position, velocity and attitude estimation of UAV'. *2007 IEEE International Conference on Robotics and Automation*, Rome, Italy, Apr 2007. pp. 2010–16

[19]   George M., Sukkarieh S. 'Tightly coupled INS/GPS with bias estimation for UAV applications'. *Australian Conference on Robotics and Automation*, Sydney, Australia, Dec 2005. pp. 1–7

[20]   Pathak P.H, Feng X., Hu P., Mohapatra P. 'Visible light communication, networking, and sensing: a survey, potential and challenges'. *IEEE Communications Surveys & Tutorials*. 2015;**17**(4):2047–77

[21]   Armstrong J., Sekercioglu Y.A., Neild A. 'Visible light positioning: a roadmap for international standardization'. *IEEE Communications Magazine*. 2013;**51**(12):68–73

[22]   Komine T., Nakagawa M. 'Fundamental analysis for visible-light communication system using LED lights'. *IEEE Transactions on Consumer Electronics*. 2004;**50**(1):100–7

[23]   Rajagopal S., Roberts R.D., Lim S. 'IEEE 802.15.7 visible light communication: modulation schemes and dimming support'. *IEEE Communications Magazine*. 2012;**50**(3):72–82

[24]   Jovicic A. 'Qualcomm LumicastTM: A high accuracy indoor positioning system based on visible light communication'. Qualcomm white paper, 2016

[25]   Acuity Brands. *Atrius navigator* [online]. 2021. Available from https://www.acuitybrands.com/products/detail/776333/Atrius/Atrius-Navigator/Atrius-Navigator-indoor-positioning-and-location-based-platform-service-SDK [Accessed 16 Aug 2021]

[26]   Chen C., Bian R., Haas H. 'Omnidirectional transmitter and receiver design for wireless infrared uplink transmission in LiFi'. *IEEE International Conference on Communications Workshops*, Kansas City, USA, May 2018, pp. 1–6.

[27]   Şahin A., Eroğlu Y., Güvenç İ., Pala N., Yüksel M. 'Hybrid 3-D localization for visible light communication systems'. *Journal of Lightwave Technology*. 2015;**33**(22);4589–99

[28]   Eroglu Y.S., Güvenç İ., Pala N., Yüksel M. 'AOA-based localization and tracking in multi-element VLC systems'. *IEEE 16th Annual Wireless and Microwave Technology Conference*, Cocoa Beach, USA, Apr 2015. pp. 1–5

[29]   pureLiFi. *LiFi integration* [online]. 2021. Available from https://purelifi.com/lifi-products/ [Accessed 16 Aug 2021]

[30]   Philips. *Carrefour in France installs Philips LED based indoor positioning system* [online]. 2021. Available from http://www.lighting.philips.com/main/cases/cases/food-and-large-retailers/carrefour-lille.html [Accessed 16 Aug 2021]

[31] Kuo Y.S., Pannuto P., Hsiao K.J, Dutta P. 'Luxapose: indoor positioning with mobile phones and visible light'. *ACM MobiCom*, New York, USA, Sep 2014. pp. 447–58

[32] Fang J., *et al.* 'High-speed indoor navigation system based on visible light and mobile phone' *IEEE Photonics Journal*. 2017;**9**(2):1–11

[33] Guan W., Chen X., Huang M., Liu Z., Wu Y., Chen Y. 'High-speed robust dynamic positioning and tracking method based on visual visible light communication using optical flow detection and Bayesian forecast'. *IEEE Photonics Journal*. 2018;**10**(3):1–22

[34] Li Y., Ghassemlooy Z., Tang X., Lin B., Zhang Y. 'A VLC smartphone camera based indoor positioning system'. *IEEE Photonics Technology Letters*. 2018;**30**(13):1171–4

[35] Grobe L., *et al.* 'High-speed visible light communication systems'. *IEEE Communications Magazine*. 2013;**51**(12):60–6

[36] Kim M., Suh T. 'A low-cost surveillance and information system for museum using visible light communication'. *IEEE Sensors Journal*. 2019;**19**(4): 1533–41

[37] Guan W., Chen S., Wen S., Tan Z., Song H., Hou W. 'High-accuracy robot indoor localization scheme based on robot operating system using visible light positioning'. *IEEE Photonics Journal*. 2020;**12**(2):1–16

[38] Zhuang Y., Wang Q., Shi M., Cao P., Qi L., Yang J. 'Low-power centimeter-level localization for indoor mobile robots based on ensemble Kalman smoother using received signal strength'. *IEEE Internet of Things Journal*. 2019;**6**(4):6513–22

[39] Jin J., Ma X., Kosonen I. 'An intelligent control system for traffic lights with simulation-based evaluation'. *Control Engineering Practice*. 2017;**58**:24–33

[40] Wei H., Zheng G., Yao H., Li Z. 'IntelliLight: a reinforcement learning approach for intelligent traffic light control'. *24th ACM SIGKDD International Conference on Knowledge Discovery & Data Mining*, London, UK, Aug 2018. pp. 2496–505

[41] Aitenbichler E., Muhlhauser M. 'An IR local positioning system for smart items and devices'. *23rd International Conference on Distributed Computing Systems Workshops*, Rhode Island, USA, May 2003. pp. 334–9

[42] Martin-Gorostiza E., *et al.* 'Infrared local positioning system using phase differences'. *2014 Ubiquitous Positioning Indoor Navigation and Location Based Service*, Corpus Christ, USA, Nov 2014. pp. 238–47

[43] Want R., Hopper A., Falcao V., Gibbons J. 'The active badge location system'. *ACM Transactions on Information Systems*. 1992;**10**(1):91–102

[44] Fujiwara T., *et al.* 'Firefly capturing method and its application to performance analysis of athlete'. *SICE Annual Conference*, Kagawa, Japan, Sep 2007. pp. 295–8

[45] States R., Pappas E. 'Precision and repeatability of the Optotrak 3020 motion measurement system'. *Journal of Medical Engineering and Technology*. 2006;**30**(1):1–16

[46]   Wong W.K., *et al.* 'An effective surveillance system using thermal camera'. *2009 International Conference on Signal Acquisition and Processing*, Kuala Lumpur, Malaysia, Apr 2009. pp. 13–17

[47]   Szajewska A. 'Development of the thermal imaging camera (TIC) technology'. *Procedia Engineering*. 2017;**172**:1067–72

[48]   Iwasaki Y., Misumi M., Nakamiya T. 'Robust vehicle detection under various environmental conditions using an infrared thermal camera and its application to road traffic flow monitoring'. *Sensors*. 2013;**13**(6):7756–73

[49]   Torabi A., Massé G., Bilodeau G. 'An iterative integrated framework for thermal–visible image registration, sensor fusion, and people tracking for video surveillance applications'. *Computer Vision and Image Understanding*. 2012;**116**(2):210–21

[50]   Leira F, Johansen T.A., Fossen T.I. 'Automatic detection, classification and tracking of objects in the ocean surface from UAVs using a thermal camera'. *2015 IEEE Aerospace conference*, Big Sky, MT, Mar 2015. pp. 1–10

[51]   Benli E., Motai Y., J. Rogers. 'Human behavior-based target tracking with an omni-directional thermal camera'. *IEEE Transactions on Cognitive and Developmental Systems*. 2017;**11**(1):36–50

[52]   Lloyd J.M. 'Thermal imaging systems'. Berlin: Springer Science & Business Media; 2013

[53]   Flir. *What's the difference between thermal imaging and night vision?* [online]. Available from https://www.flir.com/discover/ots/thermal-vs-night-vision/ [Accessed 16 Aug 2021]

Chapter 7

Security, privacy and blockchain applications in COVID-19 detection and social distancing

Dinh C. Nguyen[1], Quoc-Viet Pham[2], Pubudu N. Pathirana[1], Ming Ding[3] and Aruna Seneviratne[4]

The coronavirus (COVID-19) outbreak has posed serious challenges to healthcare systems around the world. Due to the COVID pandemic, there are a huge number of technologies implemented in our real life to prevent impacts of viruses such as contact tracing, camera surveillance and location detection. However, the use of technologies in the detection and prevention of COVID-19 also poses new problems related to the security and privacy of users. In this context, blockchain has emerged as a potential solution for supporting the prevention of the COVID-19 epidemic by providing security and privacy solutions for facilitating COVID-19 detection and social distancing. Blockchain can help combat the pandemic by offering a number of promising solutions, such as outbreak monitoring, user privacy protection, safe day-to-day operations, medical supply chain and social distancing. This chapter focuses on the applications of blockchain for the COVID-19 pandemic and discusses the key roles of blockchain for COVID-19 detection and social distancing.

7.1 Introduction

The coronavirus (COVID-19) outbreak has posed serious challenges to healthcare systems around the world. The virus spread results in the massive interruptions of various industrial sectors, from manufacturing, supply chain to transport and tourism. Due to the high infection cases, many countries have developed solutions for virus detection, disease treatment, and applied social distancing to mitigate the effects of the pandemic [1]. Given the rapid spreading of coronavirus, every attempt is necessary to address challenges raised by the pandemic, where innovative technologies such as

[1]School of Engineering, Deakin University, Waurn Ponds, Australia
[2]Korean Southeast Center for the 4th Industrial Revolution Leader Education, Pusan National University, Busan, Republic of Korea
[3]Data61, CSIRO, Australia
[4]School of Electrical Engineering and Telecommunications, University of New South Wales (UNSW), Sydney, Australia

blockchain can be useful. Blockchain can provide a wide range of solutions to combat the COVID-19 pandemic by enabling secure outbreak detection, fast-tracking drug delivery and secure social distancing.

In fact, blockchain can offer promising security solutions to assist the campaign of fighting COVID-19. Indeed, the blockchain allows for the construction of immutable ledgers of transactions for medical data exchange and management systems. The COVID-19 data, e.g., scanning images, can be recorded in the blockchain in a fashion that the records are not modified or changed by participants, such as doctors, governments, and patients in the healthcare network [2]. The blockchain, along with smart contracts, can achieve a fair and reliable data management over the distributed networks without the need for relying on a central authority that is often used in traditional data control solutions [3]. This also allows patients, hospitals, governments to obtain an effective access control over COVID-19 data usage.

The benefits of blockchain to COVID-19 fighting mostly come from its outstanding services, such as decentralization, immutability and transparency. Specifically, blockchain decentralizes the medical data system where data nodes can communicate with each other via the peer-to-peer network over the decentralized data ledger. Unlike traditional data systems that use a central server to manage the data flow, the solution with blockchain is able to provide distributed data management which means that the data is not managed by a single entity but by all entities. This feature is useful in practical COVID-19 application scenarios, e.g., decentralized data sharing and decentralized COVID-19 detection. Another motivation behind the integration of blockchain in COVID-19 healthcare systems is its immutability that makes edge data records, e.g., X-ray image data, unchangeable when they are kept on the blockchain network. In this way, thanks to using data ledgers, data clients can establish reliable communications for enabling reliable and trusted computation, such as health data processing via a trusted healthcare environment. From this fact, blockchain is able to offer transparency for health networks, where blockchain creates a shared database by allowing the data to be distributed over multiple devices and computers for authentication that also helps to improve the integrity of data stored on the blockchain [4]. This feature is particularly suitable for healthcare ecosystems in the COVID-19 context where openness and fairness are required. For example, blockchains can provide solutions to create transparent ledgers that are able to support open and fair data delivery among hospital organizations in a fashion such that patients and doctors can trace and monitor transactions.

In addition, in the intelligent COVID-19 detection systems with Artificial Intelligence (AI) [5], how to deal with privacy and security bottlenecks during the learning process, e.g., data attacks and model modification, is a critical concern. Indeed, in traditional learning architectures, one needs to rely on third parties like cloud servers for handling the data training and model turning. Accordingly, users need to upload their sensitive information and datasets to the external server for AI training, which is exposed to data attacks. For example, an attack can steal the sensitive information mixed in the dataset or model parameters, e.g., pixel resolution in imaging, which puts the data under high risks of modification or leakage. Further, the service provider can use data without the consent of users for their financial purposes or other unnecessary requirements. Without protecting user data, the AI training cannot hold a trust level to involved users and data suppliers that makes the learning less inefficient in terms of reputation and privacy. Blockchain can play important roles in secure COVID-19 data analytics where blockchain can enable decentralized data learning over distributed institutions without the need for a central server. Accordingly, data and model updates can be stored over decentralized ledgers and the training process can be recorded on the blockchain for reliable data learning.

Moreover, in the presence of contagious diseases like COVID-19, blockchain can well support social distancing, by enabling secure physical contact tracing, decentralized data management for social distancing apps, aiming to provide a high degree of security for COVID-19-related data in the COVID-19 monitoring system [6]. For example, in the public social distancing system where the information of contacts between people in public places (e.g., schools and workplaces) needs to be secure and reliable, blockchain can establish a decentralized data control mechanism for reliable data exchange and share among mobile users and healthcare authorities. This chapter presents the roles of blockchain in the COVID-19 pandemic and discusses the key roles of blockchain for COVID-19 detection and social distancing. In a nutshell, this chapter makes the following contributions:

- We present the overview of blockchain and discuss its roles to address security and privacy as well as its application to support the fighting of COVID-19.
- We extensively analyze the benefits of blockchain for combating COVID-19 via five key solutions, including outbreak monitoring, user privacy protection, safe day-to-day operations, medical supply chain and social distancing.
- We present several potential and popular projects on blockchain adaption in response to coronavirus fighting. Then, a case study using blockchain for COVID-19 detection is also given.
- We highlight a set of important challenges and directions in the applications of blockchain in the COVID-19 context.

The remainder of this chapter is organized as follows. Section 7.3 presents the background of blockchain. Section 7.4 discusses the role of blockchain for coronavirus fighting via five key solutions, including outbreak monitoring, user privacy protection, safe day-to-day operations, medical supply chain and social distancing. Section 7.5 shows several potential and popular projects on blockchain adaption in response to coronavirus fighting. Then, a case study using blockchain for COVID-19 detection is given in Section 7.6. Section 7.7 presents a set of important challenges and directions in the applications of blockchains in the COVID-19 pandemic. Finally, Section 7.8 concludes the chapter.

7.2   Security and privacy in COVID-19 detection and social distancing

Security and privacy are important issues in smart healthcare systems. The main reason for this is that health and IoT devices generate massive data, which is usually sensitive. For example, health care users are not willing to share their personal information, such as gender, sex, blood type and body temperature. Also, the architecture of healthcare systems is complex, including the first tier of IoT and end devices, the second tier of edge nodes with the Internet connection, and the third tier of cloud computing and data processing. Therefore, ensuring all the data is private and secure through the complex architecture plays an important role in any healthcare system. According to [7], there are five main requirements when implementing privacy and security solutions for smart healthcare systems, as follows.

- The sensitive data should be protected.
- The transmission of health data should be secured.
- The data should be available when it is requested.
- The healthcare system is only open to authorized users.
- The data conforms to semantic standards without unwanted manipulation.

Moreover, there are many challenges of security and privacy to smart healthcare systems, such as data confidentiality, data freshness, authentication, authorization, resiliency, nonrepudiation and fault tolerance [7].

   As there is no sign of completely stopping the COVID-19 pandemic, governments, authorities and healthcare providers have developed various approaches to mitigate the impact of the pandemic, in which collection of sensitive COVID-19 data is a must. For example, Zoom is a great video platform for remote working and meeting, but it faced many privacy and security concerns due to its rapid development during the COVID-19 outbreak in early 2020*. Another example of security and privacy in the pandemic is the collection of COVID-19 data for contact tracing applications. In particular, most approaches for contact tracing apps require users to periodically share their location with a central entity; however, periodic sharing of locations may cause serious privacy concerns and such a contact tracing app may not be accepted by most users [8]. Moreover, users need to share their data with a central server over insecure communication links for the purpose of COVID diagnoses. In such a case, users typically assume that the central server is trusted and has security approaches to protect the user data. This central data storage and learning may cause serious security concerns if the central server gets compromised. Blockchain and federated learning are two promising techniques to solve these concerns. For example, the work in [9] introduced a novel approach using blockchain with federated learning for COVID-19 detection, which is based on heterogeneous CT images from different hospitals. Blockchain is used to verify the data, and federated learning is used to aggregate verified models at a central entity (e.g., a provincial health center). Another solution in [10] was investigated by using a generative adversarial network (GAN) with federated learning, thus enhancing the COVID-19 learning performance and preserving data privacy.

   Social distancing has been a powerful approach to fight against the COVID-19 pandemic; however, there is a strong need for security and privacy issues. In a recent work [11], the authors developed a hybrid federated learning approach using blockchain and differential privacy for social distancing. In particular, IoT devices

---

*https://www.theverge.com/2020/4/1/21202584/zoom-security-privacy-issues-video-conferencing-software-coronavirus-demand-response

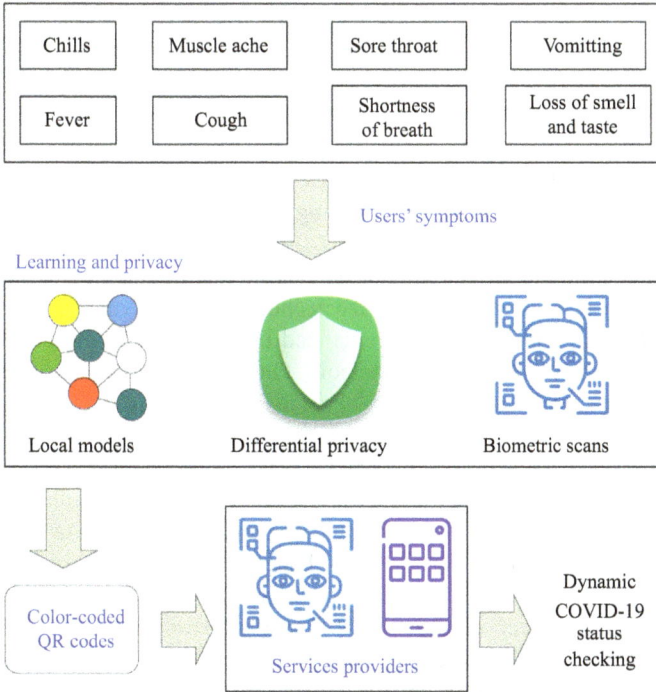

Figure 7.1   Illustration of a system for immunization certificate generation and dynamic COVID-19 status checking [11]

upload their own data with their associated edge nodes, which are equipped with computing capabilities and act as a blockchain client and federated node. The edge nodes train the local models using data from their IoT devices before performing differential privacy and sharing the encrypted model with a central cloud, e.g., Microsoft Azure, for global analysis. Moreover, the work in [11] also conducted a proof of concept that allows health users, services providers, and authorities to perform social distancing rules, as illustrated in Figure 7.1. Based on users' symptoms, the authorities can analyze the COVID statuses, store them in the blockchain, and generate immunization certificates in the form of color-coded QR codes. The services providers then use generated QR codes along with facial images to check the COVID status of their users.

7.3   Blockchain background

7.3.1   Blockchain introduction

Nowadays, banks track monetary accounts of all parties in a ledger that is confined to the public. Customers often have to rely on a bank to perform transactions, i.e., money trading. The bank verifies the user information, checks the balances of the trading parties, and updates the ledger once a transaction has occurred. This is an example of a centralized system, e.g., the bank. Blockchain is the opposite, a system that enables information transfer and service delivery among distributed participants in a fully decentralized manner without the need for any central authority [2,12]. The transaction is distributed by a public ledger network in which every network node can verify and validate with equal rights. In the decentralized banking scenario, for example, whenever a monetary transaction is engaged, the involved parties including senders and receivers can check and monitor banking accounts, agree on the transfer conditions and perform transactions. This process can be updated on the public ledger which is accessible to all network members while the role of the central bank is eliminated.

Blockchain is mostly known as the technology underlying the virtual cryptocurrency Bitcoin which was conceptualized by Satoshi Nakamoto [13] through the invention of the digital currency bitcoin. On October 31, 2008, he published a groundbreaking paper called "Bitcoin: A Peer to Peer Electronic Cash System" which outlined a new model of cash management that enables peer-to-peer cash flow without requiring a central bank or other authorities to operate and maintain the database. The key concept was laid out with a chain of blocks, each cryptographically linked to the previous one by using a hash digest, creating a chain of blocks for data storage among distributed nodes, i.e., computers [14–17]. Especially, Nakamoto sought to address the double-spending issue that plagued previous cryptocurrencies by linking every transaction to the transaction preceding it in a tamper-resistant manner. In this way, a network can examine the transaction history of an electronic coin that a participant submits for payment and can check that the coin has not already been spent, thereby avoiding the double-spending problem [18,19].

7.3.2   Potentials of blockchain for intelligent COVID-19 detection

Blockchain can be applied to develop secure frameworks for supporting intelligent COVID-19 detection enabled by AI techniques. In fact, data learning is the heart of the AI operation to enable data knowledge discovery and data mining for COVID-19 detection. There are many AI learning algorithms such as supervised, unsupervised and reinforcement learning. The traditional AI learning models often have to rely on centralized infrastructure to perform learning tasks, i.e., classification and regression. However, this AI learning architecture introduces some bottlenecks such as learning latency and privacy concerns due to single-point failures. Further, the centralized learning may be overfitting and generate outputs from memorized data since the model has to learn from the whole dataset. Decentralized data learning would be a nature choice for future AI systems. Indeed, by decentralizing learning models over the network of AI nodes and participants, AI can take advantage of learning capability of local servers for faster global learning while data is not exposed to the network for data privacy protection.

Blockchain is also used to create secure data sharing networks wherein each blockchain account (doctors or patients) can create a transaction to submit their healthcare training dataset to the blockchain [4]. Each transaction is verified by a group of miners via a consensus mechanism (i.e., Proof of Work) to confirm whether the training dataset comes from a legitimate user or a malicious node. Also, by using blockchain, incentive mechanisms enabled by smart contracts can be developed to encourage more devices to participate in the learning process, which not only accelerates the global learning process but also enhances the robustness of the AI-based COVID-19 data learning. Moreover, to support parallel data training for big data analysis in AI-based healthcare applications, shared learning models can be developed via a blockchain network. For instance, medical datasets can be trained from various computing servers hosted by different healthcare providers and patients. Blockchain creates a secure data-sharing network to exchange the learning updates among nodes and synchronize securely the global learning model. Smart contracts can be adopted to perform distributed parallel computing by self-executing programmable functions that also accelerate the training process.

One of the key objectives of AI in intelligent COVID-19 detection is to compute and optimize the training model to find the most optimal solution among all possible solutions. AI applications currently operate on edge servers and mobile devices with a high computing resources. Based on the objective of the involved system, the AI data computation can be implemented to optimize resource allocation, system throughput, network latency or QoS value. Current AI computation is executed in a centralized manner that leads to delay in updating models, in inefficient data processing and security issues. Blockchain can offer decentralized and secure solutions for AI data computation, by offering a decentralized AI solution that takes advantage of blockchain ledgers. It can be done by using all blockhain peers to authorize the accuracy of the updated model in the global AI model system without revealing the labels of datasets. In this context, anyone can contribute to the global AI model computation and provide data to enhance the model, but data manipulation is prohibited. The concept behind this idea is a smart contract-based mechanism that can provide programmed functions such as data adding, data update, and prediction functions. These would be executed automatically once the data contributors perform AI learning and submit computed updates such that no anyone can modify the contract rules, leading to a high reliability of the AI system. Another possible application of blockchain for secure AI-based COVID-19 detection is to use it to build a secure convolutional neural network (CNN) architecture. CNNs contain blocks where each block can be regarded as a convolutional layer or a combination of multilayers. The data computation process inside the block includes data preprocessing, data feature extraction, and decision making to perform a certain task (i.e., objective optimization or classification). In this context, attackers can gain malicious access to the CNN network and tamper the data flow and modify the model, making AI functions unreliable. Blockchain can come as a promising security solution by decentralizing the CNN layer operations based on immutable data ledgers. Each block of CNN models is placed on the ledger and interconnected with other blocks via a hash link. Especially, the data flow inside the CNN model is hidden using blockchain-based cryptography techniques to prevent data modification threats. Blockchain is also able to offer incentive mechanisms for addressing the optimization issues in AI algorithms for facilitating COVID-19 data analyses. By taking advantage of the computation resources of blockchain miners, the AI optimization task can be implemented in a decentralized manner while raw data is kept at local nodes for privacy preservation. In return, the miners also receive the reward as the effort of computation. This win–win model is very promising in the future wireless networks where profits of AI contributors and the AI optimization can be achieved simultaneously, thus the efficiency of AI-involved wireless applications would be improved. One can apply blockchain to train the neural networks at the local mobile devices. Each user can manage a group of devices linked together by a blockchain ledger, which transmits and exchanges the weight values of the neural network implementation. This model is feasible in wireless networks, such as IoTs where blockchain can be used to update AI model parameters and share them among mobile users for faster AI computation without AI model information leakage.

Despite the success of AI in constructing intelligent models for learning, prediction and optimization in wireless networks, the concept inside the AI algorithm in intelligent COVID-19 detection models is treated as a black box and hard to explainable. This lack of explainability is one of the most significant challenges in making AI accepted to be deployed in wireless network applications. Moreover, the decisions made by most AI systems are not verified and checked to evaluate the correctness of the outcome, which also raises trust concerns. Thus, it is vitally important to have an immutable scheme to monitor the data flow inside the AI process so that the reliability of AI models can be ensured. Blockchain can meet such requirements by using secure and immutable ledgers. Indeed, all processes of an AI algorithm, from data training, data parameter updates and model estimation can be tracked by blockchain [4]. Each behaviour in the AI model can be recorded as transactions that are secured on blocks with timestamp and signature. Blockchain also provides insights into how data is trained, learned, and optimized and detect which AI steps are compromised. Such improvements based on blockchain would improve significantly trustworthiness and immutability of AI systems, and then provide a high degree of reliability for AI-based wireless networks. A blockchain implementation is able to address the black-box AI, aiming at providing explanation and prediction of the AI algorithm outcomes. All critical decisions of the AI model are recorded, aggregated on transactions with the consensus of all AI agents or predictors. For example, the output of each step in a deep neural network architecture can be recorded on blockchain so that the detailed neural network process can be monitored closely. It also provides more insights into how data is processed and model is built with high trust. Blockchain can also be useful to provide an explanation of the AI-based decision-making process. By updating each event from AI agent interactions using transparent ledgers, blockchain can track and record the behaviour of AI agents during the learning environment interactions so that the trustor can evaluate the soundness of the agent. It also makes decision-making processes interpretable and trustworthy to both AI agents and users in AI-based use cases.

7.4   Blockchain applications in COVID-19 detection and social distancing

In this section, the applications of blockchain for COVID-19 are presented and discussed via five key solutions, including outbreak monitoring, user privacy protection,

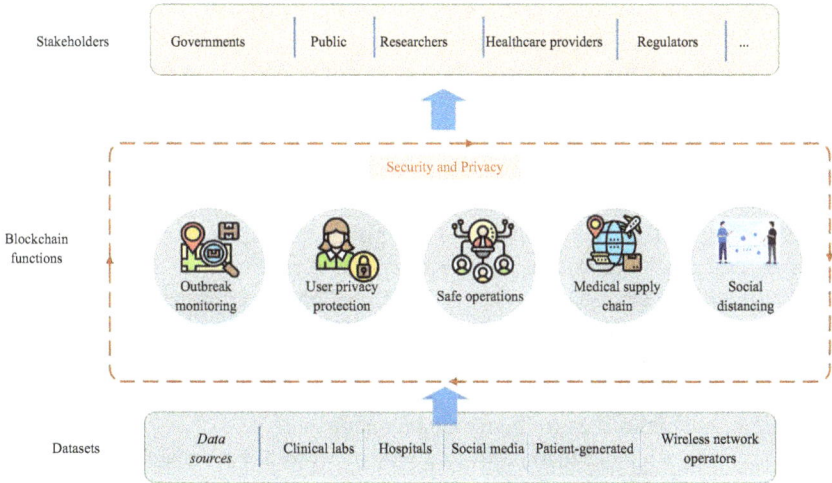

Figure 7.2   Overview of blockchain applications for the COVID-19 pandemic

safe day-to-day operations, medical supply chain and social distancing. The overview of blockchain applications in the COVID-19 context is illustrated in Figure 7.2.

7.4.1   Outbreak monitoring

Blockchain can be used as a reliable tool for monitoring the spread of the outbreak. As a distributed database ledger, blockchain is able to collect multiple updates and place securely on the ledger where blocks of data are linked together by hash values in an immutable manner. Blockchain has the potential to implement the immutable recording of infection symptoms of patients [20]. This is highly needed to keep track of the spread of the outbreak because several patients can deliberately declare their symptoms wrong for avoiding hospital visits, which can lead to the failure of patient quarantine programs. Governments and medical organizations can leverage blockchain to collect real-time data about affected areas and safe zones, aiming to support the decision-making process in the COVID-19 spread management. The data and information of safe zones such as population, location, current coronavirus outbreak status can be updated and recorded via a chain of blocks, and each block can store an update of the outbreak in each time slot. This solution can be combined with other technologies like AI and geographical information systems (GIS) for a comprehensive outbreak tracing. Enabled by information collected from blockchains, healthcare professionals can identify the free-virus zones from the infected ones to develop appropriate quarantine plans [21].

Moreover, one of the key security issues in the pandemic is the outbreak of fake news circulated through social media channels and websites which poses new challenges to the COVID-19 management plans. For example, in February 2020, Turkey and North Korea have been under scrutiny for claiming they have few diagnosed cases, but the fact is different after the actual statistic from the WHO is released [22]. Also, in this report, the accuracy of the case number given by Iran has been placed under question, where the numbers of infectious cases and deaths have been declared incorrectly. The unreliable transmission of COVID-19-related information clearly increases the public confusion about the actual situations of the pandemic in a certain area, making the fighting efforts really challenging. The spread of fake news also makes the public fear and panic due to unverified information and promotes racist forms of digital vigilantism in different regional and city areas [23]. Blockchain can offer practical solutions to address these challenges, by providing reliable data consensus mechanisms via data mining to verify the COVID-19 data records with a guaranteed ordering based on the ledger topology. This also helps provide an agreement on the data among distributed participants such as healthcare providers, medical institutions, and authorities for better data management.

Blockchain can be used to build a reliable platform for coronavirus tracking, for providing the guarantee that the information processed and stored on the blockchain is accurate, transparent, tamper-free, and trusted. Accordingly, governments and doctors can update better on the coronavirus outbreak and the spread status for better management. In summary, blockchain can support the outbreak tracking by the following solutions:

- Enforcing a high standard of shared and immutable data reports
- Establishing a single and trusted source of truth about COVID-19
- Monitoring reliably the infectious cases and deaths at home and medical facilities
- Providing immutable location tracking of the patients via secure data updating recording on blocks

Further, blockchain can be combined with other technologies like AI to facilitate outbreak monitoring for secure and intelligent COVID-19 detection. In this context, COVID-19 data can be collected from clinical labs, hospitals, social media and many other sources in the formats of time series data, radiology images, X-ray reports, virus structure sequences, and so on as illustrated in Figure 7.2. Blockchain is then exploited to form a sharing platform that allows for data exchange with AI centers for COVID-19 analytics. Particularly, blockchain can be integrated with AI for secure and intelligent data analytics. The analysed results from AI analytics enabled by blockchain will serve the stakeholders such as medical institutions and local authorities for policy-making on COVID-19 quarantine, treatment planning and insurance provision. Blockchain can also be used to build secure communication networks to provide reliable data exchange among stakeholders. This working flowchart for integrated blockchain-AI in the COVID-19 context is summarized in Figure 7.3. An illustrative use case from this concept is shown in Figure 7.4. To be clear, COVID-19 data collected from medical institutions can be exchanged securely with the data centre via blockchain. In the data centre, AI algorithms are employed to perform data training, e.g., using neural networks, for COVID-19 data analytics, e.g., classification of the positive case, outbreak estimation, etc. Blockchain is then adopted to enable secure transmissions of calculated results and necessary patient information between stakeholders such

Stage 1: Datasets can be from different sources such as hospitals, personal data, etc.

Stage 2: Blockchain is integrated for establishing secure COVID-19 analytic frameworks to provide interesting services, ranging from secure outbreak monitoring, medical supply chain to supply chain and social distancing.

Stage 3: Blockchain is also combined with AI for smart COVID-19 services, such as outbreak estimation, coronavirus detection, vaccine/drug development, etc.

Stage 4: The analysed outcomes data analytics will support the stakeholders, e.g., governments, healthcare providers, in the decision making process for COVID-19.

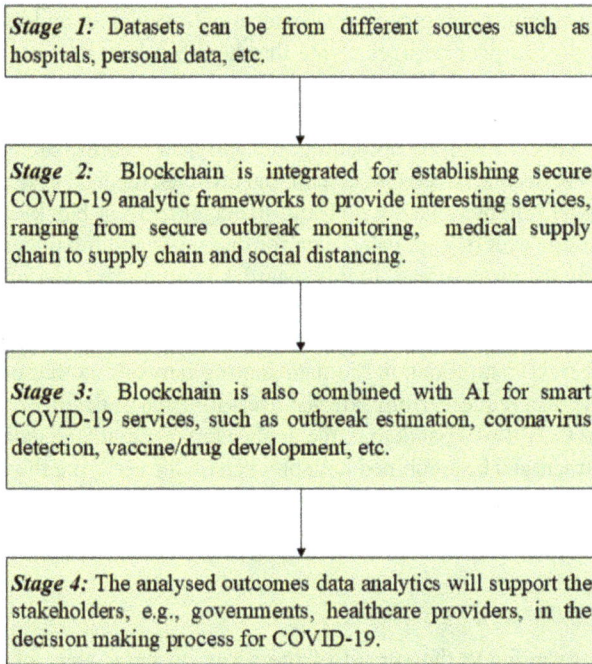

Figure 7.3   The working flowchart of blockchain-AI in the COVID-19 context

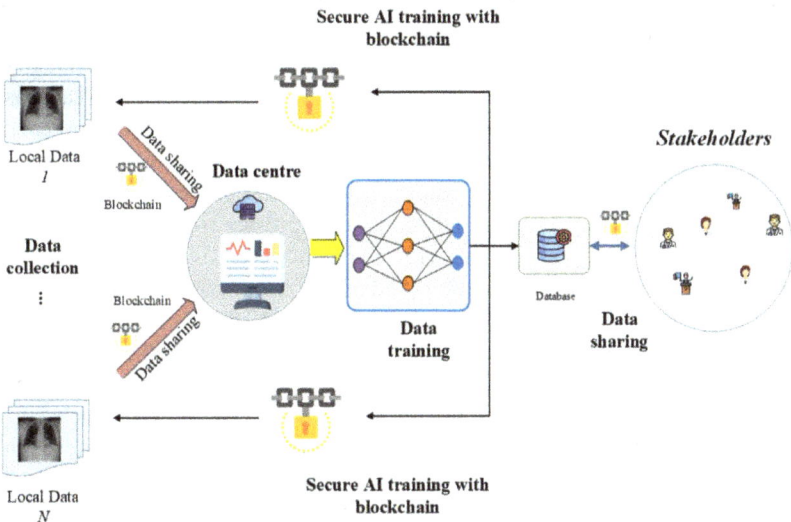

Figure 7.4   The working flowchart for integrated blockchain-AI in the COVID-19 context

as physicians, insurance companies, pharmaceutical companies and medical organizations in the peer-to-peer network. Note that here blockchain is used as a middle layer to interconnect end-users such as patients and these stakeholders in a fashion that no any third-party can hold the complete control over the data usage and data flow in the network. In fact, by uploading medical data records, such as COVID-19 image data, to the decentralized blockchain, the involved medical system can achieve a transparent and reliable data-sharing and management via block consensus over medical data usage events, e.g., using data for diagnosis and treatment or exploiting data for tracking the virus spread. In this regard, how to protect user privacy is also a critical issue that can be addressed by integrating smart contracts for authenticating the user requests and data flow control. For example, smart contracts can be deployed at different medical institutions in the data sharing network so that each entity can authenticate the data requests from external users for each of the medical tasks. To be specific, in the COVID-19 pandemic, the analysis of X-ray image data is necessary, and smart contracts can be combined with blockchain for verifying the trustfulness of data used by stakeholders when offloading the data from data sources such as clinical labs or public data entries.

7.4.2   User privacy protection

In an attempt to mitigate the spread of the virus, a wide range of solutions has been considered around the world. For instance, in South Korea, the government agencies have leveraged the personal information of mobile users for supporting the coronavirus fighting, such as using in-store purchase history and location information extracted from user mobile apps [24]. Even, the government also posted location histories of each coronavirus patient that shows where they have stayed before being tested positive and the locations of their working and study activities. Although these solutions can help the governments to trace the virus spread, they also reveal sensitive personal information to the public. This introduces new serious privacy concerns such as data leakage and information attacks, which can discourage citizens from co-working with healthcare organizations in the testing process. In the emergency situations such as the coronavirus pandemic, user privacy must be weighed against other considerations like saving lives. How to balance the values of upholding public interest in fighting mass virus infections and user privacy protection is of importance for a sustainable COVID-19 management programme. In this context, blockchain can be used as an efficient technique to address privacy issues in the pandemic. This capability is empowered by its features including accountability and transparency that blockchain can provide to enforce the COVID-19 data access. Indeed, blockchain can perform tracing for the data operation, from data collection to data transmission and data usage. Smart contracts can be used to detect any malicious data usage or illegal access behaviours without permission in a transparent manner. Each node such as a hospital can be granted permission to monitor the data flow over the network so that data records can be secure against any leakage risks.

In addition, in the coronavirus pandemic, blockchain can offer different ways to record patient information with symptoms and historical health conditions with privacy provision. This can be enabled by the trust and decentralization features of blockchain. Each COVID-19 data record is time-stamped with signature for data proof and then encrypted enabled by the mining process, e.g., Proof-of-Work. This helps prevent the possibility of modifications and changes from malicious users and adversarial attacks. As a result, the use of blockchain can preserve the data privacy for public users and patients. Also, the data block is distributed over the distributed network of healthcare professionals, clinical facilities and public users [25]. That means the COVID-19 records are managed and monitored by entities, and any modification behaviours on the recording are notified to blockchain nodes. Moreover, privacy of the surveillance networks in coronavirus tracking can be ensured by blockchain. The ubiquitous surveillance systems are often deployed in the community to monitor the virus spread, where the user privacy provision can be enabled by using distributed blockchain that allows for the data recording and storage over the distributed nodes with hashing functions. This would provide adequate transparency, reliability, and accountability on the COVID-19 data records for everyone.

7.4.3 Safe day-to-day operations

In the coronavirus crisis, blockchain has appeared as an efficient technique to safely conduct day-to-day activities via virtual environments to mitigate the risk of virus contraction. Customer services such as online education or remote healthcare can be implemented via digital blockchain networks, making it feasible to mitigate the spread of COVID-19. For example, the UAE's Ministry of Community Development has applied an online programme for using digital channels via blockchain for civil services [26]. In this way, the traditional working methods via paper works are replaced with online solutions with blockchain that can provide digital authentication of official certificates and other documents implemented by block verification and mining. Blockchain allows for eliminating the risks of infection from face-to-face contacts between workers that is very effective for reducing the spread of virus in the community.

In addition to that, digital blockchain can well support the economic activities by creating virtual platforms via data ledgers. For example, electronic payment among customers and companies can be executed via blockchain by using digital money such as coins instead of cash. This reduces the physical contacts among people via money exchange which may be a source of coronavirus infection [27]. Bitcoin [28] can be an example that has proved its potential in enabling digital transactions such as digital auction and service trading such as energy purchase. In the coronavirus crisis, electronic payment is a highly strong recommendation, where blockchain may be an ideal choice. The working concept of blockchain allows it to coordinate the transaction process in a secure and immutable interface, which is free from human interaction. Another excellent example is Ant Financial which has been launching an online bid opening platform running on blockchain to replace the cash-based bid methods [29]. In this way, customers can perform contactless bidding from remote locations when they take quarantine measures to fight against the COVID-19 outbreak. In this project, a private blockchain system has been designed which provides tamper-proof materials and supports reliable processes of bid openings, and the contactless bidding is transparent based on data ledgers. By using advanced consensus mechanisms like Delegated Proof of Stake (DPoS), the blockchain throughput can be enhanced (from 20 transactions/second to below 10 transactions/second) compared with blockchain models with Proof-of-Work that shows high delays in transaction verification and high latency in block recording in the chain.

7.4.4   Medical supply chain

Blockchain has proved its roles in supply chain applications, such as vaccine supplies, food delivery, and goods transmission [30,31]. In this pandemic context, how to achieve a reliable and trustworthy delivery of vaccines and food to vulnerable people in lockdown areas is a critical issue. In this scenario, blockchain has emerged as a promising solution to achieve a trusted supply chain from the manufacturers to end-users and medical organizations as well as other stakeholders. This is enabled by a decentralized data establishment and monitoring of data records, such as vaccine records, on the immutable chain without being modified and changed that helps the supply chain systems stable against external possible data attacks. For example, the Alipay cooperation has recently adopted blockchain to help monitor the medicine supply at remote areas in Zhejiang, China [24]. This consists of the recording and tracking of materials necessary for the COVID-19 pandemic support like masks, gloves, and other protective gear. The trials have shown that blockchain can provide high traceability for the medical supply chain enabled by data block connection and hash linking over the ledger. When an outbreak happens, a fast reaction and immediate support for the supply chain is the most important weapon that local authorities should have to prevent the virus spread. By using blockchain, the supply chain issues, such as immutability, traceability, and decentralization can be addressed, helping to save thousands of lives and billions of dollars.

7.4.5   Social distancing

Another potential application of blockchain is to provide secure social distancing for supporting the COVID-19 fighting. For example, blockchain can be used as a reliable data platform for storing information of user contacts and interactions between people by creating a historic data ledger. In a recent study in [32], blockchain can be integrated with drones to establish a flying data storage system over the quarantine areas. The contacts of users are recorded by drones and stored in the immutable blockchain. Smart contracts are also useful to create an authentication entity for performing verification on the data requesting and data retrieval from end-users. For instance, if a drone detects the contacts between two persons via image capturing enabled by drone planning, then these behaviours are updated and appended to the blockchain. In this way, the authorities can be informed of the concerning areas, and the disinfection drone joins the action by spraying disinfection around the detected man based on the recorded contact information. Blockchain can identify the trajectories of people according to captured data received from a drone, then transmit the information to healthcare professionals for appropriate actions, such as isolation forcing and quarantine provision.

Another example is in [33], where a blockchain-based system is designed to allow authorities to promote social distancing by restricting a given group of individuals in a certain area. In fact, by limiting the number of movement passes that are active at a certain time of the day, the crowd in public places can be controlled. In this way, blockchain can be exploited to establish a decentralized data announcement service by interconnecting healthcare organizations, governments, and local authorities for monitoring the social distancing implementation of the public. Blockchain can be integrated with mobile devices such as smart phones to securely trace the movement behaviours of users in every situation. For example, the restrictions to entries into grocery stores can be applied to only those with an identified movement pass in a safe place, and reliable roadside checkpoints can be implemented online and recorded on the blockchain. This helps the authorities to verify and trace the users for supporting the COVID-19 fighting.

7.5 Blockchain projects for COVID-19 fighting

7.5.1 Hashlog

One of the most popular projects is Hashlog which was created by a Georgia-based health tech startup Acoer [24]. In this project, the company built a decentralized data ledger platform called HashLog to support the governments and clinical facilities in the management of coronavirus spread. To be clear, data related to COVID-19 such as infected cases, outbreak locations, contact tracing records are appended to the blockchain, where mobile devices can be integrated into the blockchain to transform the device into a blockchain node in the COVID-19 network. Hashlog also potentially provides real-time updates on the transmission of the virus by recording the movement information on the blockchain, which is very useful to healthcare authorities to keep track of the spread of the virus.

7.5.2 Hyperchain

The next blockchain project for the fighting of coronavirus is Hyperchain [34], which aimed to construct a reliable and immutable donation tracking platform for assisting governments and healthcare organizations during the donation process in China. Hyperchain can provide a high degree of transparency of the donation, by allowing an ability to trace the records from the origins to the destinations without being modified. Moreover, the Hyperchain can well scale over the large network of millions of nodes. This feature helps to attracts more companies and individuals to donate their goods and materials to the quarantine areas and isolated regions where there have many people in need. For example, businesses can use the proposed system to reach donated goods and necessary medical equipment from the factories, for addressing the issues of facility shortage caused by the outbreak.

7.5.3   VeChain

Another potential blockchain use case is VeChain which is a data ledger-based platform built for controlling the vaccine production in China [35]. All activities for vaccine manufacturing from preparations, production planning, to mass manufacturing with materials collected from trusted sources are updated on the distributed ledger. This project also offers a secure and immutable model to mitigate the issues of potential modifications on vaccine information. VeChain also ensures that vaccine records are recorded permanently to assist the verification stage of authorities, aiming to maintain the high quality and trust of the whole vaccine production process.

7.5.4   PHBC

In addition to these projects, PHBC [36] is also a blockchain-based solution for supporting the COVID-19 detection and management. This platform is able to record information of the COVID-19 pandemic as well as applicable to other high-risk diseases caused by viruses and bacteria. PHBC can also automatically recognize the zones with authenticated incident reports by analyzing the real-time data provided by virus surveillance providers enabled by using AI techniques.

7.6   A case study on blockchain for COVID-19 detection

In this section, a case study with blockchain is illustrated for COVID-19 detection. Federated learning (FL) [37,38], has been widely used for privacy-aware COVID-19 data analytics. Although FL is able to provide privacy to data learning, security is still an issue to be addressed. In fact, in traditional FL learning architectures, one needs to rely on third parties like cloud servers for handling the data training and model turning. Accordingly, users need to upload their sensitive information and datasets to the external server for AI training, which is exposed to data attacks. For example, an attack can steal the sensitive information mixed in the dataset or model parameters, e.g., pixel resolution in imaging, which puts the data under high risks of modification or leakage. Further, the service provider can use data without the consent of users for their financial purposes or other unnecessary requirements. Without protecting user data, the FL training cannot hold a trust level to involved users and data suppliers that makes the learning less inefficient in terms of reputation and privacy [10].

Recent works such as [24] have demonstrated that blockchain can be integrated with FL for secure and privacy-aware COVID-19 data analytics. Here, blockchain can enable decentralized data learning over distributed institutions without the need for a central server. Therefore, novel decentralized FL framework by integrating a blockchain-based solution is proposed in Figure 7.5. Here, we do not use a centralized cloud server to deal with the FL aggregation; instead we adopt a decentralized solution by using a blockchain network to interconnect different servers for providing security. A set of edge nodes (ENs) is located at hospitals for local training, before uploading their models to the associated server for training synchronization via a global generative adversarial network (GAN). The ultimate goal is to produce high-quality X-ray

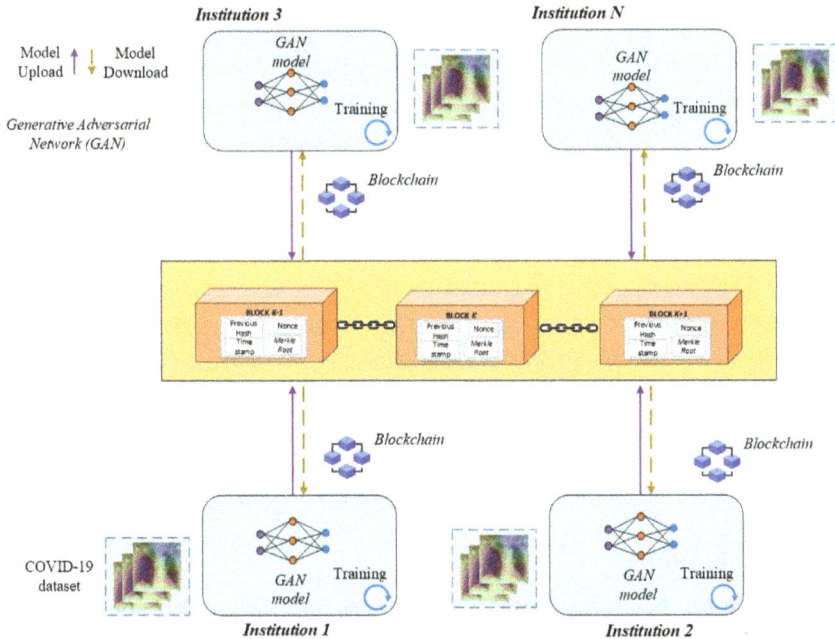

Figure 7.5 The proposed blockchain solution for COVID-19 data analytics

images for improving the quality of detection in terms of accuracy given a degree of privacy (here we can call the GAN-based FL framework as FedGAN).

7.6.1 Working procedure of blockchain-based FL for COVID-19 detection

The working procedure of the blockchain-based FL framework is explained in the following steps:

1. Each EN registers with its associated server to participate in the FL training via the blockchain. Here, we can consider each EN as a blockchain node to local training. Note that it should create an account to establish transactions that allow it for exchanging model after local training.

2. Each EN trains a GAN model (i.e., a generator and a discriminator) to determine a model gradient based on its local COVID-19 X-ray dataset. In fact, each EN can further enhance its privacy during the training by integrating privacy preservation techniques such as differential privacy that allows for protecting local gradients against external modification during the training process.

3. After local training, ENs offload their gradients to the blockchain where servers run the mining for creating a block and adding it to the blockchain.

4. By block mining, servers broadcast the local models to all ENs where they can compute the global model locally for its next round of training.

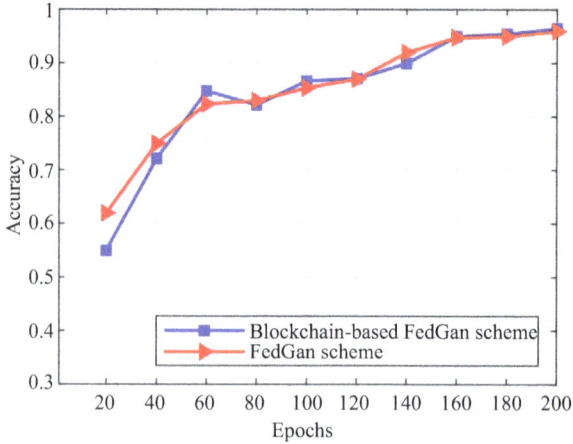

Figure 7.6   Simulation results on training 1500 COVID-19 X-ray images

7.6.2   Simulation results

We evaluate the blockchain-based FedGAN scheme via accuracy on COVID-19 detection as reported in Figure 7.6. The blockchain-based FedGAN scheme is able to produce a performance with similar accuracy rate to its FedGAN counterpart. For example, our scheme is able to achieve over 94% accuracy at the end of the training period, which is similar to the traditional FL model. Thus, the use of blockchain not only provides good detection capabilities but also ensure security via immutable ledgers.

7.7   Challenges and future directions

7.7.1   Challenges

7.7.1.1   Regulatory consideration

The application of blockchain in the healthcare sector for COVID-19 should also be considered with regulatory laws. On the one hand, blockchain is able to bring lots of benefits, such as decentralization, immutability, traceability and trust, to the involved healthcare networks for COVID-19 detection and social distancing. On the other hand, the integration of existing healthcare systems poses a legal and regulatory challenge if no party or authority is responsible and can be held accountable. Indeed, in the blockchain-based COVID-19 detection system, it is important to determine how and what laws are applied to establish data transactions, and what appropriate risk management solutions should be put into place.

7.7.1.2   People's privacy preservation

In the pandemic, how to protect the privacy of people including infected ones is highly important. The governments and healthcare professionals can exploit mobile location data to trace the outbreak spread, but it is necessary to ensure the privacy of user data, especially sensitive information, such as a home address, working office information, shopping records, etc. The governmental agencies need to impose privacy laws on mobile contact tracing apps to make sure that the safety and security of the public can be preserved [39]. Besides, nowadays many healthcare organizations and institutions are recording data from their patients via electronic data forms to serve the activities for detecting the COVID-19 disease symptoms and serve treatment [40]. In such healthcare activities, the conflict between data recording and information privacy is inevitable, and thus related laws should be introduced.

Moreover, in the blockchain-based AI systems for COVID-19 detection, e.g., blockchain-based FL, although FL can provide privacy protection for distributed data-based healthcare systems as the sharing of raw health data is not needed in the learning process, it still remains some security and privacy vulnerabilities from both health data clients and server sides. In future works, perturbation techniques such as differential privacy and secure aggregation can be considered to solve these security and privacy issues. As an example, differential privacy can be useful by inserting artificial noise into the gradients of neural network layers to protect training data. This solution would ensure that adversaries cannot retrieve the user information from exchanged messages in the blockchain-based data learning process.

7.7.1.3   Blockchain performance

Despite benefits of blockchain to COVID-19 applications, there are still several issues for blockchain performances from the perspectives of throughput, storage and networking.

- Throughput: In practical systems, blockchain has a much lower throughput compared to non-blockchain applications. Therefore, the current blockchain systems have serious performance issues in terms of scalability bottlenecks, e.g., the number of replicas, constrained throughput, and transaction latency. In many blockchain systems, a block may suffer from a long queueing time for transactions to be appended to the chain due to the block size restriction.
- Storage: When integrating blockchain into the current healthcare systems for COVID-19 prevention, a huge quantity of COVID-19 data is generated and needs to be processed by the blockchain for user transaction monitoring and data analytics. Indeed, in the conventional blockchain systems, each blockchain node must handle and hold a copy of the complete transaction data, while the scalability of such blockchains may be very high due to the distribution of multiple healthcare entities. This can pose a burden on storage and computation capabilities to resource-constrained devices to participate in the blockchain-based healthcare network.
- Networking: Networking is another critical issue that may influence the scalability of blockchain systems in the involved healthcare system. For example, blockchain with Proof-of-Work is computationally expensive and requires significant resources to perform the block mining to maintain the network which may be not satisfied in practice in lightweight healthcare systems. In fact, several

healthcare systems have limited resources, where computation and storage abilities are very limited due to the demands from healthcare devices and care service operators. This makes it impossible to meet resource requirements for blockchain to achieve scalable transaction processing in large-scale COVID-19 detection systems.

7.7.1.4   Communication issues

In blockchain-healthcare for COVID-19, communications for both uplinks and downlinks are highly sensitive due to the unbalanced health data and different transmit power of health devices in the transaction processing. Further, when the number of health entities grows exponentially, direct communications between numerous health data clients and the blockchain storage servers for data offloading become infeasible due to the increasing workload on the network channels. New designs of efficient communication protocols are desired to compress uplink and downlink communications while remaining high robustness to the increased number of health devices in blockchain-based healthcare systems. For example, clients can compress the data at each communication round before sending it to the storage servers in the data sharing process on blockchain. These properties can be achieved by using a combination of sparsification, ternarization and error accumulation for uplink compression and speeding up parallel data updating while the reliability of data records is ensured.

7.7.1.5   Economic issues in blockchain-healthcare in the COVID-19 context

In practical FL-healthcare systems for COVID-19, when an entity serves as a data collection node, how to encourage them to join the transaction process is a key challenge. This issue is particularly important in blockchain-based COVID-19 scenarios where there are no real authorities to coordinate the data management process. Incentive mechanisms such as credit-based support and revenue payment are highly useful where health users have the right to obtain profits (e.g., coins) according to their transaction processing efforts. This solution would attract more users to join the blockchain network in the blockchain-based COVID-19 data analytics and detection systems.

7.7.1.6   Lack of unified databases

One of the biggest challenges in the coronavirus fighting is the limitations of a unified database related to coronavirus such as infected cases, affected areas, spread sizes and vaccination status. Many countries may be reluctant to share their databases, thus making international healthcare organizations like WHO challenging to comprehensively evaluate the status of the epidemic in a large-scale area for global actions. In fact, WHO cannot provide the best public health guidance and advice if data and detailed local epidemic information are not shared and aggregated.

7.7.2   *Future directions*

7.7.2.1   Performance improvement of blockchain

To realize an efficient blockchain platform for COVID-19 healthcare systems, its performance should be improved in terms of scalability and network efficiency for resource consumption, throughput and network latency. These will make blockchain an ideal choice for emergency healthcare applications like the COVID-19 epidemic. For instance, scalable and lightweight blockchain designs in healthcare are needed to enhance the efficiency of data verification and transaction communication for low-latency information broadcasting [41]. In this sense, low-latency mining mechanisms with optimized block verification should be taken into account to reduce delays in the blockchain. Another way is to optimize the size of blockchain by design hybrid blockchain models by leveraging both local and private blockchain networks. In this case, each of the blockchain networks is responsible for monitoring the outbreak in a certain area for fast response. In this context, building customized ledgers that can be stored on local servers in the outbreak area can assist the minimization of latency in block broadcasting.

7.7.2.2 Security issues of blockchain

Several security issues in blockchain are still existing, such as smart contract bug vulnerabilities and double-spending threats in transaction organization within the blockchain network. In this context, developing innovative solutions such as a mining pool strategy in [42] can be helpful since it is able to provide further the security of the mining process, by eliminating the threats such as 51% vulnerability with low block processing delays. Moreover, to address issues of double spending attacks, the verification of transactions should be implemented to detect any possible attacks in the transaction processing between miners and users [43] which also increases the level of mining security.

7.7.2.3 Combination with other technologies

To bring the efficiency in solving epidemic-related issues to the next level, blockchain can be integrated with other technologies to build a comprehensive healthcare system. For example, blockchain, AI can be integrated with cloud computing for supporting secure coronavirus data analytics. The high computation capability of cloud computing can provide to facilitate AI analytics, while blockchain has the potential to create secure data storage over the virtual cloud nodes. More interestingly, China has used drones to improve the provisions of medical supplies [44]. This solution is practical in quarantine or isolated areas, where people cannot have access to foods and material resources in the pandemic. Besides, drones also help perform the contactless monitoring of the outbreak, and blockchain can be integrated to build a secure flying database interconnected by drones and terrestrial base stations.

7.8 Conclusions

In this chapter, the applications of blockchain to combat the COVID-19 epidemic have been discussed. We have first introduced the overview of blockchain and discussed its roles to address security and privacy as well as its application to support the fighting of COVID-19. We have then extensively analyzed the benefits of blockchain for COVID-19 via several key solutions and services. The potential and popular projects on blockchain adaption in response to coronavirus fighting have been then discussed. A case study using blockchain for COVID-19 detection has also been given. Finally, we highlight a couple of important challenges and directions in the applications of blockchain in the COVID-19 context.

References

[1] Pham QV, Nguyen DC, Huynh-The T, *et al*. Artificial intelligence (AI) and big data for coronavirus (COVID-19) pandemic: a survey on the state-of-the-arts. *IEEE Access*. 2020;**8**:130820–39.

[2] Nguyen DC, Pathirana PN, Ding M, *et al*. Blockchain and edge computing for decentralized EMRs sharing in federated healthcare. In: *GLOBECOM 2020– 2020 IEEE Global Communications Conference*. IEEE; 2020. pp. 1–6.

[3] Nguyen DC, Pathirana PN, Ding M, *et al*. BEdgeHealth: a decentralized architecture for edge-based IoMT networks using blockchain. *IEEE Internet of Things Journal*. 2021;**8**(14):11743–57.

[4] Nguyen DC, Pathirana PN, Ding M, *et al*. Integration of blockchain and cloud of things: architecture, applications and challenges. *IEEE Communications Surveys and Tutorials*. 2020;**22**(4):2521–49.

[5] Nguyen DC, Cheng P, Ding M, *et al*. Enabling AI in future wireless networks: a data life cycle perspective. *IEEE Communications Surveys and Tutorials*. 2020;**23**(1):553–95.

[6] Nguyen CT, Saputra YM, Van Huynh N, *et al*. A comprehensive survey of enabling and emerging technologies for social distancing-Part I: fundamentals and enabling technologies. *IEEE Access*. 2020;**8**:153479–507.

[7] Algarni A. A survey and classification of security and privacy research in smart healthcare systems. *IEEE Access*. 2019;**7**:101879–94.

[8] Ahmed N, Michelin RA, Xue W, *et al*. A survey of COVID-19 contact tracing apps. *IEEE Access*. 2020;**8**:134577–601.

[9] Nguyen DC, Ding M, Pathirana PN, *et al*. Federated learning for COVID-19 detection with generative adversarial networks in edge cloud computing. *IEEE Internet of Things Journal*. 2021.

[10] Nguyen DC, Ding M, Pham QV, *et al*. Federated learning meets blockchain in edge computing: opportunities and challenges. *IEEE Internet of Things Journal*. 2021;**8**(16):12806–25.

[11] Rahman MA, Hossain MS, Islam MS, *et al*. Secure and provenance enhanced Internet of health things framework: a blockchain managed federated learning approach. *IEEE Access*. 2020;**8**:205071–87.

[12] Gadekallu TR, Pham QV, Nguyen DC, *et al*. Blockchain for edge of things: applications, opportunities, and challenges. *IEEE Internet of Things Journal*. 2022;**9**(2):964–88.

[13] Nguyen DC, Pathirana PN, Ding M, *et al*. Blockchain for 5G and beyond networks: a state of the art survey. *Journal of Network and Computer Applications*. 2020; p. 102693.

[14] Kolb J, AbdelBaky M, Katz RH, *et al*. Core concepts, challenges, and future directions in blockchain: a centralized tutorial. *ACM Computing Surveys (CSUR)*. 2020;**53**(1):1–39.

[15] Hasselgren A, Kralevska K, Gligoroski D, *et al*. Blockchain in healthcare and health sciences—A scoping review. *International Journal of Medical Informatics*. 2020;**134**:104040.

[16] Mukherjee P, Singh D. The opportunities of blockchain in health 4.0. In: *Blockchain Technology for Industry 4.0*. Springer; 2020. pp. 149–64.

[17] Madakam S, *et al*. Blockchain technology: concepts, components, and cases. In: *Industry Use Cases on Blockchain Technology Applications in IoT and the Financial Sector*. IGI Global; 2021. pp. 215–47.

[18] Nguyen DC, Pathirana PN, Ding M, *et al*. A cooperative architecture of data offloading and sharing for smart healthcare with blockchain. *arXiv preprint arXiv:210310186*. 2021

[19] Prabadevi B, Deepa N, Pham QV, *et al*. Toward blockchain for edge-of-things: a new paradigm, opportunities, and future directions. *IEEE Internet of Things Magazine*. 2021;**4**(2):102–8.

[20] Idrees SM, Nowostawski M, Jameel R. Blockchain-based digital contact tracing apps for COVID-19 pandemic management: issues, challenges, solutions, and future directions. *JMIR Medical Informatics*. 2021;**9**(2): e25245.

[21] Saleh S, Shayor F. High-Level design and rapid implementation of a clinical and non-clinical Blockchain-Based data sharing platform for COVID-19 containment. *Frontiers in Blockchain*. 2020;**3**:51.

[22] Iran minister accuses 'some countries' of not declaring their coronavirus cases [homepage on the Internet]; 2020 [updated 2020 Dec 01; cited 2021 May 10]. Available from: https://www.cnbc.com/2020/03/06/iran-minister-accuses-some-countries-of-not-declaring-coronavirus-cases.html.

[23] Cinelli M, Quattrociocchi W, Galeazzi A, *et al*. The Covid-19 social media infodemic. *Scientific Reports*. 2020;**10**(1):1–10.

[24] Nguyen DC, Ding M, Pathirana PN, *et al*. Blockchain and AI-based solutions to combat coronavirus (COVID-19)-like epidemics: a survey. *IEEE Access*. 2021;**9**:95730–53.

[25] Dimitrov DV. Blockchain applications for healthcare data management. *Healthcare Informatics Research*. 2019;**25**(1):51–6.

[26] Blockchain and AI Amidst the Coronavirus Crisis: 'A Call to Arms' [homepage on the Internet]; 2020 [updated 2020 Dec 01; cited 2021 May 10]. Available from: https://www.cryptonewsz.com/blockchain-and-ai-amidst-the-coronavirus-crisis-a-call-to-arms/.

[27] Jain A, Sarupria A, Kothari A. The impact of COVID-19 on E-wallet's payments in Indian economy. *International Journal of Creative Research Thoughts*. 2020; pp. 2447–54.

[28] Bhatia R, Kumar P, Bansal S, *et al*. Blockchain-the technology of crypto currencies. In: *2018 international conference on advances in computing and communication engineering (ICACCE)*; 2018, pp. 372–7.

[29]   How blockchain technology is helping businesses overcome the COVID-19
       outbreak in China [homepage on the Internet]; 2020 [updated 2020 Dec
       01; cited 2021 May 10]. Available from: https://irishtechnews.ie/blockchain-
       businesses-overcome-COVID-19-China.

[30]   Juma H, Shaalan K, Kamel I. A survey on using blockchain in trade supply
       chain solutions. *IEEE Access*. 2019;**7**:184115–32.

[31]   Gonczol P, Katsikouli P, Herskind L, *et al*. Blockchain implementations and
       use cases for supply chains—a survey. *IEEE Access*. 2020;**8**:11856–71.

[32]   Gupta R, Kumari A, Tanwar S, *et al*. Blockchain-envisioned softwarized multi-
       swarming UAVs to tackle COVID-19 situations. *IEEE Network*. 2020;**35**(2):
       160–7.

[33]   Garg C, Bansal A, Padappayil RP. COVID-19: prolonged social distancing
       implementation strategy using blockchain-based movement passes. *Journal of
       Medical Systems*. 2020;**44**(9):1–3.

[34]   HyperChain-Blockchain Charity Platform to Fight Against the Coronavirus
       [homepage on the Internet]; 2020 [updated 2020 Dec 01; cited 2021 May 10].
       Available from: https://www.hyperchain.cn/en.

[35]   VeChain Announces Blockchain Vaccine Tracing Solution for China [home-
       page on the Internet]; 2020 [updated 2020 Dec 01; cited 2021 May 10]. Avail-
       able from: https://www.nasdaq.com/articles/vechain-announces-blockchain-
       vaccine-tracing-solution-China-2018-08-16.

[36]   Public Health Blockchain Consortium: PHBC [homepage on the Inter-
       net]; 2020 [updated 2020 Dec 01; cited 2021 May 10]. Available from:
       https://www.phbconsortium.org/.

[37]   Nguyen DC, Ding M, Pathirana PN, *et al*. Federated learning for internet
       of things: a comprehensive survey. *IEEE Communications Surveys Tutorials*.
       2021;**23**(3):1622–58

[38]   Brisimi TS, Chen R, Mela T, *et al*. Federated learning of predictive models
       from federated electronic health records. *International Journal of Medical
       Informatics*. 2018;**112**:59–67.

[39]   Ahmad N, Chauhan P. State of Data Privacy During COVID-19. *IEEE Annals
       of the History of Computing*. 2020;**53**(10):119–22.

[40]   Sharma T, Bashir M. Use of apps in the COVID-19 response and the loss of
       privacy protection. *Nature Medicine*. 2020;**26**(8):1165–7.

[41]   Liu Y, Wang K, Lin Y, *et al*. LightChain: a lightweight blockchain system
       for industrial Internet of Things. *IEEE Transactions on Industrial Informatics*.
       2019;**15**(6):3571–81.

[42]   Li X, Jiang P, Chen T, *et al*. A survey on the security of blockchain systems.
       Future Generation Computer Systems. 2020;**107**:841–53.

[43]   Lee H, Shin M, Kim KS, *et al*. Recipient-oriented transaction for prevent-
       ing double spending attacks in private blockchain. In: *2018 15th Annual
       IEEE International Conference on Sensing, Communication, and Networking
       (SECON)*; 2018. pp. 1–2.

[44]   Kumar A, Sharma K, Singh H, *et al*. A drone-based networked system and
       methods for combating coronavirus disease (COVID-19) pandemic. *Future
       Generation Computer Systems*. 2021;**115**:1–19.

Chapter 8

Real-time optimization for social distancing

Van-Phuc Bui[1], Trinh Van Chien[1], Eva Lagunas[1]
and Symeon Chatzinotas[1]

Optimization theories can offer unprecedented improvements in providing satisfactory quality-of-services to all people while controlling the number of people and their positions at a predetermined level in a place inside the context of social distancing. Various academic researches in the literature have employed optimization theories to design efficient algorithms under practical scenarios comprising physical space, healthcare, workforce, traffic, online service management and computer-based distance monitoring. This chapter focuses on two specific scheduling subjects: People scheduling and traffic scheduling. People scheduling can effectively exploit available space within particular workplaces or essential public centers to minimize physical contact, prevent virus spread and increase the number of people in space while still ensuring a certain safety level. Traffic scheduling can assist in reducing, for example, the peak number of pedestrians and vehicles. Furthermore, network resource optimization can help the growing demands on online services since the remote working mode has aggressively increased during the COVID-19 pandemic. We present several open issues and draw potential future research directions in the social distancing context by the optimization applications to assess fundamental properties and system performance.

8.1 Introduction

The severity of the coronavirus (COVID-19) pandemic has highlighted social distancing (or physical distancing) as one of the most critical factors for health and safety [1]. This is because when an infected person has a convulsive expulsion, such as coughing or sneezing, the virus spreads in small liquid pieces from the mouth or nose into the environment. In Figure 8.1, we collected on September 1, 2021, the COVID-19 epidemic is raging worldwide and shows no signs of abating [2,3]. The plot displays the increasing trends of the total number of infected people and fatalities over time.

[1]Interdisciplinary Centre for Security, Reliability and Trust (SnT), University of Luxembourg, Luxembourg, Luxembourg

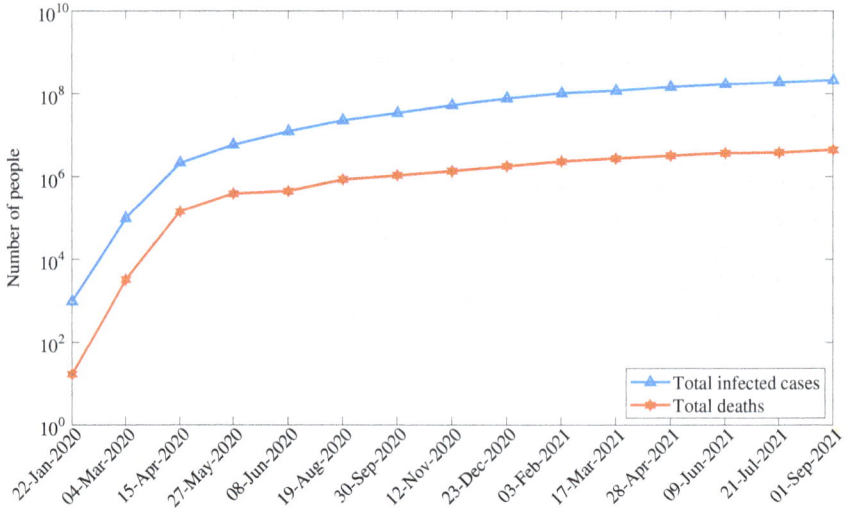

Figure 8.1   The aggregated number of infected people and deaths over time

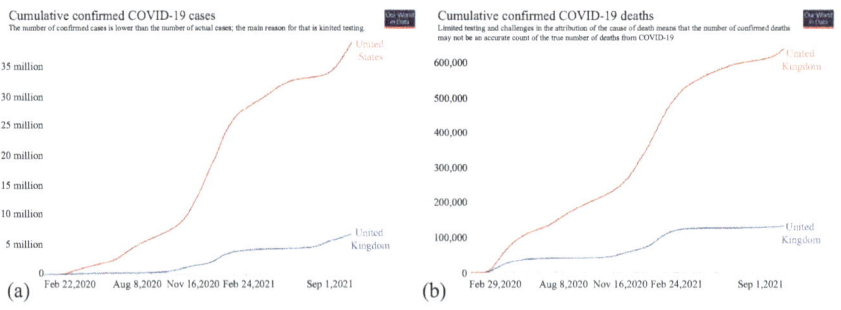

Figure 8.2   The confirmed and death cases in United States and United Kingdom

Even in the countries with high vaccination rates, such as the United States or United Kingdom, the disease mutates into new strains resistant to vaccines and shows signs of a sharp increase in the number of infections as in Figure 8.2 [3]. Up to now, vaccines for COVID-19 are known as an effective measure in fighting the pandemic and bringing normal life to citizens [4]. However, the vaccine's effectiveness gradually decreases over time, making the disease always a potential for an outbreak. Moreover, the lack of vaccine supply makes it difficult for developing countries to vaccinate the citizens. Therefore, beside effective testing and vaccinations, the development of arrangements for physical spaces has become crucial. Consequently, people's habits and regulations of conduct in society have entirely reversed to prevent the pandemic. The social distancing includes the set of principles that show how we move, how we work, how we meet and conduct meetings, and how we manage our social gatherings.

Since the pandemic's beginning, special measures have been dynamically applied to respond to the epidemic situation, such as re-planning public areas with large crowds, limiting the physical presence and close contact between people. Because of the unknown nature of the pandemic outbreak, standard operating protocols are typically issued by government authorities to match the actual situation at that time, which are changed depending on our knowledge of the virus SARS-CoV-2. In social distancing, real-time optimization plays a vital role in arranging the distribution of the maximum number of people allowed in a given place (such as a hospital, supermarket, or other public areas) to ensure that the minimum distance between people is sufficient to prevent the spreading of the virus. In addition, optimization techniques also help improving work performance and people's quality of life in the context of social distancing. Some open issues today that optimization and scheduling techniques are efficiently utilized is illustrated in Figure 8.3. The main purposes of this book chapter are summarized as follows:

- We present a survey on issues in a society where optimization techniques play an essential role in maintaining and enhancing the adoption of social distancing. Different areas of life that have been affected by the COVID-19 pandemic include physical space management, healthcare management, traffic management, online service management, workforce management, and low-cost service management. Each area is presented with a background on how the optimization technique can impact and enhance the social distancing.
- General simple examples have been illustrated to explain each area, thereby setting the stage for new problems that can be implemented within the social distancing context. It should be noted that the examples given in this chapter are intended to

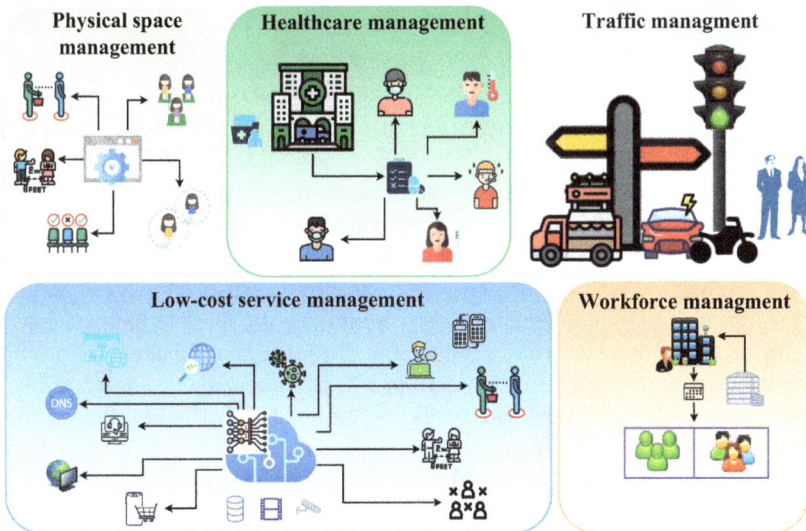

Figure 8.3   Several management activities for social distancing

Table 8.1   Mathematical notations

Notation	Description	Notation	Description
Upper bold letters ($\mathbf{X}$)	Matrices	$\mathbb{C}$	Complex field
Lower bold letters ($\mathbf{x}$)	Vectors	$\mathbb{R}$	Real field
$\mathbb{E}\{\cdot\}$	Expectation operator	$(\cdot)^T$	Regular transpose
$\|\cdot\|$	Euclidean norm	$(\cdot)^H$	Hermitian transpose

illustrate the theory of presentation only, so it is not meant to pose a new problem applicable to social distancing contexts. The examples are detailed in Sections 8.2–8.6.

- Optimization problems that can improve computation and processing speed to meet real-time requirements by applying artificial neural networks are also presented. The use of artificial neural networks provides fast processing speed based on its low complexity, thereby making the proposed solutions feasible in practice. The detail is presented in Section 8.6.

The rest of this chapter is organized as follows: Section 8.2 presents the fundamental aspects of the social distancing in the view point of the COVID-19 pandemic. We describe the workforce management which requires the minimal number of people at the office to complete a certain number of workloads in Section 8.3. Meanwhile, Section 8.4 gives an optimization framework to manage the public healthcare so that the social distancing should be maintained. The public transport management is shown in Section 8.5. Finally, Section 8.6 presents the online service managements and Section 8.7 draws the main conclusions. Mathematical notations in this chapter are given in Table 8.1.

8.2   Physical space management

One of the most important factors to consider during the ongoing pandemic is organizing and optimizing physical spaces corresponding to the appropriate level to ensure safety. The purpose of physical space optimization is to design the position of objects (objects can be people, chair positions, queue positions) in a given space to minimize the risk of infection while simultaneously optimizing the level of physical interaction among them. For example, in restaurants, the more tables that can be arranged in a restaurant, the higher the restaurant's profit. In supermarkets, conference halls, or other places where physical participation is required, the efficient exploitation of space while ensuring safety is an important goal. Social distancing, i.e., maintaining a protected distance among people to minimize the short range of interactions, has been considered the most effective method currently being widely applied worldwide to help reduce the infection rate arising from the COVID-19 epidemic. However, in reality, face-to-face communication is an influential factor necessary for human development and optimal effectiveness. Therefore, it is essential to find and develop methods for maintaining a balance between safe physical distance and effective communication. According to [5], nonverbal communication requires an environment with a relatively neutral background and sufficient light, and a short physical distance (2 m or less) between communicating objects. In addition, the authors in [6] have shown that arranging an optimal physical space environment also helps a lot in improving productivity.

There are various works in the recent literature on the transmissibility of the SARS-CoV-2 virus and the responses to overcome this COVID-19 pandemic. Based on severe consequences of the spread of infectious viruses through respiratory activities such as breathing, coughing, and sneezing [smallpox (1970), measles (1985), tuberculosis (1990), SARS (2003) and H1N1 (2009)], the authors in [7] detected and explored the distribution of droplet aerosols produced through the mouth and nose under air-conditioned room conditions. Although this work was published before the emergence of the COVID-19 pandemic, the results obtained are also very meaningful in assessing the properties of the SARS-CoV-2 virus, which is similar to those of the previous pandemics related to respiratory tract involvement. In order to prevent the spread of the virus, the authors in [8] have designed a general platform to optimize the efficient use of public space to maintain a safe distance between people. Their design can be flexible to accommodate a variety of conditions in terms of space geometry, air conditions, room air conditioner operation, or different minimum spacing requirements. The work [9] approached human placement through wind farm layout design by exploiting the similarity between the placement of turbines on an offshore area and physical space planning subject to the distance constraints between individual turbines or people. The authors proposed a method to minimize the interference/infection between close people by exploiting the fact that nearby turbines also infect each other by casting wind shadows that cause production losses.

In order to achieve an optimal solution to the problem of physical space modeling in terms of layout configuration and arrangement, several parameters need to be considered. These parameters are established and influenced by many factors, i.e., the pre-existing structure and shape of the place under consideration, the distance and number of present people required by the authorities, the influence of ventilation, lighting conditions, or other factors. Physical distancing, the first and most affecting factor in people's lives during the pandemic, is the key factor to prevent the virus from spreading during the COVID-19 pandemic. In contrast to life before the epidemic, governments and health authorities recommend that all participants keep a physical distance of at least 2 m within communal places at all times [10]. Besides, the airflow pattern and lighting also need to be considered as constraints for social distancing problems. Before the COVID-19 pandemic, the ventilation and lighting systems in buildings were optimally designed so that people could stay close to each other in order to increase communication efficiency and create closeness. However, the authors in [11] have pointed out that the airflow pattern can become a vector of transmission for the virus and need to be taken into consideration in current social distancing issues. In addition, other environmental factors such as temperature and humidity, which can affect the survival rate of the virus, also play an essential role in the transmission of infectious diseases. Therefore, these factors also need to be considered to optimize at an appropriate level in the context of social distancing [12]. Within the framework of this chapter, we would like to introduce a model with low complexity and suitable for real situations, which was presented in [8]. In this chapter, optimization techniques are applied to give the optimal solution to social distancing problems in various physical structures of the place and also consider pre-existing airflow systems. With the problem of optimally arranging objects in a given physical space, the work [8] proposed a general approach by defining the optimal configuration such that the circles of given radius to be neatly arranged within the configuration. Specifically, on a general polygonal region Ω includes K objects and each object with coordinate vector $p_i \triangleq (x_i, y_i) \in \mathbb{R}^2$ ($i \in \{1, 2, \ldots, K\}$), the problem is to is determine circles within that region such that all those circles do not overlap with each other. From there, the general optimization problem with non-negative constraints can be formatted as

$$\underset{\{(x_{i'}, y_{i'})\}}{\text{minimize}} \quad -r \tag{8.1a}$$

$$\text{subject to} \quad 2r \leq ||p_i - p_j||, \quad \forall i, j \in \{1, 2, \ldots, K\}, \tag{8.1b}$$

where $|| \cdot ||$ denotes the Euclidean distance. Herein, with different physical shapes of space, constraints limiting spatial coordinates will be added to the general problems to suit each specific design. This physical problem can be effectively solved with the interior-point method through an iterative algorithm [8]. Applying to real-world examples, the optimization problem for arranging N chairs in a square room is formulated as

$$\underset{\{(x_{i'}, y_{i'})\}}{\text{minimize}} \quad -r \tag{8.2a}$$

$$\text{subject to} \quad 2r \leq ||p_i - p_j||, \quad \forall i, j \in \{1, 2, \ldots, K\}, \tag{8.2b}$$

$$x_i \in [r, L - r], \ y_i \in [r, W - r], \quad \forall i \in \{1, 2, \ldots, K\}, \tag{8.2c}$$

$$1 \leq i < j \leq N, \quad r \in (0, 1/2), \tag{8.2d}$$

where L and W represent the length and the width of the considered room. Another example, for the triangular room, the adaptive optimization problem is

$$\underset{\{(x_{i'}, y_{i'})\}}{\text{minimize}} \quad -r \tag{8.3a}$$

$$\text{subject to} \quad 2r \leq ||p_i - p_j||, \quad \forall i, j \in \{1, 2, \ldots, K\}, \tag{8.3b}$$

$$6x_i + 2\sqrt{3}y_i \leq 3L - 4\sqrt{3}r, \quad \forall i \in \{1, 2, \ldots, K\}, \tag{8.3c}$$

$$-6x_i + 2\sqrt{3}y_i \leq 3L - 4\sqrt{3}r, \quad \forall i \in \{1, 2, \ldots, K\}, \tag{8.3d}$$

$$1 \leq i < j \leq N. \tag{8.3e}$$

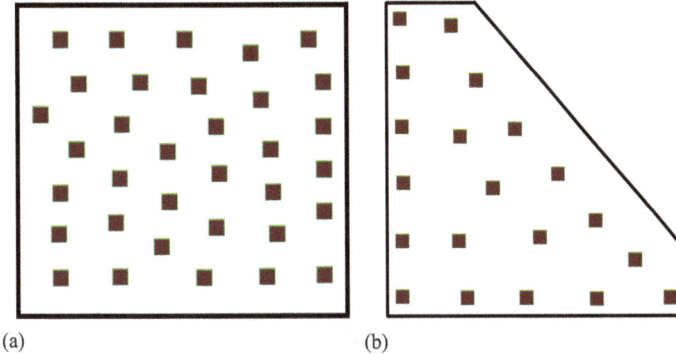

(a)   (b)

Figure 8.4   Examples of optimal arrangements versus different shapes of room [8]

Two examples of solutions for problem (8.2) and problem (8.3) are illustrated in Figure 8.4. For the sake of simplicity, we do not include visual arrange constraints, e.g., symmetric arrangement, into problems (8.2) and (8.3). Such extensions should be interested for a future work since the visual arrange constraints are of paramount importance in practice. In addition, a variety of physical constraints with different space layouts can be taken into accounts, such as space constraints, location of doors, windows, aisles, and airflow. From there, optimal solutions can be devised to optimize the placement of objects.

8.3   Workforce management

Workforce management helps to limit a minimum number of people in the workplace while completing the important work required by organizations. During the COVID-19 pandemic, although remote working is an encouraging form, some specific types of work still require employees to be present at the workplace to perform essential tasks (e.g., healthcare, transportation, and manufacturing). We emphasize that workforce management also depends on different types of constraints that depend on different types of jobs, such as dependence among tasks (precedence constraints), skill requirements (skill constraints), and limited resources usage (resource constraints), and due date (time constraints). For such scenarios, optimization approaches can be utilized to support workforce management to minimize the number of employees present at work to meet social distancing requirements. In terms of project scheduling with consideration of multi-skilled human resources, several optimization approaches have been developed aiming to reduce operation cost [13–16]. Also, the authors in [17–19] evaluated workforce scheduling in order to reduce cost. The authors in [20] proposed a hybrid genetic algorithm to optimize the service center workforce taken into account several tasks such as skill consideration, standard duration, and workload balance. Besides, the interaction between cost and service quality has also been studied. The work [21] minimized the operational cost and simultaneously maximized service level by proposing an alternative multidimensional paradigm. However, it should be noted that all of the above papers focus on reducing the cost of the workforce. In the context of social distancing, approaches to management and allocation to maximize distance and minimize physical contact among employees are critical and need to be developed.

As an example, we illustrate a workforce management problem with the objective of minimizing the total cost of agents under constraints of multiskill staffing. Indirectly, getting the job done at the minimum cost results in minimizing the number of employees present at the company. In particular, assuming there are K call types and I agent types, which an agent type I ($i \in \{1, 2, \ldots, I\}$) has the skill set $S_i \subseteq \{1, 2, \ldots, K\}$. A workday length includes P periods, and the staffing set is $\mathbf{y} \overset{\triangle}{=} \{y_{1,1}, \ldots, y_{1,P}, \ldots, y_{I,1}, \ldots, y_{I,P}\}$, where $y_{i,p}$ denotes the number of agents of type i available in period p. Then, the service level $g_{k,p}(\mathbf{y})$ in the period p and type-k is formulated as $g_{k,p}(\mathbf{y}) = \mathbb{E}[C_{g,k,p}]/\mathbb{E}[C_{g,k,p} + A_{k,p}]$, where $C_{k,p}$ represents the number of type-k calls arriving in p and getting served. $C_{g,k,p}$ and $A_{k,p}$ are the number of those calls getting served and abandoning after an acceptable waiting time $\tau_{k,p}$, respectively. Similar, the service levels per call period p, per type k, and globally for the waiting times τ_p, τ_k, τ are defined respectively as $g_p(\mathbf{y}), g_k(\mathbf{y}), g(\mathbf{y})$.

Let us denote $\{1, 2, \ldots, Q\}$ the set of all admissible shifts, $\mathbf{A}_0 \in \mathbb{R}^{P \times Q}$ the matrix whose element $a_{p,q} = 1$ if agent with shift q works in period p, and 0 otherwise. $\mathbf{c} = [c_{1,1}, \ldots, c_{1,Q}, \ldots, c_{I,1}, \ldots, c_{I,Q}]^T$ is the vector of costs, where $c_{i,q}$ is the cost of an agent of type i with shift q. The vector of schedule is defined as

$$\mathbf{x} = [x_{1,1}, \ldots, x_{1,Q}, \ldots, x_{I,1}, \ldots, x_{I,Q}]^T, \tag{8.4}$$

where $x_{i,q}$ denotes the number of agents of type i and working shift q. Then, the staffing vector $\mathbf{y} = \mathbf{A}\mathbf{x}$ corresponds to the shift vector $\mathbf{x}$ and the block-diagonal matrix $\mathbf{A}$ with I identical blocks $\mathbf{A}_0$. We also define the skill transfer variable $z_{i,j,p}$ as the number of type-i agents that are temporarily downgraded to type j during period p. The vector $\mathbf{z}$ consists of the variables $z_{i,j,p}$, and $\mathbf{B}$ is a matrix, where its' entries are in the set $\{-1, 0, 1\}$. The optimization scheduling problem can be formulated as

$$\underset{\{(x_{i'}, y_{i'})\}}{\text{minimize}} \quad \mathbf{c}^T \mathbf{x} = \sum_{i=1}^{I} \sum_{q=1}^{Q} c_{i,q} x_{i,q} \tag{8.5a}$$

$$\text{subject to} \quad \mathbf{A}\mathbf{x} + \mathbf{B}\mathbf{z} \geq \mathbf{y}, \tag{8.5b}$$

$$g_{k,p}(\mathbf{y}) \geq \ell_{k,p}, \ q \leq k \leq K, 1 \leq p \leq P, \tag{8.5c}$$

$$g_p(\mathbf{y}) \geq \ell_p, \ 1 \leq p \leq P, \tag{8.5d}$$

$$g_k(\mathbf{y}) \geq \ell_k, \ 1 \leq k \leq K, \tag{8.5e}$$

$$g(\mathbf{y}) \geq \ell, \tag{8.5f}$$

$$\mathbf{x} \geq 0, \mathbf{z} \geq 0, \mathbf{y} \geq 0 \text{ and integer}, \tag{8.5g}$$

where the levels $\ell_{k,p}$, ℓ_p, ℓ_k, ℓ are given constants. The multiskill scheduling problem (8.5) is in general NP-hard, which can be solved effectively by a simulation-based algorithm combining integer or linear programming, as presented in [14].

8.4 Healthcare management

Healthcare management focuses on managing appointments between doctors and patients to ensure the highest level of service quality and customer satisfaction while optimizing the workload for doctors as well as the number of hospital customers. Optimization techniques can also help solve appointment sequencing and scheduling problems in the context of social distancing. There are various applications of appointment management in life and are especially meaningful in healthcare management, such as managing hospital appointments, or home healthcare thus reducing unnecessary traffic and the number of patients visiting the hospital.

Several works have studied the appointment scheduling problems. A heuristic solution to the sequential scheduling problem was proposed in [22] by looking at cases where customers booked appointments but did not show up and ones who booked more than one appointment for themselves. In addition, a local search algorithm was proposed in [23] to minimize patient waiting times, doctor idle times, and lateness. To improve the efficiency of appointment scheduling, the authors in [22], [24] suggested optimization approaches by considering the uncertainty in the processing time of each appointment. Moreover, a multistage stochastic linear program was developed in [25], considering the unpredictable appointments duration and unexpected cancellations to minimize the waiting times and overtime of hospital arrivals. However, it should be noted that the above works did not consider the number of patients present simultaneously at hospitals for hospitals, which is needed of developing methods to maintain a suitable level of social distancing.

Not only managing the appointment, optimization techniques for home healthcare services are also essential to reduce pressure on hospitals in the context of hospitals being overwhelmed during the pandemic. In [26], the uncertain transportation demand of patients was addressed with a hybrid genetic algorithm to minimize transportation costs of delivering medication drugs. Similar, many perspectives such as the policy, health care facility, and logistics have been studied to improve the home health care services [27]. Nickel and Schrode [28] presented a two-phase method to solve the scheduling problem in home health care, which requires an optimal weekly (or periodical) plan. It shows the benefit of using more sophisticated planning methods in the home health care context. Besides, a branch-and-price algorithm was proposed in [29] to minimize the traveling costs and delay of services while considering stochastic service times under considering uncertainties in patient's demands. Authors in [30] proposed a multi-heuristics method to minimize traveling times and optimize the dissatisfaction level of clients and nurses while considering workload and time constraints. Different from the workforce and appointment scheduling, home healthcare scheduling techniques can be more effectively applied to social distancing scenarios because they can minimize the traveling distances while planning quality-of-service.

Assume that a treatment room is operational T time slots with length d per slot and is scheduled for a total of N patients. At time slot t, $x_t \in \{0, 1, \dots, N\}$ is the number of patients scheduled. As a general example, we develop a problem with the objective of minimizing simultaneously the mean waiting time $W(x)$, the idle time $I(x)$, and the lateness $L(x)$, which are weighted respectively with the weight factors α_W, α_I, and α_L. Then the optimization problem is then formulated as [23]

$$\text{minimize} \quad \alpha_W W(x) + \alpha_I I(x) + \alpha_L L(x) \tag{8.6a}$$

$$\text{subject to} \quad \sum_{t=1}^{T} s_t = N, \tag{8.6b}$$

$$\alpha_W + \alpha_I + \alpha_L = 1. \tag{8.6c}$$

Therein, the functions $W(x), I(x)$ and $L(x)$ are built as in [23], where a local search procedure has also been proposed to solve problem 1.5 effectively.

8.5 Traffic management

In traffic management, the objective is typically designed to minimize travel time, which can be calculated as the total of the product of the travel times and link flows in the whole traffic network. In traffic management, optimization techniques can also potentially applied in an effective manner to handle social distancing in the context of the COVID-19 pandemic. In more details, scheduling approaches are studied to accommodate traffic levels, e.g., the number of people in traffic over different time frames. The authors in [31] considered a macroscopic model to formulate a scheduling problem based on the mixed traffic demand to assess the trade-off between pedestrians and vehicles in an economic evaluation framework. After that, a mixed-integer linear programming and a discrete harmony search algorithm were proposed to produce real-time traffic light scheduling. Sharing the same concept, many model-based strategies have been proposed to solve traffic-responsive problems in real-time to obtain the optimal switching times [32,33]. In addition, there are also works that studied model-based vehicle flow scenarios by solving mixed-integer linear programming problems with high computational complexity [34,35] and some others proposing meta-heuristic algorithms that effectively solve the problems for large traffic networks [36–39]. black In addition, the work [40] evaluated criteria to select pedestrian phase patterns between the typical two-way crossing and the exclusive pedestrian phase in an economic evaluation framework in terms of both safety and efficiency factors. The authors in [41] ensured the mobility of citizens through building intelligent optimization systems and applying a harmony search algorithm to provide alternative routes when a long-term road cut occurs. Although many researches in the literature have addressed the optimization problems in traffic management that indirectly express social distancing issues, the meaningful objectives and constraints focusing on reducing the number of pedestrians and vehicles on the street at the same time still need to be researched and developed. Therefore, we can consider this direction as a potential topic for future work.

For example, the total travel time depends on several factors such as the signal timings, denoted by ψ, and the link flow patterns $q^*(\psi)$ in the network. Let us denote L the numbers of links of the network, then q_a and t_a are the flow and travel time on link a with $a \in \{1, 2, \ldots, L\}$, respectively. C is the common cycle time with the maximum C_{max} the minimum C_{min}. θ_h represents the offset at junction h. The duration of the green signal time for stage r at junction h is denoted as $\phi_{h,r}$ with the maximum and minimum are respectively $\phi_{h,r}$ max and $\phi_{h,r}$ min. Finally, the inter-green between two sequence green for the stage r and junction h is $I_{h,r}$, where S_h is the total number of stages at junction h. Following [38], the total travel time problem can be mathematically formulated as

$$\underset{\psi \in \Omega_o}{\text{minimize}} \quad \sum_{a=1}^{L} q_a t_a(\psi, q^*(\psi)) \tag{8.7a}$$

$$\text{subject to} \quad C_{min} \leq C \leq C_{max}, \tag{8.7b}$$

$$0 \leq \theta_h \leq C - 1, \tag{8.7c}$$

$$\phi_{h,r_{min}} \leq \phi_{h,r} \leq \phi_{h,r_{max}}, \tag{8.7d}$$

$$C = \sum_{r=1}^{S_h} \phi_{h,r} + \sum_{r=1}^{S_h} I_{h,r}, \quad \forall h, \tag{8.7e}$$

which can be solve effectively by using a genetic algorithm. This problem is just a general example of a traffic management problem. In the context of the pandemic, the travel time is not the top priority, but it may be the number or distance of vehicles circulating in the network.

8.6 Low-cost service management

As the COVID-19 pandemic broke out and has been happening until now, many countries have adopted social distancing in vast extensions. Many people have worked from home as a solution to reduce the virus spread. From this fact, the internet should be proper measures to efficiently explode and serve requirements such as remote work, online shopping, and virtual meeting with an extremely small delay for conveniences. As a result of the confinement period, a significant increase of Internet use and general virtual services have been observed, such as video streaming, broadcasting, and contents delivery. Therefore, online service solutions which guarantee people's quality of life becomes urgent in the social distancing scenarios. Regarding video streaming traffic, when many people are working from home, the users' demands are significantly higher than the other types of internet demands. As a consequence, emerging networking technologies can be effective solutions to the online service optimization problems. For instance, in the previous work [42], software-defined networking technology and adaptive streaming were evaluated to propose an algorithm to enable video streaming over hypertext transfer protocol (HTTP). An algorithm that allocates available people into groups was also developed to reduce communication overhead and to maximize network resources. In addition, resource usage, quality-of-service, and video stability were numerically reported to be improved by utilizing the suggested framework. With the purpose of optimizing the contents delivery process in a content delivery network semi-federation system*, the study [43] proposed an algorithm to optimally allocate the content provider's demand to different content delivery networks in the federation. Herein, a content delivery network is a globally distributed network where users can access the internet with high quality of service and low latency, and the proxy servers are deployed at the network edge to reduce the burden for the backhaul. Edge caching is potential another approach to decrease latency and network congestion (e.g., [44]). Also, two edge caching methods, coded and uncoded caching, were examined in [45]. In more detail, for each of the two caching techniques, two optimization algorithms have been proposed to minimize content delivery times. We now present in detail two practicable researches directions comprising computer-based distance monitoring and artificial intelligence based optimization model.

8.6.1 Low-cost computer-based distance monitoring

Due to the self-interested nature of human in their routines, special mechanisms and solutions are in need to monitor or encourage people to comply with social distancing under a pandemic or a disaster, especially in curfews to prevent a contagious disease. To encourage people to keep a required distance for safety from each other, computer-based models have been studied to measure the distance between any two people in a public area. The main idea is based on a particular circumstance to formulate and solve an open issue subject to some practical constraints. In the recent work [46], the authors used mobile software installed on users' phone that can track the locations and give notices to help prevent the spread of the SAR-CoV-2 virus, especially in high risk regions with many infected people. However, the application not only requires all the users to install a unique software for synchronization, but also raises a potential risk of leaking private information. Consequently, it is one of the reasons why many people might not want to use the service. To avoid such uncertainties, the authors in [47] used mobile phone users' geolocation data to track the user's location without using any mobile applications to increase privacy. Apart from this, a system for outdoor location tracking, which uses the mobile cellular network, has been developed in [48] by running entirely on the network side. A system exploiting geometry to estimate the vehicles' position and focusing on vehicle dynamics and road trajectories was presented in [49]. The above works are effectively instrumental in tracking the

---

*A semi-federation system is defined as a distributed system where multiple content delivery networks can be optimized without fully share all the information.

movement history of infected people or identifying crowded areas so that timely response solutions can be given. In addition, the data-driven approaches have been also popularly utilized in detecting objects and people in a video of a public area. Since then, algorithms have been developed to determine the distance of people in the video to help control and support social distancing [50–52]. For future works in computer-based distance monitoring, we can figure out some interesting directions as follows:

- Developing algorithms with higher accuracy and faster computation time to track infected people and localize epidemics quickly and accurately.
- Designing low cost devices that can process real-time response public camera signals, or building more accurate object detection models and distance measurement algorithms in 2D image to assist in monitoring people in public areas, thereby enhancing social distancing.

8.6.2   Low-cost optimization solution with artificial intelligence

The aforementioned classical optimization works in these areas are typically implemented in a heuristically iterative manner due to the inherent nonconvexity, which may not be optimal in general [53]. Furthermore, the proposed algorithms often require high computational complexity in many applications and will most likely not meet the requirements of real-time implementations in practice. As a potential solution to this matter, a solution to train artificial neural networks for the above mentioned optimization problems is illustrated in Figure 8.5 by using data sets to learn various features of the social distancing. It should be noticed that machine learning and deep learning do not entirely replace classical optimization algorithms, but they should be designed to support and compute faster for time-critical requirements towards online management. For supervised learning, an optimization algorithm will be treated as a black box, and its properties are learned and trained by an artificial neural network, as shown in Figure 8.5(a). The learning stage targets to exploit the relationship between the inputs and outputs of the designed optimization algorithm. The inputs and outputs of a neural network model may be coincided with the inputs and outputs of optimization problems, respectively, for example in scheduling problems. However, it can be also different. The data set is divided into several portions, which are used for training, testing, and validation. In the training stage, the training data fed into the artificial neural network is generated as the solutions to the optimization algorithms. We emphasize that these solutions are possibly either suboptimal or optimal, thus they could be pre-processed to achieve the best training performance, e.g., low mean square error. The main purpose of the training stage is to seek for good values of the weights and biases to express the input–output relations. The optimization algorithms are only utilized in the training stage and therefore we possible obtain a low cost solution for the testing state. Specifically, in the testing stage, as illustrated in Figure 8.5(b), when a new realization of data is available, the well-trained neural network will be used to predict the desired solution. The run time should be very fast in the testing stage since a neural network is only equipped with low arithmetic operators, and the prediction is done by only using the forward propagation once. Therefore, processing time is

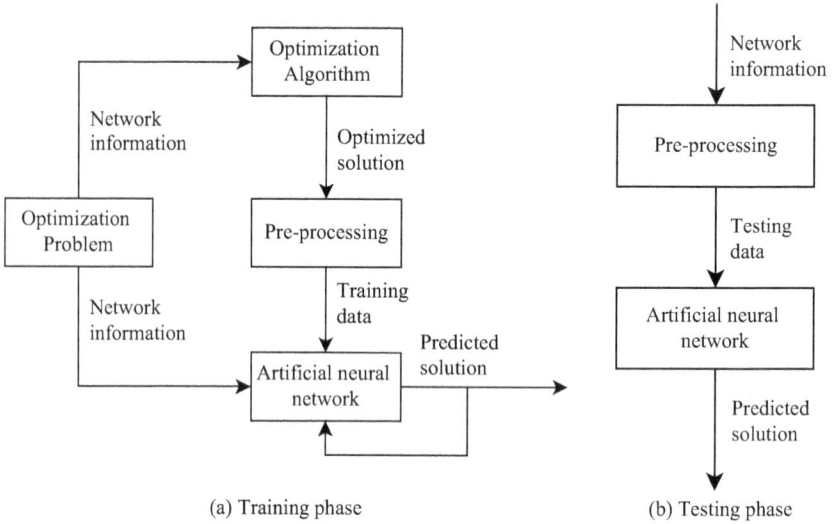

(a) Training phase   (b) Testing phase

Figure 8.5   The machine/deep learning framework applied for the optimization problems of social distancing context

significantly shortened compared to the traditional-based optimization approaches. With deep learning, we expect to meet practical problems that require fast computation. We stress that the machine learning/deep learning solutions can be exploited efficiently if the universal approximation theorem [54] is fulfilled, there always exist a basic comprehensive enough neural network to approximate a continuous function which a compact feasible set. For the applications in the social distancing, there are potential directions with the use of machine learning/deep learning:

• Reducing the computational complexity of the high-cost algorithms that produce the global optimum or a good local solution with expensive expenditure.
• Attaining a better solution than the solution produced by the traditional optimization approaches in the case that the optimal solution is not available.
• Designing a universal hardware to implement a neural network that can train and predict the solution to multiple applications and for general purposes.

With the three above potential research directions, we can formulate and solve a plenty of interesting social distancing problems, which are left for future works.

8.7   Conclusion

This chapter has examined optimization techniques from different perspectives to enhance social distancing amid the COVID-19 pandemic. The use of optimization methods not only makes sense in this context but can also be applied to respond to future infectious pandemics should they occur. In various areas, including physical space management, workforce management, healthcare management, and traffic management, general problems have been raised for illustrative purposes and hence can motivate other issues to enhance social distancing. In order to improve the feasibility of solutions in practice, the application of artificial neural networks has also been presented with the aim of providing low complexity alternatives to classical optimization methods to meet the real-time of the proposed algorithms.

References

[1]   Maier BF, Brockmann D. Effective containment explains subexponential growth in recent confirmed COVID-19 cases in China. *Science*. 2020;**368**(6492):742–6.
[2]   Coronavirus (COVID-19) cases provided by World Ometers. Available from: https://www.worldometers.info/coronavirus/.
[3]   Coronavirus (COVID-19) cases provided by Our World in Data. Available from: https://ourworldindata.org.
[4]   Unicef. Available from: https://www.unicef.org/northmacedonia/what-you-need-know-about-covid-19-vaccine.
[5]   Albert M. In: Kühnhardt L, Mayer T, editors. *Nonverbal Communication*. Cham: Springer International Publishing; 2019. pp. 453–61. Available from: https://doi.org/10.1007/978-3-319-90377-4_38.
[6]   Travers I. risk based occupational health and safety guidelines to help you prepare to re-open for business and to protect people Covid-19 Return to Working Guidance. Retrieved on 19th October 2020.
[7]   Zhang Y, Feng G, Bi Y, *et al.* Distribution of droplet aerosols generated by mouth coughing and nose breathing in an air-conditioned room. *Sustainable Cities and Society*. 2019;**51**:101721.
[8]   Ugail H, Aggarwal R, Iglesias A, *et al.* Social distancing enhanced automated optimal design of physical spaces in the wake of the COVID-19 pandemic. *Sustainable Cities and Society*. 2021;**68**:102791. Available from: https://www.sciencedirect.com/science/article/pii/S2210670721000834.
[9]   Fischetti M, Fischetti M, Stoustrup J. *Mathematical optimization for social distancing*. 2020;14:2021.
[10] Pearce K. *What is social distancing and how can it slow the spread of COVID-19*. The Hub, Johns Hopkins University. 2020.
[11] Qian H, Zheng X. Ventilation control for airborne transmission of human exhaled bio-aerosols in buildings. *Journal of Thoracic Disease*. 2018;**10**(Suppl 19):S2295.
[12] Huang Z, Huang J, Gu Q, *et al.* Optimal temperature zone for the dispersal of COVID-19. *Science of The Total Environment*. 2020;**736**:139487. Available from: https://www.sciencedirect.com/science/article/pii/S0048969720330047.

[13] Fırat M, Hurkens CA. An improved MIP-based approach for a multi-skill workforce scheduling problem. *Journal of Scheduling*. 2012;**15**(3):363–80.

[14] Avramidis AN, Chan W, Gendreau M, et al. Optimizing daily agent scheduling in a multiskill call center. *European Journal of Operational Research*. 2010;**200**(3):822–32.

[15] Li H, Womer K. Scheduling projects with multi-skilled personnel by a hybrid MILP/CP benders decomposition algorithm. *Journal of Scheduling*. 2009;**12**(3):281–98.

[16] Heimerl C, Kolisch R. Scheduling and staffing multiple projects with a multi-skilled workforce. *OR spectrum*. 2010;**32**(2):343–68.

[17] Morz M, Musliu N. Genetic algorithm for rotating workforce scheduling problem. In: *Second IEEE International Conference on Computational Cybernetics*, 2004. ICCC 2004. IEEE; 2004. pp. 121–6.

[18] Balakrishnan N, Wong RT. A network model for the rotating workforce scheduling problem. *Networks*. 1990;**20**(1):25–42.

[19] Musliu N, Gärtner J, Slany W. Efficient generation of rotating workforce schedules. *Discrete Applied Mathematics*. 2002;**118**(1–2):85–98.

[20] Valls V, Pérez Á, Quintanilla S. Skilled workforce scheduling in service centres. *European Journal of Operational Research*. 2009;**193**(3):791–804.

[21] Castillo I, Joro T, Li YY. Workforce scheduling with multiple objectives. *European Journal of Operational Research*. 2009; **196**(1):162–70.

[22] Zacharias C, Pinedo M. Appointment scheduling with no-shows and overbooking. *Production and Operations Management*. 2014;**23**(5):788–801.

[23] Kaandorp GC, Koole G. Optimal outpatient appointment scheduling. *Health Care Management Science*. 2007;**10**(3):217–29.

[24] Mak HY, Rong Y, Zhang J. Appointment scheduling with limited distributional information. *Management Science*. 2015;**61**(2):316–34.

[25] Erdogan SA, Denton B. Dynamic appointment scheduling of a stochastic server with uncertain demand. *INFORMS Journal on Computing*. 2013;**25**(1): 116–32.

[26] Shi Y, Boudouh T, Grunder O. A hybrid genetic algorithm for a home health care routing problem with time window and fuzzy demand. *Expert Systems with Applications*. 2017;**72**:160–76.

[27] Harris MD. *Handbook of home health care administration*. Jones & Bartlett Learning; 2005.

[28] Nickel S, Schröder M, Steeg J. Mid-term and short-term planning support for home health care services. *European Journal of Operational Research*. 2012;**219**(3):574–87.

[29] Yuan B, Liu R, Jiang Z. Home health care crew scheduling and routing problem with stochastic service times. In: *2014 IEEE International Conference on Automation Science and Engineering (CASE)*. IEEE; 2014. pp. 564–9.

[30] Trautsamwieser A, Hirsch P. Optimization of daily scheduling for home health care services. *Journal of Applied Operational Research*. 2011;**3**(3):124–36.

[31] Zhang Y, Gao K, Zhang Y, et al. Traffic light scheduling for pedestrian-vehicle mixed-flow networks. *IEEE Transactions on Intelligent Transportation Systems*. 2018;**20**(4):1468–83.

[32] Sen S, Head KL. Controlled optimization of phases at an intersection. *Transportation Science*. 1997;**31**(1):5–17.

[33] Henry JJ, Farges JL, Tuffal J. The PRODYN real time traffic algorithm. In: *Control in Transportation Systems*. Elsevier; 1984. pp. 305–10.

[34] Lin S, De Schutter B, Xi Y, et al. Fast model predictive control for urban road networks via MILP. *IEEE Transactions on Intelligent Transportation Systems*. 2011;**12**(3):846–56.

[35] Lo HK. A cell-based traffic control formulation: strategies and benefits of dynamic timing plans. *Transportation Science*. 2001;**35**(2):148–64.

[36] Yang Z, Benekohal RF. Use of genetic algorithm for phase optimization at intersections with minimization of vehicle and pedestrian delays. *Transportation Research Record*. 2011;**2264**(1):54–64.

[37] Ceylan H, Bell MG. Traffic signal timing optimisation based on genetic algorithm approach, including drivers' routing. *Transportation Research Part B: Methodological*. 2004;**38**(4):329–42.

[38] Teklu F, Sumalee A, Watling D. A genetic algorithm approach for optimizing traffic control signals considering routing. *Computer-Aided Civil and Infrastructure Engineering*. 2007;**22**(1):31–43.

[39] Ma W, Liu Y, Head KL. Optimization of pedestrian phase patterns at signalized intersections: a multi-objective approach. *Journal of Advanced Transportation*. 2014;**48**(8):1138–52.

[40] Ma W, Liao D, Liu Y, et al. Optimization of pedestrian phase patterns and signal timings for isolated intersection. *Transportation Research Part C: Emerging Technologies*. 2015;**58**:502–14.

[41] Salcedo-Sanz S, Manjarres D, Pastor-Sánchez Á, et al. One-way urban traffic reconfiguration using a multi-objective harmony search approach. *Expert Systems with Applications*. 2013;**40**(9):3341–50.

[42] Bentaleb A, Begen AC, Zimmermann R, *et al*. SDNHAS: An SDN-enabled architecture to optimize QoE in HTTP adaptive streaming. *IEEE Transactions on Multimedia*. 2017;**19**(10):2136–51.

[43] Wang H, Tang G, Wu K, *et al*. Speeding up multi-CDN content delivery via traffic demand reshaping. In: *2018 IEEE 38th International Conference on Distributed Computing Systems (ICDCS)*. IEEE; 2018. pp. 422–33.

[44] Hoang DT, Niyato D, Nguyen DN, *et al*. A dynamic edge caching framework for mobile 5G networks. *IEEE Wireless Communications*. 2018;**25**(5): 95–103.

[45] Vu TX, Chatzinotas S, Ottersten B. Edge-caching wireless networks: Performance analysis and optimization. *IEEE Transactions on Wireless Communications*. 2018;**17**(4):2827–39.

[46] Zhou ZJ, Qiu Y, Pu Y, *et al*. BioAider: An efficient tool for viral genome analysis and its application in tracing SARS-CoV-2 transmission. *Sustainable Cities and Society*. 2020;**63**:102466.

[47] Rahman MT, Khan RT, Khandaker MR, *et al*. An automated contact tracing approach for controlling COVID-19 spread based on geolocation data from mobile cellular networks. *IEEE Access*. 2020;**8**:213554–65.

[48]   Trogh J, Plets D, Surewaard E, *et al.* Outdoor location tracking of mobile devices in cellular networks. *EURASIP Journal on Wireless Communications and Networking.* 2019;**2019**(1):1–18.

[49]   Kaiwartya O, Cao Y, Lloret J, et al. Geometry-based localization for GPS outage in vehicular cyber physical systems. *IEEE Transactions on Vehicular Technology.* 2018;**67**(5):3800–12.

[50]   Ahmed I, Ahmad M, Rodrigues JJ, et al. A deep learning-based social distance monitoring framework for COVID-19. *Sustainable Cities and Society.* 2021;**65**:102571.

[51]   Punn NS, Sonbhadra SK, Agarwal S, *et al.* Monitoring COVID-19 social distancing with person detection and tracking via fine-tuned YOLO v3 and Deepsort techniques. *arXiv preprint arXiv:200501385.* 2020.

[52]   Ramadass L, Arunachalam S, Sagayasree Z. Applying deep learning algorithm to maintain social distance in public place through drone technology. *International Journal of Pervasive Computing and Communications.* 2020.

[53]   Van Chien T, Canh TN, Björnson E, *et al.* Power control in cellular massive MIMO with varying user activity: A deep learning solution. *IEEE Transactions on Wireless Communications.* 2020;**19**(9):5732–48.

[54]   Hornik K, Stinchcombe M, White H. Multilayer feedforward networks are universal approximators. *Neural networks.* 1989;**2**(5):359–66.

Chapter 9

Incentives for individual compliance with pandemic response measures

Balázs Pejó[1] and Gergely Biczók[1]

The common methods to fight against COVID-19 are quasi-standard measures which include wearing masks, social distancing and vaccination. However, combining these measures into an efficient holistic pandemic response instrument is even more involved than anticipated. We argue that some non-trivial factors behind the varying effectiveness of these measures are selfish decision-making and the differing national implementations of the response mechanism. In this chapter, through simple models, we analyze the impacts of individual incentives on different measures of the decisions made with respect to social distancing, mask wearing, and vaccination. We shed light on how these may result in suboptimal outcomes and demonstrate the responsibility of national authorities in designing these games properly regarding data transparency, the chosen policies, and their influence on the preferred outcome. We promote a mechanism design approach: it is in the best interest of every government to carefully balance social good and response costs when implementing their respective pandemic response mechanism; moreover, there is no one-size-fits-all blueprint when designing an effective solution.

9.1 Introduction

The current coronavirus pandemic is pushing individuals, businesses and governments to the limit. Even with the recently emerged hope of rapidly developed vaccines, people still suffer owing to reduced mobility, social life and income; complete business sectors face an almost 100% drop in revenue; and governments are scrambling to find out when and how to impose and remove restrictions. In fact, COVID-19 has turned the whole planet into a "living lab" for human and social behavior where feedback on response measures deployed is only delayed by around two weeks (the incubation period). From the 24/7 media coverage, all of us have been introduced to a set of quasi-standard measures applied by national and local authorities, including social

[1]Laboratory of Cryptography and System Security (CrySyS), Department of Networked Systems and Services (HIT), Faculty of Electrical Engineering and Informatics (VIK), Budapest University of Technology and Economics (BME), Budapest, Hungary

distancing, wearing masks, vaccination, virus testing, contact tracing, and so on. It is also clear that different countries have had different levels of success employing these measures as evidenced by the varying normalized death tolls and confirmed cases.*

We believe that apart from the intuitive (e.g., genetic differences, medical infrastructure availability, hesitancy, etc.), there are two significant factors that have not received sufficient attention. First, the *individual incentives* of citizens, e.g., "is it worth more for me to stay home than to meet my friend?" have a significant say in every decision situation. While some of those incentives can be inherent to personality type, clearly, there is a non-negligible rational aspect to it, where individuals are looking to maximize their own utility. Second, countries have differed in their specific *implementation* of response measures, e.g., providing extra unemployment benefits (affecting the likelihood of proper self-imposed social distancing), whether they have been distributing free masks (affecting the efficacy of mask wearing in case of equipment shortage), or regulating the amount of accepted vaccines (affecting the speed of reaching herd-immunity). Framing pandemic response as a mechanism design problem, i.e., architecting a complex response mechanism with a preferred outcome in mind, can shed light on these factors. What's more, it has the potential to help authorities (mechanism designers) fight the pandemic efficiently. The objective of this chapter is to show that both individual incentives and the actual design and implementation of the holistic pandemic response mechanism can have a major effect on how (ongoing and future) pandemics are going to play out.

9.1.1   Contribution

In this chapter, we model decision situations during a pandemic using game theory where participants are rational, and the proper design of the games could be the difference between life and death. This chapter is a focused version of [1], which is an extension of [2]. Our main contribution is two-fold: we elaborate on several basic decision models of social distancing, mask wearing, and vaccination, and present a pandemic mechanism design viewpoint, in which all of these games are only sub-mechanisms of the bigger picture.

We show via COVID-19 social distancing statistics that going out is only a rational choice if staying home is not impossible or it provides significant benefit, from which we derive the optimal operating point for out-of-home activities. We also present a game corresponding to mask wearing and introduce several decision models concerning vaccination, so we can detail the pandemic response from a mechanism design perspective. We show that different government policies influence the outcome of these games profoundly, and the standalone response measures of the sub-mechanisms are interdependent.

9.1.2   Organization

The chapter is structured as follows. In the remaining of this section, we recap some concepts of game theory we use throughout this chapter. In Section 9.2, we briefly

---

*Johns Hopkins Coronavirus Resource Center. https://coronavirus.jhu.edu/map.html

describe related work. In Section 9.3, we develop and analyse the Distancing Game which includes the effects of meeting duration (or size). In Section 9.4, we sketch the two-player Mask Game, while in Section 9.5 we introduce several decision models focusing on various aspects of vaccines. In Section 9.6, we frame pandemic response as a mechanism design problem using the introduced models. Finally, in Section 9.7, we outline future work and conclude the chapter.

9.1.3 Preliminaries

Here, we shortly elaborate on the main game-theoretical notions used in this chapter, to facilitate the conceptual understanding of the implications of our results.

9.1.4 Game theory

Game theory [3] is "the study of mathematical models of conflict between intelligent, rational decision-makers." Almost any multi-party interaction can be modeled as a game. In relation to COVID-19, decision-makers could be individuals (e.g., whether to wear a mask), municipalities (e.g., whether to enforce wide-range testing within the city), governments (e.g., whether to apply contact tracing within the country), or companies (e.g., whether to apply social distancing within the workplace). Potential decisions are referred to as strategies; decision-makers (players) choose their strategies rationally so as to maximize their own utility.

9.1.5 Rationality

Note that rational (in a game-theoretical context) does not necessarily mean fully and objectively informed, i.e., individuals will make their decisions based on the *perceived* utility of their actions. Such a decision can even go against scientifically proven best practices, resulting in refusing vaccination or partying carelessly. Naturally, more realistic behavioral modelling (e.g., bounded rationality, unpredictability and a large number of proven behavioral biases [4]) delves deeper into the human decision-making process. However, the simple decision models in this chapter serve more of a demonstrative purpose, illustrating (i) how (selfish) individual decisions perturb society-level behavior and (ii) how central mechanism design decisions influence the outcome of such models.

9.1.6 Nash equilibrium

The Nash Equilibrium (NE)—arguably the most famous solution concept—is a set of strategies where each player's strategy is the best response strategy. This means every player makes the best/optimal decision for itself as long as the others' choices remain unchanged. NE provides a way of predicting what will happen if several entities are making decisions at the same time where the outcome also depends on the decisions of the others. The existence of a NE means that no player will gain more by unilaterally changing its strategy at this unique state.

9.1.7 Social optimum

Another game-theoretic concept is the Social Optimum (SO), which is a set of strategies that maximizes social welfare. Note, that despite the fact that no one can do better by changing strategy, NEs are not necessarily Social Optima (we refer the reader to the famous example of the Prisoner's Dilemma [3]). In fact, it is well-studied in game theory how much a distributed outcome (e.g., a NE) is worse than a centrally planned optimum (e.g., SO); this ratio is captured by the Price of Anarchy [5] and by the Price of Stability [6].

9.1.8 Mechanism design

If one knows the NE they prefer as the outcome of a game, e.g., everybody following social distancing guidelines, and they have the power to instantiate the game accordingly, i.e., fixing the structure, game flow and any free parameters, then we talk about mechanism design [7]. In a way, mechanism design is the inverse of game theory; although a significant share of efforts within this field deals with auctions, mechanism design is a much broader term applicable to any multi-stakeholder mechanism, (e.g. optimal organ matching for transplantation, school-student allocation or, in fact, pandemic response), aimed at achieving a preferred steady-state result.

9.2   Related work

In this section, we review some well-known and/or recent epidemic response mechanisms and game-theoretic works in relation to social distancing, masks, vaccination, and pandemics in general. A comprehensive systematic literature review on COVID-19 can be found in [8,9].

Concerning social distancing, [10,11] aim to provide a comprehensive survey on how emerging technologies, e.g., wireless and networking, artificial intelligence (AI) can enable, encourage, and enforce social distancing practice.

9.2.1 Game-theoretic models

In the intersection of epidemics and game theory, a comprehensive survey was carried out in [12,13]. Behavioral changes of people caused by a pandemic and, specifically, COVID-19 were studied in [14,15], respectively. Others focused on the mobility habits of people travelling between areas affected unevenly by the disease [16]. The authors of [17] took a closer look through the lens of game theory on the effect of self-quarantine on virus spreading. In [18], an optimization problem was formalized by accommodating both isolation and social distancing. Concerning the latter, the authors of [19] considered an approach to schedule the visitors of a facility based on their importance. The impact of social distancing was also studied in [20], in combination with vaccines.

The impacts of vaccine availability on human behaviour have been well studied. For instance in [21], the authors studied personal vaccination preferences and concluded that vaccine delayers relied on herd immunity and vaccine safety information generated by early vaccinators. A similar vaccination dilemma was studied in [22], in which an incentive model was proposed for individuals to determine the best strategy taking into account risks and expenses in the epidemic campaign. The study in [23] proposed optimal use of anti-viral treatment for individuals taking into consideration both indirect and direct treatment costs.

The Centers for Disease Control and Prevention (CDC) created a policy review of social distancing measures for pandemic influenza in non-healthcare settings [24]. They identified measures to reduce community influenza transmission such as isolating the sick, tracking contacts, quarantining exposed people, closing down schools, changing workplace habits, avoiding crowds, and restricting movement. The impact of several of these (and wearing masks) was studied in [25], where authors used agent-based modelling for the pandemic, simulating actions of people, businesses and the government. Other researchers demonstrated that early school and workplace closures and the restriction of international travel are independently associated with reduced national COVID-19 mortality [26]. On the other hand, lock-down procedures could have a devastating impact on the economy. This was studied in [27] with a modified SIR model and time-dependent infection rate. The authors found that, surprisingly, in spite of the economic cost of the loss of workforce and incurred medical expenses, the optimum point for the entire course of the pandemic is to keep the strict lock-down as long as possible. Finally, the authors in [28] designed and analysed a multi-level game-theoretic model of hierarchical policy-making, inspired by policy responses to the COVID-19 pandemic. Taking the step towards making such policies explicit, the same authors have developed a novel class of games and their respective analytic solution framework in [29].

As detailed above, related work has mostly studied narrowly focused specifics of epidemic modelling such as the intricate behaviour of individuals in relation to vaccines, or the preferred actions of mechanism designers such as healthcare system operators. In contrast, our work takes a step back, and focuses on the big picture: we model decision situations during a pandemic as games with rational participants and promote the proper design of these games. We highlight the responsibility of mechanism designers such as national authorities in constructing these games properly with adequately chosen policies, taking into account their interdependent nature.

9.3 The distancing game

One of the most important concepts which got widespread due to the ongoing COVID-19 pandemic is social distancing. By definition it is a set of non-pharmaceutical interventions or measures intended to prevent the spread of a contagious disease, hence it is the first line of defence against SARS-CoV-2. Such measures influence the self-determination of individuals, restricting the freedom of mobility, minimizing social interactions outside ones' household, and potentially, threatening the livelihood of (dominantly) low-income families. Therefore, it is imperative to understand why people do or do not comply with social distancing measures, especially when strict enforcement is a (prohibitively) costly option. We study the incentives underlying the (non-)compliance of individuals via game-theoretical models. To improve readability, we summarize all corresponding parameters and variables in Table 9.1.

Table 9.1   Parameters of the distancing games

Variable	Meaning
C	Cost of staying home
B	Benefit of going out
m	Mortality rate
L	Value of Life
ρ	Probability of infection
t	Time duration

Table 9.2   Payoff matrix of the distancing game

	go	stay
go	$[B - \rho \cdot m \cdot L, B - \rho \cdot m \cdot L]$	$[-\rho \cdot m \cdot L - C, -C]$
stay	$[-C, -\rho \cdot m \cdot L - C]$	$[-C, -C]$

9.3.1   Basic distancing game

We represent the cost of going out with $\rho \cdot m \cdot L$, i.e., the probability of getting infected (i.e., the infection rate) multiplied with the mortality rate of the disease and with the player's evaluation about her own life.[†] Besides the risk of getting infected, going out and attending a meeting could benefit the player, denoted as B. In parallel to the benefit of a meeting, there is also a cost for staying home or missing a meeting, denoted as C. Of course, there are other alternative ways to capture the benefits and the cost of social distancing [30], but this simple utility function suffices for demonstration purposes.

Definition 9.1. *The Distancing Game is a tuple* $\langle \mathcal{N}, \Sigma, \mathcal{U} \rangle$, *where the set of players is* $\mathcal{N} = \{1, 2\}$, *and their actions are* $\Sigma = \{go, stay\}$. *The utility functions* $\mathcal{U} = \{u_1, u_2\}$ *are presented as a payoff matrix in Table 9.2.*

Theorem 9.1. *A trivial Nash Equilibrium of the distancing game is* **(stay, stay)**, *On the other hand* **(go, go)** *is also a NE if* $\rho \cdot m \cdot L < B + C$. *If this condition holds than* **(go, go)** *is also the social optimum, otherwise it is* **(stay, stay)**.

Proof. The strategy vector **(stay, stay)** is clearly a NE since no player have incentive to deviate from it (because $-C > -\rho \cdot m \cdot L - C$). **(stay, stay)** could be a NE as well if the same is true, however, the corresponding condition does not hold trivially except when $-C < B - \rho \cdot m \cdot L$ which is equivalent with $\rho \cdot m \cdot L < B + C$. This inequality (if true) also implies that the total payoff is greatest at **(go, go)**, but if it is false than **(stay, stay)** is the SO.   □

[†]This is an optimistic approximation, as besides dying, the infection could impose other tolls on a player.

Above we focused on analysing the pure-strategy Nash Equilibrium (e.g. either **use** or **no**), however, it is possible that the game also has mixed-strategy Nash Equilibria [i.e., **go** with probability φ and **stay** with probability $(1 - \varphi)$], which may lead to a utility increase [31]. These randomized strategies could also be easily calculated; we leave these calculations to the interested readers. We also defined this game as symmetric, however, one can easily adapt the analysis to an asymmetric payoff structure (i.e., B_1, B_2, C_1 and C_2, instead of B and C).

Example 9.1. For instance, should a rational American citizen (e.g., Alice) meet Bob based on how much they value their lives? We estimate[‡] $m = 0.0225$, because $0.021 \approx \frac{\#\{\text{deceased}\}}{\#\{\text{all cases}\}} < m < \frac{\#\{\text{deceased}\}}{\#\{\text{closed cases}\}} \approx 0.024$ and $\rho = 0.0025$ as $\rho \approx \frac{\#\{\text{active cases}\}}{\#\{\text{population}\}} \approx 0.0025$.

Using these values, Alice should go out only if she values her lifeless than $17777 \left(= \frac{1}{0.0225 \cdot 0.0025}\right)$ times the sum of the benefit of the meeting and the loss of missing out. According to [32], the value of a statistical life in the US was 9.2 million USD in 2013, which is equivalent to 11.7 million USD in 2021 (adjusted for inflation with a 0.3% rate). This means that Alice should only meet someone if the benefit of the meeting plus the cost of missing it would amount to more than 658 USD $\left(= \frac{11.7M}{17777}\right)$.

9.3.2 Extended distancing game

One way to improve the above model is by introducing meeting duration.[§] Leaving our disinfected home during a pandemic is risky, and this risk grows with the time spent in a crowd. In the original model, we captured the infection probability with $\rho = 1 - (1 - \rho)$. This ratio increases to $1 - (1 - \rho)^t$ for time t. We leave the interpretation of unit time to the reader. Moreover, the benefit of attending a meeting should depend on this new parameter, as well as the cost of isolation. For instance, staying home for a longer period might cause anxiety, which could get worse over time (i.e., increasing the cost) [33]; on the other hand, spending longer quality time with someone could significantly boost the experience (i.e., increase the benefit).

Definition 9.2. *The Extended Distancing Game is a tuple* $\langle \mathcal{N}, \Sigma, \mathcal{U} \rangle$, *where the set of players is* $\mathcal{N} = \{1, 2\}$ *and their actions are* $\Sigma = \{\textbf{go}, \textbf{stay}\}$. *The utility functions* $\mathcal{U} = \{u_1, u_2\}$ *are presented in (9.1).*

$$u(\textbf{stay}) = -C(t)$$

$$u(\textbf{go}) = \begin{cases} B(t) - (1 - (1 - \rho)^t) \cdot m \cdot L & \textit{if other plays \textbf{go}} \\ -(1 - (1 - \rho)^t) \cdot m \cdot L - C(t) & \textit{if other plays \textbf{stay}} \end{cases} \qquad (9.1)$$

A direct consequence of this extension is that the structure of the Distancing Game remained unchanged, hence, the two games share the same NEs.

[‡] Data from https://www.worldometers.info/coronavirus/ (accessed 30th April, 2021)
[§] We capture the duration of the meeting equivalently to how meeting size can be modelled, hence all our arguments about meeting duration could easily be adapted to optimal meeting size.

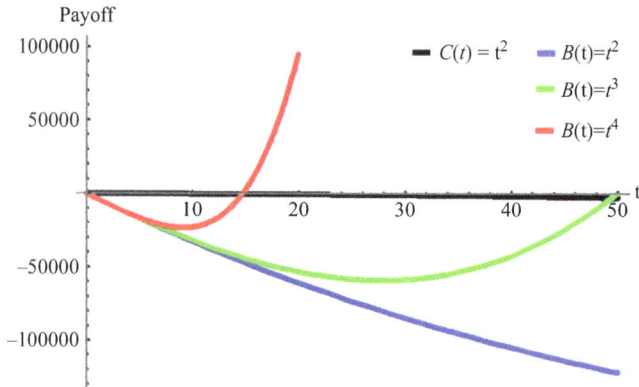

Figure 9.1   *Illustrating the payoffs of the NEs **(stay, stay)** and **(go, go)** for the extended distancing game with parameters $B = \{t^2, t^3, t^4\}$, $C(t) = t^2$, $m = 0.0225$, $\rho = 0.0025$, and $L = 11,700,000$*

Corollary 9.1. *Similarly to the basic Distancing Game, the Extended Distancing Game have the same trivial NEs **(stay, stay)** and **(go, go)** if $(1 - (1 - \rho)^t) \cdot m \cdot L < B(t) + C(t)$. If this condition holds than **(go, go)** is also the Social Optimum, otherwise it is **(stay, stay)**.*

Example 9.2. In Figure 9.1, we illustrate the payoffs for several polynomial benefit functions (e.g., $B(t) = \{t^2, t^3, t^4\}$) and a cost function $C(t) = t^2$. The rest of the parameters are defined as before, i.e., $\rho = 0.0025$, $m = 0.0225$, and $L = 11,700,000$. It is visible that the change of the cost is insignificant compared to the benefit, consequently the illustration and the reasoning would be similar if C would be linear or constant. It is visible that for small t **(stay, stay)** is the SO as the utility is higher. On the other hand, as t grows **(go, go)** becomes the SO. Another clearly visible take-away message is that the threshold of t where the SO changes is lower as the benefit is higher.

9.4   The Mask Game

Another visible effect most people has experienced during the current COVID-19 pandemic is masks: before, their usage was mostly limited to some Asian countries, hospitals, constructions, and banks (in case of a robbery). Nowadays an unprecedented spreading of mask-wearing can be seen around the globe. Policies have been implemented to enforce their usage in some places, but in general, it has been up to the individuals to decide whether to wear a mask or not, based on their own risk assessment. In this section, we model this decision situation via game theory: we introduce a simple[‖] Mask Game to be played in sequence with the previously intro-duced Distancing Game: once a player decided to meet up with friends she can decide whether to wear a mask for the meeting by playing the Mask Game. We assume that there are several types of masks, providing different levels of protection.

- **No** corresponds to the behavior of using no masks during the COVID-19 (or any) pandemic. Its cost is consequently zero; however, it does not offer any protection against the virus.
- **Out** is the most widely used mask (e.g., cloth mask or surgical mask). They are meant to protect the environment of the individual using them. They work by filtering out droplets when coughing, sneezing or simply talking, therefore they limit the spreading of the virus. They do not protect the wearer itself against an airborne virus. The cost of deciding for this protection type is noted as $C_{out} > 0$.
- **In** is the most protective prevention gear designed for medical professionals (e.g., FFP2 or FFP3 mask with valves). Valves make it easier to wear the mask for a sustained period of time, and prevent condensation inside the mask. They filter out airborne viruses while breathing in; however, the valved design means they do not filter the while air breathing out. The cost of this protection type is $C_{in} > C_{out}$.

Besides which mask they use (i.e., the available strategies), the players are either susceptible or infected, i.e., we are using a basic SI model. Being infected has some undesired effects; hence, we model it by adding a cost C_i to these players' utility (which is magnitudes higher than the cost of wearing any mask, i.e., $C_i \gg C_{in} > C_{out}$). Consequently, in the Mask Game we minimized the costs instead of maximizing the payoff as with the Distancing Game. We summarize all the parameters and variables used for the Mask Game in Table 9.3. Using these states and masks, we can present the basic game's payoffs where two players with known health status meet, and decide which mask to use.

Definition 9.3. *The basic Mask Game is a tuple* $\langle \mathcal{N}, \Sigma, \mathcal{U} \rangle$*, where the set of players is* $\mathcal{N} = \{1, 2\}$ *and their actions are* $\Sigma = \{\textbf{\textit{no}}, \textbf{\textit{in}}, \textbf{\textit{out}}\}$*. The utility functions* $\mathcal{U} = \{u_1, u_2\}$ *are presented as a cost matrix in Table 9.4. In details, Table 9.4 corresponds to the*

Table 9.3   *Parameters of the various Mask Games*

Variable	Meaning
C_{out}	Cost of playing **out**
C_{in}	Cost of playing **in**
C_i	Cost of being infected
C_{use}	Cost of playing **use**

[‖]The interested reader can follow up on the various extensions of this basic game in [1].

Table 9.4   Payoff matrices of the Mask Game

	No	Out	In
No	$[0,0]$	$[0, C_{out}]$	$[0, C_{in}]$
Out	$[C_{out}, 0]$	$[C_{out}, C_{out}]$	$[C_{out}, C_{in}]$
In	$[C_{in}, 0]$	$[C_{in}, C_{out}]$	$[C_{in}, C_{in}]$

(a) Payoff matrix when both players are susceptible

	No	Out	In
No	$[C_i, C_i]$	$[0, C_{out} + C_i]$	$[C_i, C_{in} + C_i]$
Out	$[C_{out} + C_i, C_i]$	$[C_{out}, C_{out} + C_i]$	$[C_{out} + C_i, C_{in} + C_i]$
In	$[C_{in}, C_i]$	$[C_{in}, C_{out} + C_i]$	$[C_{in}, C_{in} + C_i]$

(b) Payoff matrix when exactly one player is susceptible

case when both players are susceptible, while Table 9.4(b) corresponds to the case when one player is infected while the other is susceptible. Note that when both players are infected, the payoff matrix would be as when both are susceptible, with an additive constant cost C_i.

Theorem 9.2. *When perfect knowledge is available about the states of the players, then if both players are of the same type, both the pure strategy Nash Equilibrium and the Social Optimum of the Mask Game are **(no, no)**; while if exactly one is susceptible (e.g., player 1) then the NE is **(in, no)** and the SO is **(no, out)**.*

Proof. From Table 9.4(a) it is trivial that both players' cost is minimal when they do not use any masks, i.e., the Nash Equilibrium of the game when both players are susceptible is **(no, no)**. This is also the social optimum, meaning that the players' aggregated cost is minimal. The same holds in case both players are infected, as this only adds a constant C_i to the payoff matrix.

When only one of the players is susceptible as represented in Table 9.4(b), using no mask is a dominant strategy for the infected player[¶], since it is a best response, independently of the susceptible player's action. Consequently, the best option for the susceptible player is **in**, i.e., the NE is **(in, no)**. On the other hand, the social optimum is different: **(no, out)** would incur the least burden on the society since $C_{out} \ll C_{in}$.   □

In social optimum, susceptible players would benefit, through a positive externality, from an action that would impose a cost on infected players; therefore it is

[¶]Note that the payoffs does not take into account the legal consequences of a deliberate infection such as in https://www.theverge.com/2020/4/7/21211992/coughing-coronavirus-arrest-hiv-public-health-safety-crime-spread.

not a likely outcome. In fact, such a setting is common in man-made distributed systems, especially in the context of cybersecurity. A well-fitting parallel is defence against Distributed Denial of Service Attacks (DDoS) attacks [34]: although it would be much more efficient to filter malicious traffic at the source (i.e., **out**), Internet Service Providers rather filter at the target (i.e., **in**) owing to a rational fear of free-riding by others.

9.5 Vaccination models

The most recent virus spreading prevention mechanism against the COVID-19 is vaccination. Since researching and developing a vaccine takes time, it could not be utilized as rapidly as the rest of the techniques detailed in this work (e.g., social distancing and masks). On the other hand, this protection mechanism is considered to be the most efficient and has proven its strength several times in the past [35]. Concerning the rapidly developed COVID-19 vaccines, most governments and international organizations agree that all vaccines are safe to use and protect (to an extent) against COVID-19 for the general population. Yet, there are various aspects in which these vaccines differ, so individuals could have preferences.

Here, we introduce several optimization models, where—in contrast to multi-player games—the utility of an individual does not depend on other players' actions.**

The decision we model originates from the choice among multiple specific vaccines. Instead of focusing on whether to be vaccinated or not, as several previous works [21–23] did, we compare two hypothetical vaccines, differing along 6 different dimensions as summarized in Table 9.5. *Technology* refers to the working mechanism of the vaccine (e.g., using dead/weakened virus, mRNS, etc) [36]. *Availability* means the point in time when the vaccines are at the actual disposal of individual decision-makers. It is reasonably expected that vaccines based on traditional technologies could be mass manufactured and transported with ease, while vaccines based on new technologies could be delayed for many reasons [37]. A similar difference corresponds to the potential *side-effect*: vaccines based on older technologies were utilized in the past around the globe, hence the rare side-effects are either known or nonexisting. On the other hand, side-effects concerning modern vaccines are only based on tests with a limited number of participants [38]. The *efficiency* and *duration* of the vaccines

Table 9.5   *The two vaccines and their properties: technology, availability, side-effect, efficiency, duration, and usability*

	Tech	Availability	Side-effect	Efficiency	Duration	Usability
α	Old	Now	No	Low	Long	Limited
β	New	Soon	Maybe	High	Short	Wide

**The interested reader can see a game-theoretic extension of these basic decision models in [1].

(e.g., the probability of mitigating the severe consequences of an infection and the length of the response of the body triggered by the vaccine, respectively) also differ, favouring the newer technology [39]. Finally, the *usability* of a vaccine refers to the portion of individuals who could/should get it, e.g., there are vaccines which were associated with severe side effects which affects various demographic groups differently [40]. These differences between the two vaccines considered by the individuals are formalized in Table 9.6 with the corresponding cost and benefit variables.

In the following optimization models, we select 2–3 of the dimensions above, and present the utility/objective function for which the individuals optimize by selecting the vaccine with the higher payoff. We do not provide formal theorems and proofs as the results are trivial corollaries of the exact definitions.

9.5.1 Duration-efficiency decision

As defined in Table 9.6(a), we assume vaccine α provides protection for duration d_α with protection level e_α. On the other hand, we assume that Vaccine β protects for a shorter duration d_β but with a stronger protection level e_β. This is also illustrated in Figure 9.2(a).

Definition 9.4. *The duration-efficiency decision problem is a tuple $\langle \Sigma, \mathcal{U} \rangle$, where the actions are $\Sigma = \{\alpha, \beta\}$ and the corresponding utility functions $\mathcal{U} = \{U(\alpha), U(\beta)\}$ are presented in Equation (9.2):*

$$U(\alpha) = \int_0^{d_\alpha} e_\alpha \mathrm{d}t = e_\alpha \cdot d_\alpha \qquad\qquad U(\beta) = \int_0^{d_\beta} e_\beta \mathrm{d}t = e_\beta \cdot d_\beta \quad (9.2)$$

It is clear that the optimal decision depends on the exact values of e_α, e_β, d_α, and d_β: if $e_\alpha \cdot d_\alpha > e_\beta \cdot d_\beta$ then Vaccine α is the optimal choice, otherwise it is

Table 9.6   *The parameters concerning the Vaccines and the*
                  Vaccination Models

Vaccine	α	β
Protection efficiency	e_α	e_β
Effect duration (time)	d_α	d_β
Availability (from time)	0	t_0
Side-effect probability	0	ϵ
Benefit of being vaccinated	B_α	B_β

(a) Vaccine specific variables

Variable	Meaning
C_i	Cost of being infected
C_s	Cost of the side-effect
p	Vaccine preference

(b) Costs & benefits of the vaccination models

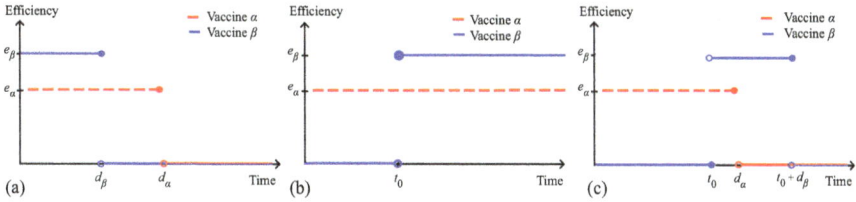

Figure 9.2 Illustration of vaccine properties: (a) Duration and efficiency of the vaccines, (b) Availability and efficiency of the vaccines and (c) Availability, duration, and efficiency of the vaccines

Vaccine β. For instance if we set $e_\alpha = 0.76$, $e_\beta = 0.95$, $d_\alpha = 49$ and $d_\beta = 35$ then $U(\alpha) \approx 37 > 33 \approx U(\beta)$.[††]

9.5.2 Availability-efficiency decision

Following Table 9.6(a), we assume vaccine α is available now (i.e., at $t = 0$), but it only provides protection level e_α. On the other hand, vaccine β will only become available at t_0, but with a stronger protection level e_β. This is also illustrated in Figure 9.2(b). Even without taking the duration into account, we have to introduce time-based discounting for the utility via the factor δ, as generally treated in the economics literature [41].

Definition 9.5. *The availability-efficiency decision problem is a tuple $\langle \Sigma, \mathcal{U} \rangle$, where the actions are $\Sigma = \{\alpha, \beta\}$, and the corresponding utility functions $\mathcal{U} = \{U(\alpha), U(\beta)\}$ are presented in (9.3):*

$$U(\alpha) = \int_0^\infty e_\alpha \cdot \delta^t \mathrm{d}t = \frac{-e_\alpha}{\log \delta} \qquad U(\beta) = \int_{t_0}^\infty e_\beta \cdot \delta^t \mathrm{d}t = \frac{-e_\beta}{\log \delta} \delta^{t_0} \quad (9.3)$$

Again, the optimal decision trivially depends on the exact values of e_α, e_β, t_0, and δ: if $e_\alpha < e_\beta \cdot \delta^{t_0}$ then Vaccine α is the optimal choice, otherwise it is Vaccine β. For instance, with $e_\alpha = 0.76$, $e_\beta = 0.95$, $t_0 = 28$, and $\delta = 0.999$, the utilities are $U(\alpha) \approx 1749$ and $U(\beta) \approx 2126$, respectively.

9.5.3 Duration-efficiency-availability decision

It is possible to combine the previous two decision models as illustrated in Figure 9.2(c).

[††]The values used through all our examples within this section are serving only illustrative purposes and do not correspond to any existing vaccines.

Definition 9.6. *The Duration-Efficiency-Availability decision problem is a tuple* $\langle \Sigma, \mathcal{U} \rangle$, *where the actions are* $\Sigma = \{\alpha, \beta\}$, *and the corresponding utility functions* $\mathcal{U} = \{U(\alpha), U(\beta)\}$ *are presented in (9.4):*

$$U(\alpha) = \int_0^{d_\alpha} e_\alpha \cdot \delta^t \, dt = (\delta^{d_\alpha} - 1) \cdot \frac{e_\alpha}{\log(\delta)}$$

$$U(\beta) = \int_{t_0}^{t_0 + d_\beta} e_\beta \cdot \delta^t \, dt = (\delta^{d_\beta} - 1) \cdot \delta^{t_0} \cdot \frac{e_\beta}{\log(\delta)} \tag{9.4}$$

The optimal decision depends on the exact values of e_α, e_β, d_α, d_β, t_0, and δ: if $\frac{e_\alpha}{e_\beta} \cdot \delta^{t_0} > \frac{\delta^{d_\alpha} - 1}{\delta^{d_\beta} - 1}$ then Vaccine α is the optimal choice, otherwise it is Vaccine β. For instance, with $e_\alpha = 0.76$, $e_\beta = 0.95$, $d_\alpha = 49$, $d_\beta = 35$, $t_0 = 28$, and $\delta = 0.999$, the utilities are $U(\alpha) \approx 84$ and $U(\beta) \approx 73$, respectively.

9.5.4   Side-effect decision

Suppose Vaccine α is based on a traditional vaccination technology, hence, it provides protection level e_α with a negligible risk of any undesired side-effect. On the other hand, Vaccine β is a product of the most advanced technological improvements, consequently, it offers a stronger protection level e_β but with a small likelihood ϵ of serious undesired consequences. In the following, instead of defining the utility purely based on the vaccine parameters as previously, we utilize explicit costs and benefits variables. B_α and B_β are the benefits of the corresponding vaccines. These might differ due to regional diversity of acceptance: one could be accepted worldwide, while the other may be accepted by only specific national authorities. Concerning the costs, we capture the cost of infection with C_i while C_s corresponds to the cost of the side-effect which occurs with probability ϵ. We assume the individual is exposed to the virus, hence non-efficient protection corresponds to infection.

Definition 9.7. *The Side-Effect decision problem is a tuple* $\langle \Sigma, \mathcal{U} \rangle$, *where the actions are* $\Sigma = \{\alpha, \beta\}$, *and the corresponding utility functions* $\mathcal{U} = \{U(\alpha), U(\beta)\}$ *are presented in (9.5):*

$$U(\alpha) = B_\alpha - (1 - e_\alpha) \cdot C_i \qquad U(\beta) = B_\beta - (1 - e_\beta) \cdot C_i - \epsilon \cdot C_s \tag{9.5}$$

The optimal decision depends on the exact values of e_α, e_β, B_α, B_β, C_i, and C_s: if $B_\beta - B_\alpha > \epsilon \cdot C_s - (e_\beta - e_\alpha) \cdot C_i$ then Vaccine α is the optimal choice, otherwise it is Vaccine β. For instance, with $e_\alpha = 0.76$, $e_\beta = 0.95$, $b_\alpha = b_\beta = 100$, $C_i = C_s = 1000$, and $\epsilon = 0.001$, the utilities are $U(\alpha) \approx -140$ and $U(\beta) \approx 49$, respectively.

9.6   Pandemic mechanism design

The three counter-COVID mechanisms (social distancing, mask wearing, vaccination) modeled above are only parts of the bigger picture. Here we analyse the impact of specific policies on data transparency, social distancing, mask wearing, testing and contact tracing, and vaccination.

9.6.1 The Government as mechanism designer

We refer to the collection (and interplay) of measures implemented by a specific government fighting the epidemic in their respective country as *mechanism*. Consequently, decisions made with regard to this mechanism constitutes *mechanism design* [7]. In its broader interpretation, mechanism design theory seeks to study mechanisms achieving a particular preferred outcome. Desirable outcomes are usually optimal either from a social aspect or maximising a different objective function of the designer.

In the context of the coronavirus pandemic, the immediate response mechanism is composed of, e.g., social distancing, wearing a mask, testing and contact tracing, among others, followed by vaccination. Note that this is not an exhaustive list: financial aid, creating extra jobs to accommodate people who have just lost their jobs, declaring a national emergency and many other conceptual vessels can be utilized as sub-mechanisms by the mechanism designer, i.e., usually, the government; we do not discuss all of these in detail. Instead, we shed light on how government policy can affect the sub-mechanisms, how sub-mechanisms can affect each other and, finally, the outcome of the mechanism itself. We illustrate the importance of mechanism design applying different policies to our three games, and adding testing and contact tracing to the mix.

9.6.2 Data quality and transparency

It is well-known that inaccurate reporting of epidemic data can potentially decrease the efficacy of forecasting, and thus, response measures [42]. A less understood aspect of the data quality problem is the deliberate distortion of such reports. While not specific to handling the COVID-19 situation, a government's decision to be fully transparent or to partially conceal information from its citizens could have a profound impact on the success of the pandemic response. It is fairly straightforward to see that if people make their individual decisions based on deliberately manipulated, coarse-grained or gappy data, the results will be suboptimal and, potentially even more detrimental, unpredictable. If there is no unanimously trusted source of information available, people's beliefs will be heterogeneous, as if they were playing different games altogether. As a simple example, take the Distancing game in Section 9.3: individuals will make their assessments whether to meet based on ρ, the probability of getting infected. If media reports on this parameter are altered or varying across different channels, people may (a) meet up when it is not in their best interest, or (b) stick to staying home even if it is no longer sensible. While the detrimental effect of data concealment seems rather indirect and hard to piece together, there exist quantitative reports aiming to shed light on such issues, e.g., on data concealment and COVID-19 mortality [43].

9.6.3 Social distancing

Within the Extended Distancing Game in Section 9.3.2, the time parameter t captures the duration of a meeting. This could have another interpretation as well, as meeting size could be captured the same way as time. Consequently, if the government imposes an upper limit T for the size of congregations, this will put a strict upper bound on the "optimal meeting size" t^*, and the resulting group size will be min (T, t^*), instantiating a decreased benefit, and, therefore, promoting staying at home.

Social distancing can be a strong measure in good hands. However, the need for individual (dis)incentives for adhering to distancing policy is clear; especially, after the novelty of the pandemic has worn out. Governments and municipalities could encourage home offices, compensate workers whose jobs would demand physical presence, promote open-air cultural activities, and educate citizens on the benefits of social distancing. Schools and universities could enforce a hybrid system, where only half of the students are present physically at the same time, with weekly (or daily) shifts. Furthermore, indoor venues, such as restaurants, movie theaters, museums, etc., could restrict their capacity to, e.g., 50% to enable proper distancing. Each of these policies, when enforced, has an effect on the outcome of distancing games presented in Section 9.3.

On the other hand, if the chosen restrictive measure is a total lock-down, both the Distancing Game and the Mask Game are rendered moot, as people are not allowed to leave their households.

9.6.4   Mask wearing

If the government declares that wearing a simple mask is mandatory in public spaces (such as shops, mass transit, etc.), it can enforce an outcome (**out, out**) that is indeed socially better than the NE. The resulting strategy profile is still not SO, but it (i) allocates costs equally among citizens; (ii) works well under the uncertainty of one's health status; and (iii) may decrease the first-order need for large-scale testing, which in turn reduces the response cost of the government. By distributing free masks, the government can reduce the effect of selfishness and, potentially, help citizens who cannot buy or afford masks owing to supply shortage or unemployment.

9.6.5   Vaccination

By far, vaccination policy is the most complicated and scrutinized among all sub-mechanisms, owing to its direct relation to control over one's own body, a pillar of human rights.

The availability of multiple, high efficacy vaccines enables governments to contain and suppress the pandemic. It is clear that, even if herd immunity is never reached, the more people are vaccinated, the less problem COVID-19 will cause in the near future. As mandatory vaccination is not feasible even in semi-democracies, the design of an efficient carrot-and-stick system is sensible. Therefore, countries have started to introduce vaccination passports [44], which give to its holders benefits over their non-vaccinated countrymen, such as attending indoor venues, mass events like concerts or football matches, and traveling internationally without continuous testing. Sensibility notwithstanding, even the vaccine passport concept is under heavy legal and ethical scrutiny. Note, that some EU countries have used many types of vaccines, including ones developed in China and/or Russia, currently not recognized by the

European Medicines Agency (EMA); citizens who had received such a vaccine are not entitled to a vaccine passport[‡‡].

As vaccines have so far been a scarce resource, government decisions on which vaccines to purchase in what quantities can be crucial. Exacerbated by incomplete trial documentation, the lack of trust between countries, being in different stages of the pandemic, and having greatly varying financial and healthcare means available, national governments have followed different strategies. In a country, where the pandemic is fairly well-contained with mild restrictive measures, playing it safe makes perfect sense.[§§] However, it is in the best interest of a country with high mortality and collapsing healthcare to grab any available, perhaps under-documented or lower efficacy vaccine in significant quantities. In the latter scenario, there might be 5–6 different types of vaccines in a national vaccination program.[‖‖]

Adding to the set of available vaccines, the proposed order of vaccination is another important control lever. Most implemented policies agree on prioritising medical staff and emergency first responders but can differ on prioritising the elderly (demographic segment with the highest risk of death/severe symptoms) or the actively working people (segment with the highest risk of transmission) [45]. Combining this aspect with the individual preference for a certain type of vaccine, the part of the population that does not want to be vaccinated, and the uncertainty of to which extent vaccines prevent transmission, realistically, the mechanism designer can only aim for an approximately optimal policy design. Adding to this, the right policy for relaxing restrictive measures as the vaccination progresses constitutes an issue of its own [45], and has an effect on all the submechanisms and games mentioned above.

9.6.6 Testing and contact tracing

It is clear that the Distancing and the Mask Games are not played in isolation: people deciding to meet up invoke the decision situation on mask wearing. On the other hand, so far we have largely ignored two other widespread pandemic response measures: testing and contact tracing.

With appropriately designed and administered coronavirus tests, medical personnel can determine two distinct features of the tested individual: (i) whether she is actively infected spreading the virus and (ii) whether she has already had the virus, even if there were no or weak symptoms. (Note that detecting these two features require different types of tests, able to show the presence of either the virus RNA or specific antibodies, respectively.) In general, testing enables both the tested person and the authorities to make more informed decisions. Putting this into the context of our games, testing reduces the uncertainty, enables the government to impose mandatory quarantine thereby removing infected players, and identifies individuals who are

---

[‡‡]https://www.schengenvisainfo.com/news/all-details-on-eu-covid-19-passport-revealed-heres-what-you-need-to-know/
[§§]https://www.fhi.no/en/id/vaccines/coronavirus-immunisation-programme/
[‖‖]https://abouthungary.hu/news-in-brief/coronavirus-heres-the-latest

temporarily immune, and thus, can be vaccinated at a later stage without imposing greater risk on them.

Even more impactful, mandatory testing (as in Wuhan[¶]) render the situation to a full information game: it serves as an exogenous "health oracle" imposing no monetary cost on the players. To sum it up, the testing sub-mechanism outputs results that serve as inputs to the distancing and Mask Games as well as to the Vaccination Decision Models.

Naturally, a "health oracle" does not exist: someone has to bear the costs of testing. From the government's perspective, mandatory mass testing is extremely expensive.[***] (Similarly, from the concerned individual's perspective, a single test might be unaffordable.) Contact tracing, whether traditional or mobile app-based, serves as an important input submechanism to testing [46]. It identifies the individuals who are *likely* affected based on spatial proximity, and informs both them and the authorities about this fact. In game-theoretic terms, for such players, the benefit of testing outweighs the cost (per capita) with high probability. From the mechanism designer's point of view, contact tracing reduces the overall testing cost by enabling *targeted testing*, potentially by orders of magnitude, without sacrificing proper control of the pandemic. Another potential cost of contact tracing for individuals could be the loss of privacy. Note that mobile OS manufacturers are working on integrating privacy-preserving contact tracing into their platform to eliminate adoption costs for installing an app.[†††]

9.6.7 The big picture

As far as pandemic response goes, the mechanism designer has the power to design and parametrize the games that citizens are playing, taking into account that submechanisms affect each other. It is vital to observe and profit from the interdependence of the submechanisms; not even a strong weapon such as social distancing can stand against the pandemic on its own. If done properly, submechanisms can strengthen each others' effectiveness, e.g., selective testing based on contract tracing can change the framing for social distancing. If done poorly or without acknowledging the interdependence, the submechanisms may undermine each other, resulting in a suboptimal pandemic response with potentially catastrophic consequences.

After (i) games have been designed and parameterized, (ii) games have been played by selfish individuals, (iii) outcomes have been determined, and (iv) the cost for the mechanism designer itself is realized (see Figure 9.3). The corresponding cost function is very complex incorporating factors from ICU beds through civil unrest and affected future election results to a drop in GDP over multiple time scales [47]. Therefore, governments have to carefully balance the—very directly interpreted—social optimum and their own costs; this indeed requires a mechanism design mindset.

[¶]New York Times. https://www.nytimes.com/2020/05/26/world/asia/coronavirus-wuhan-tests.html
[***]But not without precedence, e.g., in Slovakia (https://edition.cnn.com/world/live-news/coronavirus-pandemic-10-18-20-intl/h_beb93495fe9b83701023eafd5f28e39d)
[†††]Apple. https://covid19.apple.com/contacttracing

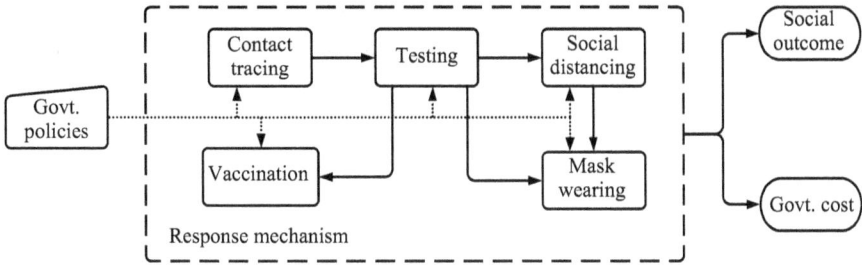

Figure 9.3   *Pandemic response mechanism as influenced by government policy (dotted lines) and the interplay of submechanisms (solid lines)*

9.7 Conclusion

In this chapter, we have made a case for treating the pandemic response as a mechanism design problem. Through simple games and decision models modeling interacting selfish individuals, we have shown that it is necessary to take individual incentives into account during a pandemic. First, we have shown how individual decisions (and, therefore, social impact) concerning social distancing depend on the perceived benefits of meeting up and the cost of missing out. Second, we have shown how individual incentives impact mask wearing and illustrated how individuals could optimize when selecting between two hypothetical vaccines taking into account availability, efficacy and duration of immunity.

We have also reported a significant impact of specific government policies on these games performance, and collaboration possibilities among various response measures (submechanisms). We demonstrated these effects via the influence that contact tracing can have to improve the certainty of individual decision-making (social distancing and wearing masks) via targeted testing. Furthermore, we have discussed the notoriously complex nature of the vaccination policy; designing such in an even approximately optimal way has to take into account medical, behavioral, economic and legal factors. We have also argued that sharing high-quality and truthful pandemic data with the public promotes better individual decision-making, and thus, more efficient handling of the pandemic. Governments have significantly more power than traditional mechanism designers in distributed systems; therefore, it is even more crucial for them to carefully study the trade-off between social good and the cost of the designer when implementing their pandemic response mechanism.

9.7.1 Limitations

The work presented here has several limitations from a policy-making standpoint. First, although the mechanism designer can directly influence the payoff functions and thus the outcome of the games presented (e.g., by imposing fines on non-compliant citizens or giving benefits to the vaccinated), and the factors currently used in the payoffs—without doubt—do play a part in individual decisions making, the utility functions themselves are—of course—simplified: behavioral decision-making aspects are out of scope for this chapter. Second, at this level of abstraction, the games and their respective designs cannot form a practical guidebook for governments. In fact, complex simulation studies and the analysis of already existing real historical data have to be undertaken in order to make real-world decisions affecting human lives. The objective of this study is to *illustrate the impact of individual decision-making* on social distancing and other common pandemic measures, and *advocate for a mechanism design mindset* for policy-makers.

9.7.2   Future work

The current models mostly focus on demonstrative performance based on simplified assumptions, which are not sufficient to deal with complex pandemic situations. However, these starting points open up several promising research directions. One possibility is to take into account temporal aspects together with epidemic models. Social graphs are then be used to model them as multi-agent games. Therefore, any future pandemic and be beneficial by importing real data, e.g., user mobility, policy changes, into the parameterized models. Another direction is to tolerate the rational decision-making aspect in the games by considering other-regarding preferences, risk-taking and behavioral modeling. Moreover, a formal treatment of the mechanism design problem constitutes important future work, incorporating hierarchical designers (WHO, EU, nations, municipality, household), an elaborate cost model, and analyzing optimal policies for different time horizons. Finally, special attention should be given to sustainable pandemic response measures, such as milder forms of social distancing, which could be used for prolonged times as COVID-19 seems to be staying with us for years to come. These aspects are expected to enable an extensible mechanism design framework which is helpful in future pandemic treatment.

References

[1]   Balázs Pejó and Gergely Biczók. Games in the time of COVID-19: promoting mechanism design for pandemic response. *arXiv preprint arXiv:2106.12329*; 2021.

[2]   Balázs Pejó and Gergely Biczók. Corona games: masks, social distancing and mechanism design. In *Proceedings of the 1st ACM SIGSPATIAL International Workshop on Modeling and Understanding the Spread of COVID-19*, 2020, pp 24–31.

[3]   John C Harsanyi, Reinhard Selten *et al.* A general theory of equilibrium selection in games. *MIT Press Books*, 1988.

[4]   Dan Ariely and Simon Jones. *Predictably irrational*. Harper Audio New York, NY, 2008.

[5]   Elias Koutsoupias and Christos Papadimitriou. Worst-case equilibria. In *Annual Symposium on Theoretical Aspects of Computer Science*, pp. 404–13. Springer, 1999.

[6]   Elliot Anshelevich, Anirban Dasgupta, Jon Kleinberg, Eva Tardos, Tom Wexler, and Tim Roughgarden. The price of stability for network design with fair cost allocation. *SIAM Journal on Computing*, 2008;**38**(4):1602–23, 2008.

[7]   Andreu Mas-Colell, Michael Dennis Whinston, Jerry R Green, et al. *Microeconomic theory*, vol. 1. Oxford University Press New York; 1995.

[8]   Dinesh Kumar Rajendran, Varthini Rajagopal, S Alagumanian, T Santhosh Kumar, SP Sathiya Prabhakaran, and Dharun Kasilingam. Systematic literature review on novel corona virus SARS-COV-2: a threat to human era. *VirusDisease*, 2020;**31**(2):161–73.

[9]   Praveen Kumar Khosla, Mamta Mittal, Dolly Sharma, and Lalit Mohan Goyal. *Predictive and Preventive Measures for Covid-19 Pandemic*. Springer, 2021.

[10] Cong T Nguyen, Yuris Mulya Saputra, Nguyen Van Huynh *et al*. A comprehensive survey of enabling and emerging technologies for social distancing – part i: fundamentals and enabling technologies. *IEEE Access*, 2020;**8**:153479–507.

[11] Cong T Nguyen, Yuris Mulya Saputra, Nguyen Van Huynh *et al*. A comprehensive survey of enabling and emerging technologies for social distancing — part ii: emerging technologies and open issues. *IEEE Access*, 2020;**8**:154209–36.

[12] Sheryl L Chang, Mahendra Piraveenan, Philippa Pattison, and Mikhail Prokopenko. Game theoretic modelling of infectious disease dynamics and intervention methods: a mini-review. *arXiv preprint arXiv:1901.04143*, 2019.

[13] Yunhan Huang and Quanyan Zhu. Game-theoretic frameworks for epidemic spreading and human decision making: a review. *arXiv preprint arXiv:2106.00214*, 2021.

[14] Piero Poletti, Marco Ajelli, and Stefano Merler. Risk perception and effectiveness of uncoordinated behavioral responses in an emerging epidemic. *Mathematical Biosciences*, 2012;**238**(2):80–9.

[15] Martin Brüne and Daniel R Wilson. Evolutionary perspectives on human behavior during the coronavirus pandemic: insights from game theory. *Evolution, Medicine, and Public Health*, 2020;**2020**(1):181–6.

[16] Shi Zhao, Chris T Bauch, and Daihai He. Strategic decision making about travel during disease outbreaks: a game theoretical approach. *Journal of The Royal Society Interface*, 2018;**15**(146):20180515.

[17] Murat Özkaya and Burhaneddin İzgi. Effects of the quarantine on the individuals' risk of Covid-19 infection: game theoretical approach. *Alexandria Engineering Journal*, 2021;**60**(4):4157–65.

[18] Anupam Kumar Bairagi, Mehedi Masud, Do Hyeon Kim *et al*. Controlling the outbreak of Covid-19: a noncooperative game perspective. *arXiv preprint arXiv:2007.13305*;2020.

[19] Deepesh Kumar Lall, Garima Shakya, and Swaprava Nath. Prior-free strategic multiagent scheduling with focus on social distancing. *arXiv preprint arXiv:2104.11884*; 2021.

[20] Wongyeong Choi and Eunha Shim. Optimal strategies for vaccination and social distancing in a game-theoretic epidemiologic model. *Journal of Theoretical Biology*, 2020;**505**:110422.

[21]   Samit Bhattacharyya and Chris T Bauch. "wait and see" vaccinating behaviour during a pandemic: a game theoretic analysis. *Vaccine*, 2011;**29**(33):5519–25.

[22]   Chris T Bauch and David JD Earn. Vaccination and the theory of games. *Proceedings of the National Academy of Sciences*, 2004;**101**(36):13391–4.

[23]   Michiel van Boven, Don Klinkenberg, Ido Pen, Franz J Weissing, and Hans Heesterbeek. Self-interest versus group-interest in antiviral control. *PLoS One*, 2008;**3**(2):e1558.

[24]   Min W Fong, Huizhi Gao, Jessica Y Wong et al. Nonpharmaceutical measures for pandemic influenza in nonhealthcare Settings—social distancing measures. *Emerging Infectious Diseases*, 2020;**26**(5):976.

[25]   Petrônio CL Silva, Paulo VC Batista, Hélder S Lima, Marcos A Alves, Frederico G Guimarães, and Rodrigo CP Silva. Covid-abs: an agent-based model of Covid-19 epidemic to simulate health and economic effects of social distancing interventions. *Chaos, Solitons & Fractals*, 2020, p. 110088.

[26]   Dimitris I Papadopoulos, Ivo Donkov, Konstantinos Charitopoulos, and Samuel Bishara. The impact of lockdown measures on Covid-19: a worldwide comparison. *medRxiv*; 2020.

[27]   Sung-Po Chao. Simplified model on the timing of easing the lockdown. *arXiv preprint arXiv:2007.14072*; 2020.

[28]   Feiran Jia, Aditya Mate, Zun Li *et al* A game-theoretic approach for hierarchical policy-making. *arXiv preprint arXiv:2102.10646*; 2021.

[29]   Zun Li, Feiran Jia, Aditya Mate *et al*.
Solving structured hierarchical games using differential backward induction. *CoRR*, abs/2106.04663; 2021.

[30]   Linda Thunström, Stephen C Newbold, David Finnoff, Madison Ashworth, and Jason F Shogren. The benefits and costs of using social distancing to flatten the curve for Covid-19. *Journal of Benefit-Cost Analysis*, 2020;**11**(2): 179–95.

[31]   Martin J. Osborne. *An introduction to game theory*. Oxford University Press, New York, NY; 2004.

[32]   Polly Trottenberg and RS Rivkin. Guidance on treatment of the economic value of a statistical life in US department of transportation analyses. *Revised departmental guidance, US Department of Transportation*; 2013.

[33]   Ashwin Venkatesh and Shantal Edirappuli. Social distancing in Covid-19: what are the mental health implications? *BMJ*, 2020;bf369.

[34]   MHR Khouzani, Soumya Sen, and Ness B Shroff. Incentive analysis of bidirectional threat filtering in the internet. In *Workshop on Economics of Information Security*. Citeseer; 2013.

[35]   Stanley A Plotkin. Vaccines: past, present and future. *Nature Medicine*, 2005;**11**(4):S5–11.

[36]   T Thanh Le, Zacharias Andreadakis, Arun Kumar *et al*. The Covid-19 vaccine development landscape. *Nature Reviews Drug Discovery*, 2020**19**(5):305–6.

[37]   Roxanne Khamsi. If a coronavirus vaccine arrives, can the world make enough. *Nature*, 2020;**580**(7805):578–80.

[38]   Søren Dinesen Østergaard, Morten Schmidt, Erzsébet Horváth-Puhó, Reimar Wernich Thomsen, and Henrik Toft Sørensen. Thromboembolism and

the oxford – astrazeneca Covid-19 vaccine: side-effect or coincidence? *The Lancet*, 2021.

[39] Jia Wei, Nicole Stoesser, Philippa C Matthews *et al*. The impact of SARS-COV-2 vaccines on antibody responses in the general population in the United Kingdom. *medRxiv*, 2021.

[40] Elisabeth Mahase. Astrazeneca vaccine: Blood clots are "extremely rare" and benefits outweigh risks, regulators conclude. *BMJ*, 2021;**373**:n931.

[41] Shane Frederick, George Loewenstein, and Ted O'donoghue. Time discounting and time preference: a critical review. *Journal of Economic Literature*, 2002;**40**(2):351–401.

[42] Hyokyoung G Hong and Yi Li. Estimation of time-varying reproduction numbers underlying epidemiological processes: a new statistical tool for the Covid-19 pandemic. *PLoS One*, 2020;**15**(7):e0236464.

[43] Istvan Janos Toth. Data concealment and mortality in Covid-19 pandemic. an illustrative figure and a hypothesis. Technical report, Corruption Research Center Budapest, April 2021. https://bit.ly/2S5cNG4.

[44] Alexandra L Phelan. Covid-19 immunity passports and vaccination certificates: scientific, equitable, and legal challenges. *The Lancet*, 2020; **395**(10237): 1595–8.

[45] Xia Wang, Hulin Wu, and Sanyi Tang. Assessing age-specific vaccination strategies and post-vaccination reopening policies for Covid-19 control using seir modeling approach. *medRxiv*; 2021.

[46] Luca Ferretti, Chris Wymant, Michelle Kendall, Lele Zhao, Anel Nurtay, Lucie Abeler-Dörner *et al*. Quantifying sars-cov-2 transmission suggests epidemic control with digital contact tracing. *Science*, 2020;**368**(6491).

[47] M McDonald, KB Scott, WJ Edmunds, P Beutels, and RD Smith. The macroeconomic costs of a global influenza pandemic. In *Global Trade Analysis Project 11th Annual Conference on Global Economic Analysis, Future of Global Economy, Helsinki, June*, 2008.

Chapter 10

Open issues and future research directions

Ngoc Tan Nguyen[1], Thang Xuan Vu[2], Dinh Thai Hoang[1],
Diep N. Nguyen[1] and Eryk Dutkiewicz[1]

Vaccination is considered as the most effective solution to fight against the COVID-19 epidemic as well as other contagious and infectious diseases to bring the world to a "new normal" lifestyle. This lifestyle is defined as a new way of living our work, routines, and interactions with other people to adapt with COVID-19 [1]. With the ambition to open up the economy, many countries such as United Arab Emirates, Portugal, and Singapore have achieved the coverage rate of COVID-19 vaccines for the 2nd dose above 80% [2]. However, when the vaccine has not been evenly distributed to all countries worldwide, it means that COVID-19 cannot be ended. This is because fully vaccinated people can still be positive with COVID-19, and the effectiveness of the vaccine also decreases significantly after 6 months. Therefore, protective measures like social distancing, wearing mask, and frequent handwashing must also be practiced simultaneously to enable the "new normal" lifestyle. In this chapter, we discuss the open issues of social distancing implementation such as pandemic mode, hybrid technology solutions, security and privacy concerns, social distancing encouragement, real-time scheduling, and negative effects. Furthermore, potential solutions to these issues are also discussed.

10.1 Pandemic mode using social distancing

To avoid the crisis that occurs when a pandemic breaks out, as well as to reduce the mortality rate and mitigate economic impacts, we need a set of rules/protocols for digital infrastructures to respond, i.e., pandemic mode. The ongoing COVID-19 pandemic has driven the mobile service providers, e.g., Google and Apple, to build up a pandemic mode application for smartphones. This application represents a comprehensive framework utilizing the current pandemic situation, i.e., infected movement data, to help the mobile users stay aware of the contagious diseases and perform cautious actions to slow down the spread of the diseases through implementing social

[1]School of Electrical and Data Engineering, University of Technology, Sydney, Australia
[2]Interdisciplinary Center for Security, Reliability and Trust, University of Luxembourg, Luxembourg

distancing. To this end, like the disaster mode applications in mobile phones [3], a built-in pandemic mode application on smart devices can be the very first step, yet crucial in early responding to future pandemics. When a contagious disease outbreak is imminent, the government can first broadcast an urgent notification to activate the official pandemic mode to national digital infrastructures (e.g., databases, communications systems, intelligent transportation systems, including users' mobile phones, etc.). Then, based on the current infected movement data, e.g., the current reported number of infected people and currently infected areas, from the government officials, the service providers can determine the risk levels of the pandemic and activate or prepare suitable responding measures. As an example, considering the risk level, the smartphones can leverage the existing sensors and wireless connections to perform effective contact tracing activity for contagious disease containment.

10.1.1   Infected movement data

To determine the risk levels of the pandemic mode, the authorities first need to monitor the current infected movement information, i.e., infected areas and the number of infected people. Based on this observation, the authorities then can orchestrate the pandemic mode risk levels and notify mobile users so that they can avoid the areas where the highly likely infection exists according to the current risk level. In [4], the authors introduce an identification framework to observe the spatial infection spread based on the arrival records of infectious cases in subpopulation areas. Considering susceptible and infectious people movement in metapopulation networks, the framework first splits the whole infection spread into disjoint subpopulation areas. Then, a maximum likelihood estimation is applied to predict the most likely invasion pathways at each subpopulation area. Using a dynamic programming-based algorithm, the framework can finally reconstruct the whole spread by iteratively assembling the invasion pathways for each subpopulation to produce the final invasion pathways. Then, the authors in [5] present a spatial–temporal technique to locate real-time influenza epidemics utilizing heterogeneous data from the Internet. In particular, the technique constructs a multivariate hidden Markov model through aggregating influenza morbidity data, influenza-related data from Google, and international air transportation data. This aims to identify the spatial–temporal relationship of influenza transmission which will be used for surveillance application. Through experimental results, the technique can predict an influenza epidemic ahead of the actual event with high accuracy. Recently, Google and Apple also create a framework to demonstrate the community mobility trend with respect to the COVID-19 outbreak [6–8]. In particular, this framework is generated based on the regions of mobile users and changes in visits monitoring at various public places, e.g., groceries, pharmacies, parks, transit stations, workplaces, and residential areas.

Motivated by the above works, the authorities can first collect the spatio-temporal infectious disease-related information from the Internet and official reports. Using the aforementioned methods, the authorities can then extract meaningful information about the spread locations/pathways and time of the infectious diseases, which lead to various spatio-temporal disease spread levels. Based on these disease spread levels,

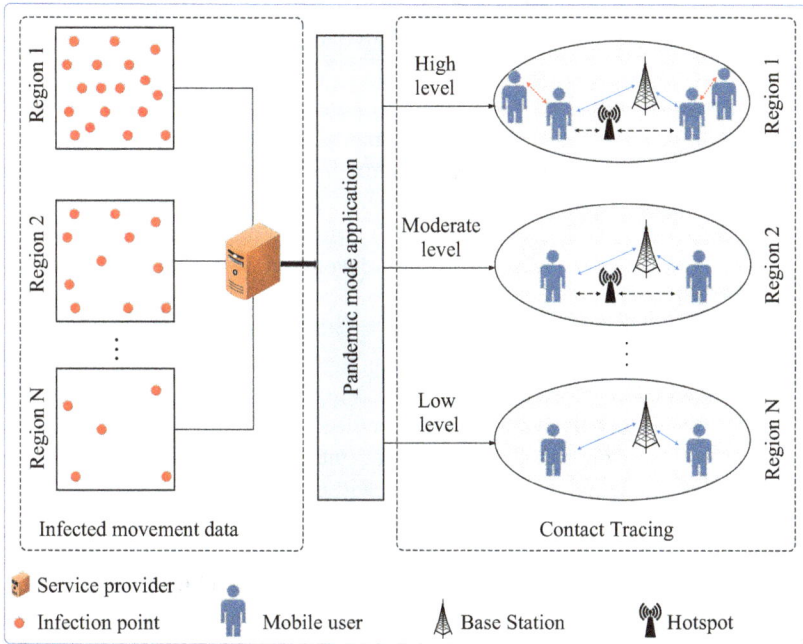

Figure 10.1 Pandemic mode in future infrastructures to support social distancing. In the infected movement data scenario, the locations of infected cases can be used to determine the pandemic mode risk level of a region. Based on this risk level, the authorities can allow different types of technologies for contact tracing, e.g., low-risk regions only use cellular for contact tracing, whereas high-risk regions can utilize cellular, Wi-Fi, and Bluetooth for contact tracing

the authorities can customize the pandemic mode risk level for different regions, e.g., states, cities, and provinces, at different times. For example, if the disease spread level, e.g., the density of infected people, at a particular city is high, the authorities can set the pandemic mode into a high-risk level for a week (as shown in Figure 10.1). Otherwise, the pandemic mode level can be set at a low-risk level.

10.1.2 Contact tracing

After determining the risk levels of the pandemic mode based on the infected movement data, the authorities can broadcast the risk level notification through smartphones' pandemic mode application. Afterward, the smartphones can perform contact tracing to help quickly discovering infected people for efficient outbreak containment [9]. Based on the risk level of the pandemic mode, the smartphones can automatically trace contacts using certain sensors and wireless connections. For example, Google and Apple currently collaborate to develop a contact tracing application utilizing Bluetooth technology, aiming to quickly detect past contacts among mobile users in close proximity [10]. In this case, the Bluetooth is used to exchange beacon signals containing unique keys between two smartphones prior to storing these keys to the cloud server for infected people notification. In another work, an epidemiological data collection scheme utilizing users' smartphones is described in [11]. Specifically, a user's smartphone can be used as a sensor platform to collect high accurate information including the user's location, activity level, and contact history between the user and certain locations.

Inspired by the aforementioned works, smartphones can be utilized as crucial tools to implement contact tracing considering the current risk level of the pandemic mode activated by the authorities (as illustrated in Figure 10.1). In particular, if the authorities activate low-risk levels, i.e., the current number of infected people and areas are small, smartphones can trace close contacts using cellular networks only. In this case, the pandemic mode application will disable certain sensors, Bluetooth, and Wi-Fi by default. However, if the high-risk level pandemic mode, i.e., the current number of infected people and areas are large, is activated, the pandemic mode application will enable all of the wireless connections including Bluetooth, Wi-Fi, and cellular network, as well as relevant sensors automatically to trace contacts faster.

Besides smartphone's built-in sensors, wearable sensors such as physiological (e.g., respiration rate, body temperature, etc.), audio, video, and inertial sensors, as well as wearable devices (GoPro, smartwatch), can all provide meaningful information for contact tracing. For example, when two persons wearing body sensors networks (BSNs), i.e., sets of wearable sensors attached to the body, making contact with each other, a collaborative BSNs system can be utilized to extract information from the contact. In [12], a framework for computing and data fusion from multiple sensors of different BSNs is proposed. To allow the collaboration between the two BSNs, the authors develop novel mechanisms including inter-BSN data communication, BSN Proximity Detection, BSN mutual service discovery and activation, inter-BSN high-level protocols, and cooperative multisensor data fusion. As a result, the framework can detect physical interactions such as handshakes between two persons. Although these wearable systems can provide meaningful and accurate data for contact tracing, they pose a threat to people's privacy. Therefore, the data from these wearable devices should only be used when a pandemic mode is in effect.

10.2   Hybrid technologies

Although the wireless and emerging technologies presented in the previous chapters provide highly accurate location information for real-life scenarios of social distancing in a certain environment, the accuracy can decrease rapidly when people move from one place to another, or are in interference environment. Therefore, it is necessary to combine available technologies to improve the accuracy of positioning in social distancing. For example, Wi-Fi signal can be interfered by surrounding Wi-Fi

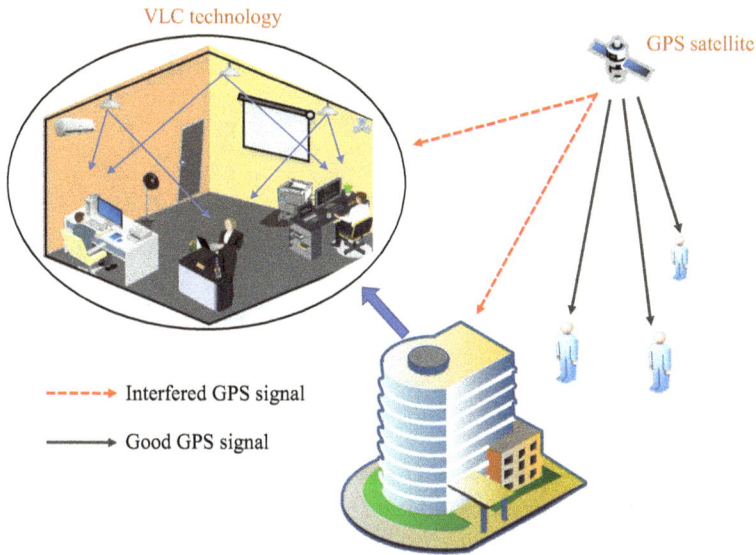

*Figure 10.2   Auto-switching between wireless technologies for different
environments. For example, when people move from outside into the
building, the GPS accuracy can be decreased dramatically. Thus, the
VLC technology can be used to maintain localization accuracy*

networks, or prevented physical obstacles, that reduces the accuracy of Wi-Fi based
localization systems. To overcome this, we can combine Wi-Fi based solution with
other proximity technologies like ultra-wideband and Bluetooth (if they are available).
Another example is illustrated in Figure 10.2 when people go from outside (using GPS
for localization) to inside of a building where GPS may not provide good positioning
accuracy. Instead, the Visible Light Communications (VLC) technology is utilized to
enhance the localization performance in indoor scenarios. On the other hand, for the
social distancing monitoring application in public places (e.g., supermarkets), computer vision approach may not be feasible in hidden corners behind shelves. Then, we
can use other technologies such as visible light communication, or cellular networks
to assist for public place monitoring as demonstrated in Figure 10.3.

To tackle aforementioned scenarios, policymakers need to develop a scalable,
reliable software platform, or a super app for individuals, factories/organizations, and
governments that links all available technologies at the specific location to collect the
mobility data. This allows the super app to intelligently select a positioning data source
from the highest accurate candidate at the time, or combine multiple data sources from
multiple technologies [13]. Furthermore, beyond the location tracking/monitoring,
other utilities such as healthcare assistance for infected people, navigation system with
mobility and infection information, management tools for factories/organizations,
etc., should also be integrated into the super app. Moreover, when offering social
distancing measures, governments also need to consider other factors such as the

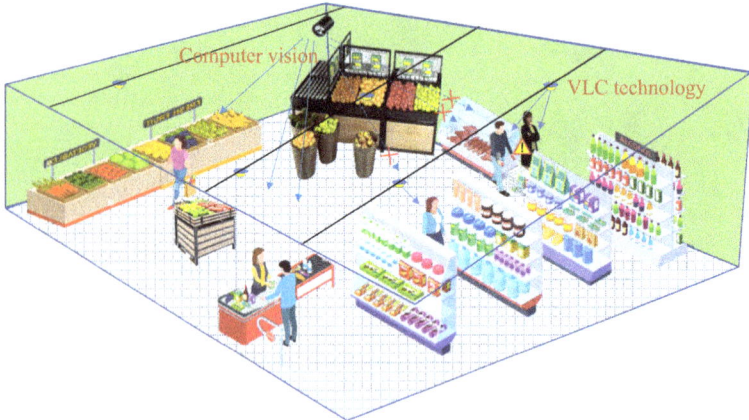

Figure 10.3   Multiple wireless technologies supporting social distancing. For example, computer vision can not monitor the area in the right side of the figure which can be supported by VLC communications

economy, the ready of health care system, and the spread of infection, together with mobility pattern [14]. These create the heterogeneity of data sources including World Health Organization (WHO), social networking, edge devices, private and public hospitals, patients, and academic institutes, which eventually enables better spatial mapping of mobility and increases the number of contexts [15].

With a huge voluminous amount of aforementioned complex data, an effective big data analytics mechanism is needed to manage this data proficiently. First, although positioning data collected from different sources is rich in spatial–temporal characteristics, they must be cleaned and processed before being extracted. State-of-practice data cleaning methods are removing outliers, checking for potential consistency issues in the data (e.g., unreasonable high-speed records), detecting duplicate observations from the same device, and merging data. To guarantee the reliability and real-time for the system, a distributed approach is preferred in which end devices with recent advances in hardware such as wireless sensors, wearable devices, or mobile phones can be used to easily process these raw data. Then, only preprocessed data or extracted features are sent to cloud/edge servers for further processing. Finally, sophisticated technologies such as AI, deep learning, and blockchain, can be applied to be able to predict/monitor the spread of infection and the mobility, or provide data inputs for other services.

10.3   Security and privacy-preserving

Most aforementioned social distancing scenarios require people's private information, to a different extent, ranging from their face/appearance to location, travel records, or health condition/data. These data, if not protected properly, attract cyber attackers and can turn users into victims of financial, criminal frauds, and privacy violation [16]. Users' data like health conditions can also adversely impact people's employment opportunities or insurance policy. Given that, to enable technology-based social distancing, it is critical to develop privacy-preserving and cybersecurity solutions to ensure that users' private data are properly used and protected.

The general principle of users' privacy-preserving is to keep each individual user's sensitive information private when the available data are being publicly accessed. To do so, data privacy-preserving mechanisms including data anonymization, randomization, and aggregation can be utilized [17]. For example, Apple, Google, and Facebook have developed people mobility trend reports while preserving users' privacy during the COVID-19 outbreak. In particular, Apple utilizes random and rotating identifiers to preserve mobile users' movements privacy [8]. Meanwhile, Google aggregates and uses anonymized datasets from mobile users who turn on their location history settings in their Android smartphones. In this case, a differential privacy approach is applied by adding random noise to the location dataset with the aim to mask individual identification of a mobile user [6]. Similarly, Facebook utilizes aggregated and anonymized user mobility datasets and maps to determine the mobility trend in certain areas including the social connectedness intensity among nearby locations [7]. In addition to the Apple's, Google's, and Facebook's latest privacy-preserving implementation, in the following, we will thoroughly discuss how the latest advances in security and privacy-preserving techniques can help to facilitate social distancing without compromising users' interest/privacy.

10.3.1 Location information protection

To protect the exact location/trajectory information of participating mobile users in social distancing, some advanced location-based privacy protection methods can be adopted. Specifically, we can anonymize/randomize/obfuscate/perturb the exact location of each mobile user to avoid malicious attacks from the attackers using the following mechanisms. For example, the authors in [18] develop a privacy-preserving location-based framework to anonymize spatio-temporal trajectory datasets utilizing machine-learning-based anonymization (MLA). In this case, the framework applies the K-means machine learning algorithm to cluster the trajectories from real-world GPS datasets and ensure the K-anonymity for high-sensitive datasets. Using the K-anonymity, the framework can collect location information from K mobile users within a cloaking region, i.e., the region where the mobile users' exact locations are hidden [19]. In [20], the use of K-anonymity is extended into a continuous network location privacy anonymity, i.e., KDT-anonymity, which not only considers the average anonymity size K but also takes the average distance deviation D and the anonymity duration T into account. Leveraging those three metrics, the mobile users under realistic vehicle mobility conditions can control the changes of anonymity and distance deviation magnitudes over time. The authors of [21] propose a mutually obfuscating paths method which allows the vehicles to securely update accurate real-time location to a location-based service server in the vehicular network. In this case, the vehicles first hide their IP addresses due to the default network address translation operated by mobile Internet service providers. Then, they generate fake path segments that separate from the vehicles' actual paths to prevent the location-based service server from tracking the vehicles. Exploiting dedicated short-range communications (DSRC) among vehicles and road navigation information from the GPS, the vehicles can mutually generate made-up location updates with each other when they communicate with the location-based service server (to obtain spatio-temporal-related information).

In addition to the anonymization and obfuscating methods, randomization and perturbation are the methods that can protect user's location privacy in social distancing scenarios. In [22], a location privacy-preserving method leveraging spatio-temporal events of mobile users in continuous location-based services, e.g., office visitation, is investigated. Specifically, an ε-differential privacy is designed to protect spatio-temporal events against attackers by adding random noise to the event data. In [23], the authors present a location privacy protection mechanism using data perturbation for smart health systems in hospitals. In particular, instead of reporting the patient's real locations directly, a processing unit attached to a patient's body can adaptively produce perturbed locations, i.e., the relative change between different locations of the patient. In this case, the system considers the patient's travel directions and computes the distance between the patient's current locations and the patient's sensitive locations (i.e., patient's predefined locations which he/she does not want to reveal to anyone, e.g., patient's treating room). Using this dynamic location perturbation, the need for a trusted third party to store real locations can be removed. Leveraging the aforementioned methods, we can also prevent the service provider from accessing mobile users' and vehicles' exact locations/trajectories/paths when they implement social distancing for crowd/traffic density and movement detection. Specifically, a platoon of mobile users/vehicles in a certain area can collaborate together to mix their real locations/trajectories/paths anonymously (Figure 10.4(a)). In this way, the service provider will only obtain the aggregated location/trajectory/path information of the platoon instead of each individual's exact location/trajectory/path for its location privacy.

10.3.2   Personal identity protection

In addition to protecting mobile users' location-related information, preserving their personal identities is of importance to improve users' acceptance of the latest technologies to social distancing. Specifically, we can exchange or anonymize personal identities among trusted mobile users to avoid the attackers identifying the actual identity of each individual user. In [24], the authors develop a pseudo-identity exchanging protocol to swap/exchange identity information among mobile users when they are at the same sensitive locations, e.g., hospital and residential areas. In particular, when a mobile user receives another trusted user's identity and private key, the mobile user will verify if the encryption of another user's identity hash function and public key is equal to the encryption of the received private key. If that condition holds, the mobile user will change his/her identity with that user's identity and vice versa.

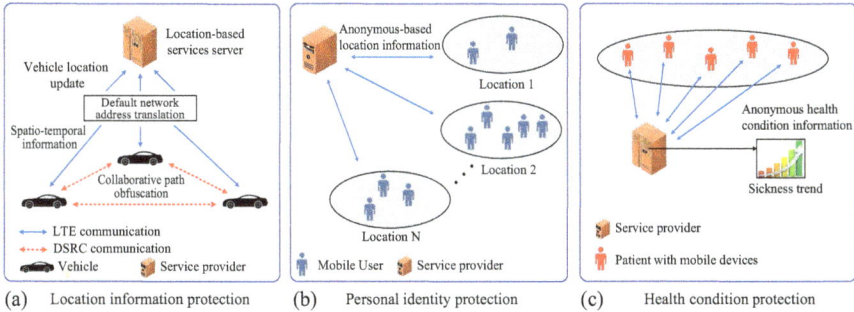

Figure 10.4   Location-based privacy preserving for social distancing scenarios. In (a) location information protection, the exact location of a vehicle can be obfuscated to protect people's privacy. To protect (b) personal identity, a user can exchange its identity with nearby trusted users in each location, and thus that user cannot be identified by the attackers. For (c) health-related information protection, the health information can be anonymized.

Another method to protect personal identity in social distancing scenarios is individual information privacy protection through indirect- or proxy-request as proposed in [25]. In particular, instead of directly submitting a request to the server, a mobile user can have his/her social friends through the available social network resources, i.e., trusted social media, to distribute his/her request anonymously to the server. The request result can be returned to his/her social friends and then forwarded to the requested mobile user, thereby preserving the requested mobile user's identity. In fact, there may exist some malicious friends who expose the identity of the mobile user. Therefore, the authors in [26] investigate a user-defined privacy-sharing framework on social networks to choose his/her particular friends who are trusted to obtain the mobile user's identity information. In this case, the mobile user only shares his/her identity information with the particular friends whose pseudonyms match the mobile user's identity through the authorized access control. Using the same approaches from the above works, we can use local wireless connections, e.g., Bluetooth and Wi-Fi Direct, to anonymously exchange actual location information in a mobile user group, i.e., between a mobile user and his/her trusted nearby mobile users, in an ad hoc way. As shown in Figure 10.4(b), when the service provider requires to collect location-related information for the current crowd density detection, a representative mobile user from the group can send the group's anonymous location information to the service provider, aiming at preserving the personal identity of each mobile user in the group.

Moreover, Apple and Google have recently introduced a key schedule for contact tracing to ensure the privacy of users [10]. Specifically, there are three types of key: (i) tracing key, (ii) daily tracing key, and (iii) rolling proximity identifier. The tracing key is a 32-byte string that is generated by using a cryptographic random number generator when the app is enabled on the device. The tracing key is securely stored on the device. The daily tracing key is generated for every 24-h window by using the SHA-256 hash function with the tracing key. The rolling proximity identifier is a privacy-preserving identifier which is sent in Bluetooth advertisements. This identifier is generated by using the SHA-256 hash function with the daily tracing key. Each time the Bluetooth MAC address is changed, the app can derive a new identifier. When a positive case is diagnosed, its daily tracing keys are uploaded to a server. This server then distributes them to the clients who use the app. Based on this information, each of the clients will be able to derive the sequence of the rolling proximity identifiers that were broadcasted from the user who tested positive. In this way, the privacy of the users can be protected because, without the daily tracing key, one cannot obtain the user's rolling proximity identifier. In addition, the server operator also cannot track the user's location or which users have been in proximity.

With outstanding performance in data integrity, decentralization, and privacy-preserving, blockchain technology can be an effective solution to preserve privacy to enable technology-based social distancing scenarios. A blockchain is a distributed database shared among users in a decentralized network. This decentralized nature of blockchain ensures its immutability property, i.e., the data stored within cannot be altered without the consensus of the majority of network users [27]. Another advantage of blockchain technology is that the users' anonymity is ensured due to the public-private keys pair mechanism [28]. As a result, blockchain technology can effectively address the personal identity issue in social distancing scenarios where people have to share their movement and location information but not their exact identities. For example, in the infected movement data scenario, we only need to know the movement path of a person, and whether or not that person is infected. In this case, the person anonymity can be ensured with the public-private keys pair mechanism, since there is no way to link the public key to that person's true identity.

10.3.3   *Health-related information protection*

To monitor the sickness trend in a certain place, e.g., the hospital, for the social distancing purpose (i.e., to inform the upcoming mobile users not to enter a high-risk area/building), the health-related condition information of visiting mobile users has to be shared to provide reliable learning dataset. To protect this highly sensitive information, the authors in [29] propose a differential privacy-based protection approach to preserve the electrocardiogram big data by utilizing body sensor networks. In particular, non-static noises are applied to produce sufficient interference along with the electrocardiogram data, thereby preventing the malicious attackers to point out the real electrocardiogram data.

To provide secure health-related information access for authenticated users, a dynamic privacy-preserving approach leveraging the biometric authentication process is introduced in [30]. Specifically, when a user wants to access the medical server containing his/her health condition, a secure biometric identification at the server for the user's validity is employed where the exact value of his/her biometric template remains unknown to the server. In this way, the personal identity of the authenticated user can be preserved. To further enhance the anonymity of his/her medical information, the random number that is used to protect the biometric template is updated after every successful login. Then, the authors in [31] propose a secure anonymous authentication model for wireless body area networks (WBANs). Specifically, this framework enables both patients and authorized medical professionals to securely and anonymously examine their legitimacies prior to exchanging biomedical information in the WBAN systems. Motivated by the above works, we can utilize mobile devices, secure service provider, and the aforementioned privacy-preserving approaches to anonymously collect people's health condition information for illness monitoring in the hospital/medical center (Figure 10.4(c)). In this way, the social distancing through monitoring the sickness trend can be implemented efficiently while preserving the sensitive information of the people in the illness areas.

10.4 Adverse effects of social distancing

Social distancing has been proven to be an effective method during the time of the COVID-19 epidemic which helps to reduce both the infectivity and mortality rates [32]. According to the age-structured compartmental Susceptible-Exposed-Infected-Recovered (SEIR) model designed by the Imperial College London COVID-19 Response Team [32], the infectivity rate can be dropped by 15% and the mortality rate can be reduced by 8.3% at least in USA when social distancing methods are practiced. However, social distancing also brings the dangers of rising social rejection, increasing impersonality and individualism, and the loss of community consciousness [33]. Socializing, which is the most fundamental human need, is prohibited during social distancing. In some extreme cases, it might cause mental health matters such as anxiety, depression, sleep disruptions, and thoughts of suicide [34], as well as impacts on the economic growth. In the following, we deeply present negative effects of social distancing on mental health and economics. We also discuss some measures to mitigate or leverage these effects to improve mental health and economics.

10.4.1 Economic impacts

According to the United Nations World Economic Situation and Prospects (WESP) mid-2020 report [35], most developed economies experienced a drop of 5% of their gross domestic product (GDP) and the global economy was reported a reduce by 3.2% which is almost 2.8 trillion U.S. dollars in 2020 due to the COVID-19 pandemic and lockdown. Clearly that social distancing measures has ceased many economic activities such as manufacturing, logistics, tourism, etc., which directly leads to a dramatic decline in occupation. A survey [36] conducted in the U.S. points out that the employment-to-population ratio witnessed a 7.5% decrease in April 6, 2020, i.e., nearly 20 million jobs lost corresponding to an adult American population.

Besides, due to the COVID-19 pandemic and social distancing measures, a fall in demand causes the largest damage for the economics. There are less consumers willing to purchase the goods and services available. This can be clearly seen in the most affected industries such as logistics and tourism. For example, in an attempt to contain the spreading of the COVID-19, the European Union closed all its external borders on March 17, 2020, following by most countries over the world. The whole year 2020 witnessed the crisis of the airline industry with more than 1 billion passengers cancelling their flights, corresponding to an enormous drop of 64.6% of global passenger traffic [37]. Similarly, the tourism industry has also been hit hard by social distancing measures, with the number of international tourist arrivals falling by 73% in 2020 and 80% in 2021, respectively [38].

Another reason for the decline in consumption spending is a noticeable degradation in household income. Economic reports show that the percentages of participants who saying a reduction in their family income due to the COVID-19 are 31% in G7 countries and 45.6% in India, respectively [39,40]. As a result, people are willing to spend less on consumption such as goods and services. A study in [41] shows that there is a dramatic decrease of 32% or 18.57 million RMB per city in daily offline consumption in China, for 12 consecutive weeks from January to April 2020. Similarly, Singapore also witnessed a decline of 22.8% in total household consumption spending in April 2020. Consumers intend to shift most of their spending to essential items, such as groceries and household supplies, while cutting back on unnecessary categories. People also tend to save money for emergencies or invest, for example, personal savings in the US has doubled at the end of 2020 compared to December 2019 [42].

However, the impacts of COVID-19 on the economy also bring great opportunities to speed up the digital transformation of manufacturing. In the time of crisis, industry 4.0 (I4.0), which is the convergence of a number of emergence technologies such as additive manufacturing, Internet of Things (IoT), blockchain, advanced robotics and artificial intelligence, is considered as the major key to improve the manufacturing performance while ensuring social distancing measures. Digital technologies enable remote work and collaboration to eliminate the need for noncritical employees to leave their homes. Computer vision and wearable technologies are also helping maintain safe distancing for manufacturing operations. Delivering the finished goods to customers also rises a challenge for producers with traditional transportation modes, e.g., last-mile contact-less delivery. To overcome this, digital and analytics solutions such as drone delivery, digital fleet management, route optimization, and carrier analytics, can increase the performance of logistics services in both demand and supply, improving real-time performance [43].

The COVID-19 pandemic also changes the consumer behavior from in-store to online shopping. Due to mobility restriction caused by social distancing measures and fear of infection, people tend to shop online instead of off-line buying. As a result, there is an unprecedented increase in e-commerce throughout the world. For example, revenue of e-commerce in U.S. was $105 billion in 2020 and the online sales was reported an increase of 32.4% from $598.02 in 2019 to $791.70 in 2020 [44]. This new customer behavior is encouraged to implement social distancing measures.

To recover the economy, governments should gradually reopen it by relaxing business closures as well as service facilities when the second dose of vaccine are covered at 70% [45], the spread of infection sustainably reduced, and public health and health care systems are ready to face future outbreaks. Together with that, at the same time, governments also need provide instructions and policies of appropriate social distancing measures and wearing mask as a prerequisite for reopening shops, restaurants, or factories.

10.4.2 Mental health effects

Due to social distancing, many fundamental human activities are restricted or banned such as working, doing exercise, or meeting friends. This leads to sudden distur-bance in people's daily life, which in turn, causes adverse effects on mental health like depression, generalized anxiety disorder (GAD), insomnia, intrusive thoughts, acute stress, etc., [46,47]. In [46], a research is conducted with 435 U.S. adults about the association of stay-at-home orders and individuals' personal distancing behavior with mental health symptoms. In March 2020, people, who stay at home, get higher depression, GAD symptoms, insomnia, and acute stress, but they do not have intrusive thoughts. Meanwhile, personal distancing behavior is linked with higher depression, GAD symptoms, intrusive thoughts, and acute stress, but not with insomnia. Similarly, authors in [47] study mental health symptoms including depression examined by the 9-item Patient Health Questionnaire (PHQ-9), anxiety measured by the 7-item Gener-alized Anxiety Disorder questionnaire (GAD-7), trauma-related symptoms assessed by the 20-item PTSD Checklist (PCL-5), and alcohol use evaluated by the 3-item AUDIT-C, in relevant to the COVID-19 pandemic and social distancing. The report shows that the percentage of participants have depressive, generalized anxiety disor-der, and trauma symptoms are 29%, 16.58%, and 5.38%, respectively. In addition, 33.59% of women and 33.08% of men are reported for alcohol abuse.

Losing job and financial difficulties are also causes of bad effects on mental health [48,49]. A recent study [48] reveals that the European labor force bears sev-eral important insights relevant between economic hardships and mental health issues including feelings of depression, loneliness, and anxiety. Specifically, people losing job and/or income during the lockdown have almost doubled the risk of feelings of depression compared to job retention. By contrast, due to social distancing, work-load has been increased which causes significant effects on health anxiety for lowest and middle occupational prestige (International Socio-Economic Index—ISEI), e.g., manual workers, clerks.

Other studies focus on the impacts of social distancing on mental health of spe-cific research subjects such as students, parents and children, etc. In [50], authors study the negative effects of closing universities and distance learning on the men-tal health of undergraduate students in Greece. The non-parametric Friedman test (with $\chi^2 = 369.051$ and degree of freedom $df = 8p < 0.001$) is used to analyze the collected data on 181 students about their emotions during lockdown. Most of respon-dents vote for concern and angry with mean ranks of 7.08 and 6.33, respectively. The study also provides an interesting analysis that male students experienced fear, panic, concern, and despair at lower level than their female counterparts, because they are more optimistic about the pandemic. The work in [49] investigates the psychological impact of financial difficulties on the mental health of college students. This study points out that students living in urban area (where can get more help and support of the society), family income stability, or living with parents, have less anxiety compared to other students. Different to mature students, children are the most vulnerable to psychological impact of the COVID-19, due to their limited understanding of the pandemic. Thus, the authors in [51] conduct a research on mental health of children and adolescents during the COVID-19 pandemic. Children are highly affected by family and community conditions (e.g., family conflict related to finances), school closure, and increasing time in front of screens.

With the above indisputable evidence about certain negative effects of social distancing on people's mental health, policymakers who are considering introducing social distancing measures to contain COVID-19 may benefit from understanding such health and wellbeing implications. In the different cities/countries, social distancing measures must be introduced gradually based on studies and evaluations on their socioeconomic and cultural aspects, the characteristics of their political and healthcare systems. Moreover, social support, i.e., receiving advice, assistance, or caring, or financial support for people losing job or low-income families, is positively associated with mental health. Thus, governments should also provide more social support as well as social media campaigns to improve mental health and reduce distress besides offering social distancing measures.

Nevertheless, recent virtual reality (VR) and augmented reality (AR) technologies enables people to live within a digital universe, i.e., metaverse which is both claimed by Facebook and Microsoft [52]. In metaverse, people can work, play, and stay connected with friends/colleagues through everything from conferences and classes to virtual trips around the world. This creates a sense of "virtual presence". We can communicate with other people virtually, these are completely new experiences of interactions that traditional communication methods like phone, instant message apps, cannot provide.

References

[1] The Star, "Covid-19: What does the 'new normal' mean?". Accessed: Nov. 19, 2021. [Online]. Available: https://www.thestar.com.my/lifestyle/health/2020/05/21/covid-19-what-does-the-039new-normal039-mean.

[2] Our World in Data, "Coronavirus (COVID-19) Vaccinations". Accessed: Nov. 19, 2021. [Online]. Available: https://ourworldindata.org/covid-vaccinations.

[3] Z. Lu, G. Cao, and T. La Porta, "TeamPhone: networking smartphones for disaster recovery" *IEEE Transactions on Mobile Computing*, vol. 16, no. 12, pp. 3554-3567, Dec. 2017.

[4] J. Wang, L. Wang, and X. Li, "Identifying spatial invasion of pandemics on metapopulation networks via anatomizing arrival history," *IEEE Transactions on Cybernetics*, vol. 46, no. 12, pp. 2782-2795, Dec. 2016.

[5] X. Zhou et al., "A spatial-temporal method to detect global influenza epidemics using heterogeneous data collected from the Internet," *IEEE/ACM Transactions on Computational Biology and Bioinformatics*, vol. 15, no. 3, pp. 802-812, Apr. 2018.

[6] Google, "Google COVID-19 community mobility reports." Google. Accessed: Apr. 20, 2020. [Online]. Available: https://www.google.com/covid19/mobility/

[7] Facebook, "Our work on COVID-19." Facebook Data For Good. Accessed: Apr. 20, 2020. [Online] Available : https://dataforgood.fb.com/docs/covid19/

[8] Apple, "Mobility trends reports." Apple. Accessed: Apr. 20, 2020. [Online]. Available: https://www.apple.com/covid19/mobility

[9] H. Chen, B. Yang, H. Pei, and J. Liu, "Next generation technology for epidemic prevention and control: data-driven contact tracking," *IEEE Access*, vol. 7, pp. 2633–2642, Jan. 2019.

[10] T. Romm, D. Harwell, E. Dwoskin and C. Timberg, "Apple, Google debut major effort to help people track if they've come in contact With Coronavirus." Washington Post, Apr. 11, 2020. Accessed: Apr. 20, 2020. [Online]. Available: https://www.washingtonpost.com/technology/2020/04/10/apple-google-tracking-coronavirus/

[11] M. S. Hashemian, K. G. Stanley, D. L. Knowles, J. Calver and N. D. Osgood, "Human network data collection in the wild: the epidemiological utility of micro-contact and location data," in *ACM SIGHIT*, Miami, FL, Jan 28-30, 2012, pp. 255–264.

[12] G. Fortino, S. Galzarano, R. Gravina and W. Li, "A framework for collaborative computing and multisensor data fusion in body sensor networks," *Information Fusion*, vol. 22, no. 1, pp. 50-70, Mar. 2015.

[13] S. Callery and K. Meehan, "Contact tracing, proximity safety, data correlation and visualization for global corporations," *IEEE 6th International Forum on Research and Technology for Society and Industry (RTSI)*, pp. 480-485, 2021.

[14] L. Zhang et al., "Interactive COVID-19 Mobility Impact and Social Distancing Analysis Platform", *Transportation Research Record.*, September 2021.

[15] A. Sadowski et al., "Big data insight on global mobility during the Covid-19 pandemic lockdown", *J Big Data*, vol. 8, no. 78, 2021.
G. Fortino, S. Galzarano, R. Gravina and W. Li, "A framework for collaborative computing and multisensor data fusion in body sensor networks," *Information Fusion*, vol. 22, no. 1, pp. 50-70, Mar. 2015.

[16] T. Menzies, E. Kocaguneli, L. Minku, F. Peters and B. Turhan, *Sharing Data and Models in Software Engineering*. Burlington, MA: Morgan Kaufmann, 2015.

[17] H. -Y. Tran and J. Hu, "Privacy-preserving big data analytics a comprehensive survey," *J. Parallel Distrib. Comput.,* vol. 134, pp. 207–218, Dec. 2019.

[18] S. Shaham, M. Ding, B. Liu, S. Dang, Z. Lin and J. Li, "Privacy preserving location data publishing: a machine learning approach," *IEEE Transactions on Knowledge and Data Engineering*, early access, doi: 10.1109/TKDE.2020.2964658.

[19] J. Y. Koh, G. W. Peters, D. Leong, I. Nevat and W. Wong, "Privacy-aware incentive mechanism for mobile crowd sensing," in *IEEE ICC*, Paris, France, May 21–25, 2017.

[20] G. P. Corser, H. Fu and A. Banihani, "Evaluating location privacy in vehic-
      ular communications and applications," *IEEE Transactions on Intelligent
      Transportation Systems*, vol. 17, no. 9, pp. 2658-2667, Sept. 2016.

[21] J. Lim, H. Yu, K. Kim, M. Kim and S. Lee, "Preserving location pri-
      vacy of connected vehicles with highly accurate location updates," *IEEE
      Communications Letters*, vol. 21, no. 3, pp. 540-543, Mar. 2017.

[22] Y. Cao, Y. Xiao, L. Xiong, L. Bai and M. Yoshikawa, "Protecting spa-
      tiotemporal event privacy in continuous location-based services," *IEEE
      Transactions on Knowledge and Data Engineering*, early access, doi:
      10.1109/TKDE.2019.2963312.

[23] I. Natgunanathan, A. Mehmood, Y. Xiang, L. Gao and S. Yu, "Location privacy
      protection in smart health care system," *IEEE Internet of Things Journal*, vol.
      6, no. 2, pp. 3055-3069, Apr. 2019.

[24] G. Sun, D. Liao, H. Li, H. Yu and V. Chang, "L2P2: A location-label
      based approach for privacy preserving in LBS," *Future Generation Computer
      Systems*, vol. 74, pp. 375-384, Sep. 2017.

[25] M. Han, et al., "Cognitive approach for location privacy protection," *IEEE
      Access*, vol. 6, pp. 13466-13477, Mar. 2018.

[26] G. Sun, Y. Xie, D. Liao, H. Yu and V. Chang, "User-defined privacy location-
      sharing system in mobile online social networks," *Journal of Network and
      Computer Applications*, vol. 86, pp. 34-45, May 2017.

[27] C. T. Nguyen, D. T. Hoang, D. N. Nguyen, D. Niyato, H. T. Nguyen, and
      E. Dutkiewicz, "Proof-of-Stake consensus mechanisms for future blockchain
      networks: Fundamentals, applications and opportunities," *IEEE Access*, vol.
      7, pp. 85727–85745, June 2019.

[28] A. Dorri, M. Steger, S. S. Kanhere and R. Jurdak "Blockchain: A dis-
      tributed solution to automotive security and privacy," *IEEE Communications
      Magazine*, vol. 55, no. 12, pp.119-125, Dec. 2017.

[29] C. Lin, Z. Song, H. Song, Y. Zhou, Y. Wang and G. Wu, "Differential privacy
      preserving in big data analytics for connected health," *Journal of Medical
      Systems*, vol. 40, no. 97, pp. 1-9, Feb. 2016.

[30] L. Zhang, Y. Zhang, S. Tang and H. Luo, "Privacy protection for e-health
      systems by means of dynamic authentication and three-factor key agreement,"
      IEEE Transactions on Industrial Electronics, vol. 65, no. 3, pp. 2795-2805,
      Mar. 2018.

[31] P. Vijayakumar, M. S. Obaidat, M. Azees, S. H. Islam, and N. Kumar,
      "Efficient and secure anonymous authentication with location privacy for
      IoT-based WBANs," *IEEE Transactions on Industrial Informatics*, vol. 16,
      no. 4, pp. 2603-2611, Apr. 2020.

[32] Z. Barnett-Howell, O. Watson, A. Mobarak, "The benefits and costs of social
      distancing in high- and low-income countries, *Trans. R. Soc. Trop. Med. Hyg.,*
      vol. 115, no. 7, pp. 807-819, 2021.

[33] K. Sikali, "The dangers of social distancing: How COVID-19 can reshape
      our social experience." *J Community Psychol.*, vol. 48, no. 8, pp.2435-2438,
      2020.

[34] KFF, "The Implications of COVID-19 for Mental Health and Substance Use". Accessed: Nov. 19, 2021. [Online]. Available: https://www.kff.org/coronavi rus-covid-19/issue-brief/the-implications-of-covid-19-for-mental-health-and-substance-use/.

[35] United Nations, "COVID-19 to slash global economic output by $8.5 trillion over next two years". Accessed: Nov. 19, 2021. [Online]. Available: https://www.un.org/en/desa/covid-19-slash-global-economic-output-85-trillion-over-next-two-years.

[36] O. Coibion, Y. Gorodnichenko, and M. Weber, "Labor Markets During the COVID-19 Crisis: A Preliminary View", 2020. Accessed: Nov. 19, 2021. [Online]. Available: https://www.nber.org/papers/w27017.

[37] Airport Council International, "The impact of COVID-19 on the airport business and the path to recovery". Accessed: Nov. 19, 2021. [Online]. Available: https://aci.aero/2021/03/25/the-impact-of-covid-19-on-the-airport-business-and-the-path-to-recovery/.

[38] UNWTO, "International tourism and COVID-19". Accessed: Nov. 19, 2021. [Online]. Available: https://www.unwto.org/international-tourism-and-covid-19.

[39] S. Keelery, "Impact on household income due to the coronavirus (COVID-19) in India from February to April 2020". Available online at: https://www.statista.com/statistics/1111510/india-coronavirus-impact-onhousehold-income/. Accessed November 19, 2020.

[40] E. Duffin, "Opinion of Adults in G7 Countries of the Expected Impact of the COVID-19 Pandemic on their Household Income as of March 2020". Accessed: Nov. 19, 2021. [Online]. Available: https://www.statista.com/statistics/1107322/covid-19-expected-impact-household-income-g7/.

[41] H. Chen, W. Qian, and Q. Wen, "The impact of the Covid-19 pandemic on consumption: Learning from high frequency transaction data", 2020. Accessed: Nov. 19, 2021. [Online]. Available: https://ssrn.com/abstract=3568574.

[42] A. Barua, "A spring in consumers" steps: Americans prepare to get back to their spending ways". Accessed: Nov. 19, 2021. [Online]. Available: https://www2.deloitte.com/us/en/insights/economy/us-consumer-spending-after-co vid.html.

[43] M. Agrawal, K. Eloot, M. Mancini, A. Patel, "'Industry 4.0: Reimagining manufacturing operations after COVID-19". Available at https://www. mckin-sey.com/ /media/McKinsey/Business%20Functions/Operations/Our%20Insi ghts/Industry%204%200%20Reimagining%20manufacturing%20operations/ industry-4-0-reimagining-manuacturing-ops-after-covid-19.pdf. Accessed November 19, 2020.

[44] Digital Commerce 360, "Coronavirus adds $105 billion to US ecommerce in 2020". Accessed: Nov. 19, 2021. [Online]. Available: https://www.digitalcommerce360.com/article/coronavirus-impact-online-retail/.

[45] NSW Government, "Reopening started at 70% full vaccination". Accessed: Nov. 19, 2021. [Online]. Available: https://www.nsw.gov.au/covid-19/easing-covid-19-restrictions/70-percent.

[46]   B. Marroquín, V. Vine, and R. Morgan, "Mental health during the COVID-19 pandemic: Effects of stay-at-home policies, social distancing behavior, and social resources," *Psychiatry Research,* vol. 293, 2020.

[47]   A. C. Shermana, M. L. Williamsb, B. C. Amickc, T. J. Hudsond, E. L. Messias, "Mental health outcomes associated with the COVID-19 pandemic: Prevalence and risk factors in a southern US state," *Psychiatry Res.,* vol. 293, 2020.

[48]   D. Witteveen and E. Velthorst, "Economic hardship and mental health complaints during COVID-19," in *Proceedings of the National Academy of Sciences of the United States of America,* vol. 117, no. 44, 2020.

[49]   W. Cao et al., "The psychological impact of the COVID-19 epidemic on college students in China," *Psychiatry Research,* vol. 287, 2020.

[50]   E. Karasmanaki, G. Tsantopoulos, "Impacts of social distancing during COVID-19 pandemic on the daily life of forestry students," *Children and Youth Services Review,* vol. 120, 2021.

[51]   D. M. De Miranda, B. Da Silva Athanasio, A. C. De Sena Oliveira, and A. C. S. Silva, "How is COVID-19 pandemic impacting mental health of children and adolescents?" *Int. J. Disaster Risk Reduct.,* 2020.

[52]   Venture Beat, "Why Microsoft may beat Zuckerberg to the metaverse". Accessed: Nov. 19, 2021. [Online]. https://venturebeat.com/2021/11/12/why-microsoft-may-beat-zuckerberg-to-the-metaverse/.

Index

www.ingramcontent.com/pod-product-compliance
Lightning Source LLC
Chambersburg PA
CBHW050511190326
41458CB00005B/1501